Beyond Borders

Beyond Borders
Fresh Perspectives in History of Science

Edited by

Josep Simon and Néstor Herran
with Tayra Lanuza-Navarro, Pedro Ruiz-Castell
and Ximo Guillem-Llobat

Cambridge Scholars Publishing

Beyond Borders: Fresh Perspectives in History of Science, Edited by Josep Simon and Néstor Herran with Tayra Lanuza-Navarro, Pedro Ruiz-Castell and Ximo Guillem-Llobat

This book first published 2008 by

Cambridge Scholars Publishing

15 Angerton Gardens, Newcastle, NE5 2JA, UK

British Library Cataloguing in Publication Data
A catalogue record for this book is available from the British Library

ISBN (10): 1-84718-483-9, ISBN (13): 9781847184832

TABLE OF CONTENTS

INTRODUCTION

JOSEP SIMON AND NÉSTOR HERRAN

In the last fifty years the history of science has experienced a profound renovation in relation to its methods, its subject of study and its place in the map of scientific and humanistic knowledge. These changes, while reinforcing its institutional and disciplinary identity, have also produced an undesired fragmentation. Perception of disciplinary crisis is apparent behind calls for the search of new "big pictures" and their implementation in teaching and communicating the history of science to wider audiences.[1]

From the perspective of scholars entering the discipline, this situation is both a cause for concern and an opportunity. It is a matter of concern because it affects the growth and health of our favourite intellectual subject, as well as its capability to excite wider interest. But it is an opportunity because it allows for critical creativity and fresh intellectual challenges. In this perspective, this book stands as a contribution made by postgraduate students – many of whom have completed their PhDs in the course of its preparation – in their attempt to identify historiographical problems in the course of their research, and to contribute in their own right with potential solutions and fresh perspectives based on their growing historical expertise.

As the reader can infer from an inspection of the table of contents, the collection of papers selected engage with a diversity of subjects and periods, and a wide range of approaches. From studies of astrological texts in the sixteenth century to contextual analysis of X-ray spectroscopy in the twentieth century, the contributions to this volume reflect the rich variety of themes and questions characteristic of our discipline. It has been far from our pretensions to be comprehensive or even uniform in our approach to creating this collection. However, its conception has sought to go beyond a simple harvest of recent contributions to the field. Rather, we aim to create synergies connecting particular case studies to the questions we perceive as the most fertile in reacting to the current major challenges of our discipline.

This introduction provides the background to the papers and essay reviews which are presented in the book. Accordingly, we first highlight the major problems that, in our opinion, are afflicting history of science and causing its fragmentation. Subsequently, we propose solutions and connect them with the structure, contents and major arguments of the book. As we will explain in the following paragraphs, in our opinion the highest priority involves providing space for international – and also transnational – approaches, privileging well-defined comparative studies and conceptualizing the role of communication in science in a systematic manner.

A fragmented and parochial discipline?

In the last two decades, historians of science, technology and medicine have increasingly expressed their concern over the perceived lack of a well-defined and consistent, over-arching framework in our field. From the late 1980s, scholars from various traditions have communicated their displeasure about the confrontation that philosophers and sociologists of science have caused in the history of science playground. Instead of disputing about whether science is a cognitive or a social fact, they have pleaded for a genuinely historical approach and some reflection on the making of history of science narratives.[2]

In parallel, the debates on how history of science should be written have translated into general agreement about the need for a larger picture in our discipline. This concern is still felt today.[3] The need for a large-scale synthesis arises not only for conceptual reasons, but also for practical motivations in relation to the teaching of history of science in universities, and the role that historians of science could and should play in the "popular science" publishing market. Considering the fragmentation of the discipline by temporal and spatial discontinuity, synthesis, introductory texts and companions are fundamental in defining the place of history of science in society and to shape future research directions undertaken by prospective students.[4] Writing a synthetic account requires the selection of facts and the construction of historiographical themes and arguments, as well as a narrative style and an intended readership. Thus, the intersection of practical necessity and fundamental reflection is a crucible for general historiographical debates.

In the 1990s, John Christie illustrated the forced pragmatism which historians of science had to exploit in defining the reading lists for their courses. A combination of "older big pictures" with more recent "little and middle sized ones" has arguably been the most commonly adopted formula. Furthermore, as one of the editors of the *Companion to the History of Modern Science*,[5] Christie recalled that this reference work was motivated by the lack of more up-to-date big pictures. At the same time, however, he confessed to the awkward practicalities that such project would have had to face, and admitted the temporal, thematic and conceptual heterogeneity that did finally characterize this reference work.[6] The "Leeds" Companion was presented by its editors as a complement to Charles Gillispie's *Dictionary of Scientific Biography*. Gillispie – the author of one of the major older big pictures [7] – accepted this characterization and gave it a positive review in *Isis*. The Companion has now been a useful reference work for several generations of students and researchers.[8]

Even so, this epitome of the history of science was also criticized at the time of its publication. José M. López Piñero expressed his profound discontent about the general orientation of works such as the Companion and Gillispie's Dictionary. [9] López Piñero was a privileged and experienced observer of the development of our discipline. Trained as a historian of medicine in 1960s Germany, he had subsequently played a major role in the articulation of history of science as a professional discipline in Spain.[10] His critique was built on what he considered were two of the major failures in the prevalent historiography of science at that moment: on the one hand, the lack of integration between history of science and history of medicine, and on the other, the loss of international perspective.[11] López Piñero, like Martin Rudwick, Robert Fox and other scholars, had an international training and knew various national historiographies in different languages.[12] This background had convinced him that a major force in the constitution of history of science as an international academic discipline had been the migration and circulation of scholars between Europe and America between the two World Wars.[13]

As an illustrative example of these two historiographical aspects, López Piñero commented on the "discovery" of Ludwik Fleck's work by the English-reading historian of science. Fleck's *Entstehung und Entwicklung einer wissenschaftlichen Tatsache* was published in 1935 and translated into English in 1979 with a preface by Thomas S. Kuhn. According to López Piñero, it was a sad episode in the historiography of science that Fleck's work had been presented simply as an early precursor of Kuhn's *The Structure of Scientific Revolutions* and of social construction.[14] Fleck's book had had an important impact in the German-speaking world at the time of its publication, and in the 1960s it was still a recommended text in history of medicine programmes. The loss of internationality and of integration between history of science and history of medicine had led many scholars to ignore the rich historiographical legacy of central European history of medicine – the context for Fleck's work – and its international significance for the development of history of science as a discipline.[15] The misrepresentation of Fleck's work was also pinpointed by other scholars at the time, although they did not attempt to investigate its causes, nor did they see in it an illustration of historiographical and professional crisis.[16]

Several decades later, it is fair to say that Fleck's work has found its place in history of science,[17] its intrinsic value has been recognized,[18] and the use of its English translation by historians of science has contributed to produce original historiographical work. [19] However, López Piñero's criticism is still compelling. The lack of integration between history of science and history of medicine can now be seen as a subset of the general problem of specialization that has fragmented our professional context into compartments defined by historical periods and sciences. [20] This fragmentation goes together with a large imbalance in the distribution of historians among periods and subdisciplines. [21] In addition, as recently recalled by Robert Fox,[22] there is also fragmentation based on geography, nation and language. As we shall argue in the following paragraphs, our discipline has increasingly become national and even local in its approaches and writing.

Today, most history of science and medicine theses are, in general, case studies of national character. David Kaiser has recently shown how, from the 1980s, the geographical focus of theses written in North America has become less international and more parochial.[23] If we look at the list of theses written in Britain since 1999, the picture is analogous or even more insular. Three-quarters of all British theses in the history of science and medicine deal strictly with British cases. Most theses on other countries are devoted to Africa and India, in relation to the existing institutional programmes on the history of British colonial medicine, and they are typically undertaken by students born in these areas. Furthermore, the number of theses dealing with more than one national context is very low (less than one tenth). These works typically deal with two national contexts (more than two is rare), where the combination of Britain with France or the USA is the most usual, and they are often undertaken by non-British students.[24] The mononational character of history of science theses is similarly very pronounced in many other European countries.[25]

The turn towards the local has often been associated with greater historiographical sophistication that – in spite of the important concerns raised in this introduction – has benefited the discipline in various ways. This phenomenon is certainly also connected to the importance that archival research has acquired in our field,[26] which tends to be almost unaffected by the rise of online digitization projects.[27] It is a matter of fact that digitization projects are not equally distributed among different countries due to economical, political and cultural factors, and are, in general, constructed under explicit national agendas. They are usually based around major library collections, thus constrained by the relative cosmopolitanism of their holdings but also by selection criteria that in general prioritize a single language and a national cultural heritage. The rhetorical myth of universalism combined with real cultural parochialism surfaces even in major private initiatives with transnational pretentions such as Google Books, which is, in fact, highly biased towards Anglo-American sources. In this sense, the diversification of the field by integration of other projects such as the digitization programme of the French National Library and its promotion of a European digital library could be a valuable counterweight. Surprisingly, the heated debates on digitization and Google Books taking place in Europe have apparently not aroused much interest in the Anglo-American community of historians of science.[28]

Caught in a shrinking cultural web

If the development of history of medicine and science as a discipline during the first half of the twentieth century was predominantly driven by the work of German, French and Italian-speaking scholars, it is indisputable that in the last few decades the discipline has been led by Anglo-American scholarship.[29] Hence, the American and British national history of science societies have attracted a large number of foreign scholars, and they are arguably the most *international* national societies for this discipline. However, the overnationalization of history of science, while securing a prominent position for American and British culture in history of science, has also contributed to the obscuration of the international character of many historical events. While the loss of internationality affects the discipline everywhere with more or less particular emphasis,[30] solving the problem in the Anglo-American axis is particularly crucial. On the one hand, there is solid evidence for the considerable dimensions of the problem in this academic context. On the other, the Anglo-American contribution to the field is currently the largest and best communicated.[31] For this reason, in this introduction we use the Anglo-American focus as the illustration of an exemplary case.

The turn towards the local, brought on by socio-constructivism,[32] has been pinpointed by scholars with different agendas as one of the main reasons for this state of affairs.[33] As a reaction to positivism, socio-constructivists converted the making of any generalization in a national or international scale into anathema. In challenging science's universality, the field has moved towards the production of microhistories that, while illuminating the role of social processes in the construction of scientific knowledge in local settings, have also obscured the relevance of macrohistorical explanations.[34] Paradoxically, socio-constructivists, like their predecessors, gave an implicit status of universality to a set of categories, in their case stressing the locality of knowledge.[35] Microhistories have had a longer tradition in our field, in particular in the study of the early modern period. But microhistorical analysis is only useful if integrated with macrohistorical parameters and vice versa. Accuracy and generalization require the connection of narratives of the local with the global through comparative analysis and the study of communication in an international perspective.[36] Indeed, in spite of the rise of nation states, from the nineteenth century onwards international communication and transnational comparison were equally central to scientific activity. Assessing internationality is therefore a task that historians cannot evade.[37]

Dealing with internationality, both as a characteristic of science in certain periods, and – in a reflexive way – of the community of historians of science today, is certainly not an easy task. It requires knowledge of several languages, several national historiographies, and the ability to engage in international projects with foreign colleagues. A recent report requested by the European Union has revealed that 56% of Europeans think they can engage (at conversational level) in other language than their native one. Eighteen countries of a group of twenty nine are above this average. English is the most spoken foreign language (38%). The countries with the lowest rate of multilingualism are Turkey (67%), Ireland (66%), and the UK (62%).[38]

In Britain, the critical role of language skills in research has indeed been stressed by scholars such as Martin Rudwick and institutions such as the British Academy.[39] In 2006, concerned about the inadequacy of language training in secondary schools and its consequences for postgraduate research, the British government appointed two commissioners to investigate and report on this issue. The British Academy endorsed the government commission, stating that research on the humanities and social sciences "is becoming increasingly insular in outlook, because PhD students do not have language skills, or the time to acquire them".[40] In contrast with the major American postgraduate programmes in history of science, British postgraduate students are in general not required to learn foreign languages. Furthermore, the pressure to publish – especially affecting postgraduate students and young scholars

– also contributes to hinder educational opportunities providing professionals with tools to go beyond mono-linguistic and mono-cultural studies.[41] In addition, despite the process of European construction and the funding projects promoted as a part of it, European historians of science are far behind the transnational initiatives exploited and promoted by professionals in other disciplines. Thus, overall it would be desirable that postgraduate programmes and scholarly journals[42] take care of preparing students for the challenges and opportunities of a future professional life, with the international outlook required by scholarship.

Beyond the local: Comparison, communication and appropriation

In proposing solutions to the problems perceived to afflict history of science, scholars have rather focused on conceptual and historiographical issues. In many cases their proposals have mainly suggested to put the methodology and approaches characterizing their own current work at the very heart of the discipline. For example, Frederic L. Holmes proposed studying individual investigative pathways and integrating them into a network constituting the moving boundaries of a well defined set of research problems within a scientific field. Casper Hakfoort envisaged the possibility of synthesis through an iterative methodology based on the historical investigation of large epistemological categories such as "scientific knowledge" and their change in time. John Christie suggested pragmatism and building around narratives of power. Robert Köhler has considered drawing on the socio-constructivist legacy by using basic categories such as practice, social role, or credibility and trust. David Kaiser has advocated bringing pedagogy to the centre.[43] All of these approaches provide interesting elements in building a successful programme of historiographical renovation. However, in their current form, they perhaps lack a more ambitious and general force.

In dealing with the aforementioned problems, the most ambitious and innovative proposals have been, in our opinion, raised by James Secord, Margaret Jacob and Lewis Pyenson. By raising the idea of science as "Knowledge in Transit", Secord has seriously suggested an alternative to previous frameworks: the study of practices of communication of knowledge could be used – if consistently problematized – to build narratives potentially useful for a wide range of scholars and contexts.[44] His previous research on literary replication and on the multifarious ways in which scientific knowledge was appropriated through reading in nineteenth-century Britain,[45] and his most recent emphasis on the role of conversation in nineteenth-century science,[46] provide good examples of how communication practices can constitute central elements of the historical analysis of science.

Secord's proposal considers that making communication central to our analysis and narratives allows for a better understanding of production, circulation and acquisition of scientific knowledge in local and international contexts. Making communication central and moving the focus from origins and production to audiences and uses[47] allows the historian to blur the distinction between the making and the communication of knowledge and helps us to link practices such as science in the laboratory, science in the field, reading and pedagogy.[48] As has recently been argued by Jonathan Topham, this approach can also be fruitful in rethinking the historiography of science popularization.[49] However, as Secord has recognized, his programme would only be effectively boosted through a better defined conceptualization of geographical and disciplinary boundaries.[50] Indeed, it is still necessary to define more precisely what is "communication", and elucidate how "communication" in local and national contexts relates to international "communication" (is this a transnational phenomenon or can it only be characterized and constrained by local and national parameters?). Therefore, a clear conceptualization of the national and the international is still necessary in order to strengthen the powerful potential of this approach. Indeed, this is an important driving question in this book.

Communication is not a newcomer in history of science. Centre-periphery models and reception studies, despite restricting the focus to producers, placed communication at the centre of historical analysis.[51] What distinguishes new approaches from these old and powerful models is that they give important agency to audiences. An important inspiration here comes from cultural history. In his study of popular culture, Roger Chartier concluded that the "popular" could not be defined through objects or texts, but through the active ways in which those were "appropriated" by different kinds of readers.[52] In the same period, in a now classic paper, Abdelhamid Sabra used the concept of "appropriation" in order to challenge traditional narratives considering that Greek science had been passively received in Medieval Islam.[53] More recently, the concept of appropriation has been used in historical studies on science in the so-called scientific "European periphery", with the collective STEP as an important force behind this movement.[54] The study of the processes of mutual communication between different national contexts has also been central in the study of colonial science,[55] and in the study of cultural transfers in general history. In the latter it has been contended that the study of local appropriation and international mediation is fundamental to overcome the constriction imposed by the national character of most historiographical frameworks.[56] "Appropriation" thus offers a useful tool allowing analysis of historical phenomena in both local and international contexts, and to produce more balanced and accurate historical accounts by shifting the focus from production to use. In this book, we have deliberately promoted the use of this concept, which in our opinion should be complemented with extensive use of the comparative approach.

This active promotion of comparative history departs from the original "Secordian" approach, as Secord dismisses this method prompted by his perception that it has led to defective results in history of science. His brief survey of the field highlights the fact that many works presented as comparative are in fact compilations of national case studies rather than international synthesis.[57] These studies would therefore not contribute to solving the historiographical problems already described. However, this is only a partial view as there are many other genuine comparative histories highlighting the usefulness of this approach. Its virtues have been stressed by many scholars, including historians of science and medicine such as Margaret Jacob, Lewis Pyenson and Ilana Löwy. Jacob has seen comparative history as the fundamental tool allowing the integration of precise microhistories into the larger scope of macrohistorical frameworks and to formulate more relevant and larger historical questions.[58] In a recent paper providing an analytical overview of twenty years of the journal *Social History of Medicine*, Löwy has stressed the heuristic power of comparison, and its fundamental role in taking properly into account the transnational dimension of medicine, its actors, practice and objects of study.[59] Furthermore, as argued by Pyenson, the comparative approach has in fact led to major historiographical innovations in the history of science.[60] Among the examples raised by Pyenson we can cite, for example, Jack Morrell's idea of the research school, originating in a comparative study of Thomson and Liebig's laboratories; Thomas P. Hughes' concept of technological network, arising from comparison of European and North American electrical supply networks, or even Joseph Needham's comprehensive – and, to some extent, comparative – survey of Chinese science and technology.[61]

As put by Marc Bloch's ground-breaking analysis, the comparative method allows history to be intelligible by "establishing explanatory relationships between phenomena".[62] According to William H. Sewell, the logic of hypothesis testing underlies the comparative method, and provides it with unmatchable explanatory power.[63] Comparison allows the historian to discriminate, to synthesize, to challenge and finally to produce an original perspective as a more satisfactory answer to the problem thus confronted. The challenge to a large amount of current scholarship is therefore to give a wider range to comparative study, in order to go beyond the local and to accurately conceptualize the national and the international.

Comparison has been a central methodological tool in many disciplines since at least the nineteenth century, contributing to create well-defined subfields within subjects such as philology, law, education, anatomy, geology, archaeology, religion studies, anthropology, linguistics, sociology, literature and their associated histories.[64] In general history and literature, comparison was originally seen as a tool that allowed historians to move their historical research beyond national frameworks. In using a comparative approach in history of science for this purpose, we could greatly benefit from the experience accumulated in these disciplines. Needless to say, intellectual and disciplinary pluralism has been and still is one of the best virtues that history of science has to offer.[65] Thus, for instance, we should certainly take into account the criticisms that have already been raised in history and literature against the comparative method, based on the idea that, paradoxically, it can in fact contribute to reinforce national singularities. Positioning a national culture in relation to the others has often rather resulted in stressing differences than commonalities.

In order to solve this problem, the diachronic study of cultural transfers was proposed in the 1980s.[66] Other historians have drawn on anthropology, and proposed an explanation of cultural change based on the study of cultural encounters, in which all interacting parts matter. Although, echoing anthropology, this approach has traditionally been applied to the confrontation between the European metropolis and its colonies, it might prove useful in a wider range of contexts including European states themselves.[67] More recently, a new proposal termed the "transnational turn" has shaken academic literature and history in the USA, and is progressively making its way in Europe and on other continents. Influenced in part by awareness of the effects of globalization and multiculturality, this approach contends that many historical phenomena transcend the boundaries of the nation state. Transnational history thus offers a further challenge to the analytical framework of the nation. Hence, the intention is to focus on transnational historical objects, phenomena and actors, which are considered to have been left out by the paradoxical reification of national boundaries effected by international comparison, even when complemented with the study of cultural transfers. In spite of its deconstruction and dismissal of the nation, writing transnational history still involves the use of international comparison and the assessment of the circulation of knowledge, objects and actors in an international perspective. In addition to its critical contribution to the recasting of the nation as just another historical parameter, one of its major values resides in its promotion of internationality and interdisciplinarity.[68]

In brief, a major aim of the editors of this book has been to problematize the local, the national and even the international through comparison and through the assessment and analysis of communication practices within these contexts and across them. This has a double intention in relation to the current configuration of history of science as outlined in this introduction. On the one hand, it intends to enrich its historiography by diversifying the national character impressed through its compilation of case studies. Through comparison, the different case studies presented in this book aim at providing the tools for a more fruitful integration and diversification of national case studies in our field. The use of sources belonging to different national historiographies and published in different languages and media platforms express our conviction of the need of promoting internationality in history of science as a requisite of outstanding scholarship. On the other hand, the five parts constituting this book pay important attention to the study of the practices of communication in different periods and in local, national and international contexts. Studying and conceptualizing these practices, their agency in linking local and global contexts and cultures, and their contribution to the making of scientific knowledge, is a first step towards the writing of a big-picture history of science that will satisfactorily assess the complex diversity of science as a human practice.

A fresh perspective in history of science

The structure of this book – as one of the results of a generalist international postgraduate conference – emerged through a combination of pragmatism and intellectual debate. [69] The meeting in València in November 2005 brought together a group of postgraduate students and young researchers from different countries and with various educational experiences, whose work dealt with different scientific disciplines, historical periods and national contexts. This diversity is reflected in the range of topic and period across this monograph. The time scope is broad, from the early modern period to the twentieth century, although nineteenth- and twentieth-century case studies are predominant, which is a particular reflex of participation in the València conference, but also a reflex of larger international developments of the discipline at the moment.[70] In spite of the generalist character of the meeting, the generous and dynamic participation of all its contributors coalesced in the identification of critical topics and historiographical challenges to our discipline. The subsequent configuration of a historiographical project, aimed at debating these issues in the context of the research of each of the authors, has resulted in the publication of this book.

In an effort to transgress boundaries and connect with topical issues in the discipline, the editors of this book have worked in three parallel directions. Authors were encouraged to use the tools of comparative history in their analyses, and to connect their local case studies with issues relevant in an international or transnational level. Furthermore, a conscious effort was made to diversify the use of literature in linguistic, cultural and historiographical terms. Finally, the book was structured through a series of central themes and questions in the discipline, defined by an essay preceding the papers in each of the parts. The five parts of this book are thus devoted respectively to the early modern map of knowledge, the pedagogy of science, the popularization of science, the tension between science and nation, and the geography of scientific centres and peripheries. A brief introduction to the major narrative lines and themes articulating the book and its parts occupies the next paragraphs as a closure to this introduction.

The first part of this book is devoted to the early modern period, contrasting in various ways with the rest of the book, which is focused on the modern period. The practices of science and its communication have changed enormously since the "Scientific Revolution", a core event in the original conceptualization of our discipline, for which early positivistic agendas considered science as universal knowledge circulating without constraints.[71] In fact, as exposed in this book, in spite of the more fluid boundaries of the early modern state, mathematics and astrology were practiced in different ways in various places in this period. Nonetheless, pedagogical programmes had a privileged role in the communication of scientific knowledge throughout Europe. But the purposes and modes of this communication varied in different institutions which contributed to encourage and at the same time to constrain the practice of science across different kingdoms, through their political, social, cultural and pedagogical aims. Pedagogical tools such as school books and teaching treatises were fundamental to ensure the communication of new outlooks that contributed to changing the place of subjects such as mathematics, its subdisciplines, and its practitioners, within European culture and society. In this process, translation had a fundamental role, displaying the tension between the global and the local through the dialogue between Latin and vernacular languages such as French.

The second part of this book precisely stresses the role of translation (into vernacular) and pedagogy as driving agents of the international communication of science during the eighteenth and nineteenth centuries. Textbooks played a fundamental role in the organization of national systems of formal education and the integration of science in the school curriculum taking place in this period. As privileged communicators of scientific and pedagogical knowledge, their circulation contributed to determine the configuration of educational and scientific practices, and consequently of disciplines within and across nation states. Through the international circulation of booksellers and students, and the spread of scientific education, textbooks emerged in the nineteenth century as a well-defined genre. The national character of educational systems in Europe was defined through mutual observation, comparison and appropriation of other cultures in an international perspective. In their communicative agency, textbooks intersected and fruitfully interacted with other genres of communication also contributing to the definition of science as a discipline in this period such as journalism, public lecturing and social conversation.

The third part of this book is especially devoted to study this diversity of communicational genres gathered under the umbrella of the phenomenon conventionally termed "science popularization". These modes of communication and appropriation of knowledge were increasingly developed in the nineteenth century with purposes not restricted to this period but accentuated in the action of science and its practitioners in society. Disciplinary formation, professional aspirations, political dominance, social control and national construction intersected in the shaping of discourses and programmes aimed at communicating science to a wide range of audiences. The communication of science across society pervaded cultural practices in different countries, and the increasing centrality of science in society shaped science popularization as a powerful communicational tool that supported programmes of local, national and international range.

The fourth part of this book focuses on the tension between the national and the international in history of science. In spite of the nineteenth-century rise of the nation state and the associated construction of national identities, as shown in this part, the national unit is too often taken for granted by historians of science. Nations are not homogeneous entities and scientific practices were often more defined by local institutional, disciplinary and pedagogical cultures. Furthermore, a national spirit is always built across time in relation to other nations. The national character of science is thus fragmented in its unit and pervaded by the international in its essence. On the other hand, scientific internationalism can be seen as a powerful ideological force aimed at increasing communication over national barriers, through disciplinary, technological, economical, intellectual and moral interests. These interests coalesce on certain occasions to form transnational entities whose presence in the local and the international sphere is not essentially driven by the nation state.

The fifth and last part of this book stresses again the importance of the conceptualization of communication in national, international and transnational contexts in order to map the structure of science and its practice and to build the big picture. In this context it is particularly important to diversify the number of national cases configuring the constructed archive of evidence in our discipline. As shown in this part, countries conventionally considered peripheral in the development of science in fact offer original insights into the constitution of scientific disciplines and the role of communicational practices. Furthermore, the overarching notion of periphery and centre fails to account for national heterogeneity and for the diachronous character of historical phenomena. Its unidirectional conception of communication is inaccurate as communication always involves interaction and mutual dependence between at least two active agents, whether in teaching, in popularizing, in writing and reading, or in national and international construction, and in science practice in general.

In February 2006, at a roundtable discussion of Martin Rudwick's *Bursting the Limits of Time* at Leeds, Jack Morrell, with his characteristic humour, emphasized the advantages of the condition he shared with Rudwick – being a septuagenarian and thus being free of academic affiliations – offered to their work in terms of intellectual freedom. At the other end of the line, postgraduate students are often subjected to various pressures related to prospective professional opportunities and to their integration into and their interaction with the established community of scholars. This book – with all its failures and work-in-progress virtues – is nonetheless an indication that postgraduate students and young researchers can engage in independent, original and productive research ventures.

Acknowledgements

This book was possible thanks to the generous efforts of all its contributing authors, to whom we would like to offer our sincere gratitude. We are also indebted to the Generalitat Valenciana, who awarded us a grant towards the linguistic revision of the manuscript. We wish to thank

Leucha Veneer, a history of science postgraduate student at Leeds, who carried out an excellent revision of the whole manuscript. The cover of the book was designed by Virginia Vinagre, a design student in València, whom we would like to thank for her generous and enthusiastic collaboration. This project has also benefited from sincere encouragement by several scholars, and we therefore wish to express our heartfelt thanks to José Ramón Bertomeu, Jon Topham, Pepe Pardo, Antonio García-Belmar, Agustí Nieto-Galan, Robert Fox, James Sumner and Greg Radick.

Notes

[1] Instead, for Steven Shapin and other scholars the problem does not lie in the lack of "big pictures", but in the impoverishment of our narratives, writing styles and readerships, as a result of the constitution of an "hyperprofessionalized" discipline that has become almost self-referential. Although a major focus of this book is cutting across national and ideological boundaries within the discipline, we agree with Shapin that it is equally important to cut across social and cultural boundaries within local contexts. Shapin, S. (2005). "Hyperprofessionalism and the Crisis of Readership in the History of Science". *Isis* 96 (2): 238-243.

[2] Rossi, P. (1986). *I ragni e le formiche: Una apologia della storia della scienza.* Bologna: Il Mulino; Garber, E. (1990). "Introduction". In Garber, ed. *Beyond History of Science. Essays in Honor of Robert E. Schofield.* Bethlehem: Lehigh University Press, pp. 7-20; Hakfoort, C. (1991). "The Missing Syntheses in the Historiography of Science". *History of Science* 29: 209-216, on pp. 211-3; Holmes, F. L. (1993). "Justus Liebig and the Construction of Organic Chemistry". In Mauskopf, S. H., ed. *Chemical Sciences in the Modern World.* Philadelphia: University of Pennsylvania Press, pp. 119-35, on pp 119-21; López Piñero, J. M. (1993). "La tradición de la historiografía de la ciencia y su coyuntura actual: los condicionantes de un congreso". In Lafuente, A.; Elena, A., and Ortega, M. L., eds. *Mundialización de la ciencia y cultura nacional.* Madrid: Ediciones Doce Calles, pp. 23-49, on p. 47; Harris, S. J. (1998). "Introduction: Thinking Locally, Acting Globally". *Configurations* 6 (2): 131-9; Jacob, M. C. (1999). "Science Studies after Social Construction: The Turn toward the Comparative and the Global". In Bonnell, V. E., and Hunt, L., eds. *Beyond the Cultural Turn: New Directions in the Study of Society and Culture.* Berkeley: University of California Press, pp. 95-120, on pp. 95, 99-101; Heilbron, J. L. (November 2002). "Science as a subject of history". [Conference at the Palau de la Generalitat]. *VII Trobada d'Història de la Ciència i de la Tècnica.* Barcelona: SCHCT; Warner, J. H. (2004). "Grand Narrative and its Discontents: Medical History and the Social Tranformation of American Medicine". *Journal of Health Politics, Policy and Law* 29 (4-5): 757-80; Fox, R. (2006). "Fashioning the Discipline: History of Science in the European Intellectual Tradition". *Minerva* 44: 410-432, on pp. 426-7.

[3] Holmes. "Justus Liebig"; Jacob. "Science Studies after Social Construction"; Hakfoort. "The Missing Syntheses"; Secord, J., ed. (1993). "The Big Picture". Special issue. *British Journal for the History of Science* 26: 387–483; Köhler, R., ed. (2005). "Focus: The Generalist Vision in the History of Science". *Isis* 96 (2): 224-51; Pickstone, J. V. (2007). "Working Knowledges Before and After circa 1800: Practices and Disciplines in the History of Science, Technology, and Medicine". *Isis* 98 (3): 489-516, on p. 490.

[4] Hakfoort. "The Missing Syntheses"; Secord. "The Big Picture"; Christie, J. (1993). "Aurora, Nemesis and Clio". *British Journal for the History of Science* 26: 391-405; Harris. "Thinking Locally, Acting Globally", pp. 136-8.

[5] Olby, R., Cantor, G. N., Christie, J. R. R., and Hodge, M. J. S. (1990). *Companion to the History of Modern Science*. London: Routledge.

[6] Christie. "Aurora, Nemesis and Clio", pp. 401, 394.

[7] Gillispie, C. C. (1960). *The Edge of Objectivity: An Essay in the History of Scientific Ideas*. Princeton: Princeton University Press.

[8] Gillispie, C. C. (1991). "Review: Scholarship Epitomized". *Isis* 82 (1): 94-8; Basalla, G. (1992). " [Review of *An Encyclopaedia of the History of Technology* by Ian McNeil]". *Isis* 83 (1): 115-7; Fyfe, A. (2005). " [Review of *The Oxford Companion to the History of Modern Science*, by J. L. Heilbron, ed.]". *British Journal for the History of Science* 38 (3): 349-73.

[9] López Piñero, J. M. (1992). "La historia de la ciencia como disciplina". *Saber Leer* 55: 8-9; on p. 9.

[10] The constitution of history of science as a professional discipline in 1970s Spain, is a phenomenon that still needs research, and especially taking into account the promising development of the subject in the 1930s, broken by the Spanish Civil War. See for example, Bujosa Homar, F. (1990). "Histoire de la médecine en Espagne". In Bernes, A. C., eds. *Nouveaux enjeux de l'histoire de la médecine*. Strasbourg: Institut Pasteur, pp. 7-27, on pp. 21-4; López Piñero, J. M. (1988). "The History of Science Unit of the Institute of Science Studies (Valencia, Spain)". *Nuncius* 3 (2): 193-207; Nieto Galan, A. (2008). "La memoria histórica de la ciencia en España". *Claves de Razón Práctica* [forthcoming]; Asociación Española de Historiadores de la Ciencia Española. (1935). *Estudios sobre la ciencia española del siglo XVII*. Madrid: Gráfica Universal, pp. vii-xiv.

[11] The first aspect has also been in the last decade recurrently stressed by John Pickstone, and the second aspect has recently been emphasized by James Secord, Martin Rudwick and Robert Fox. See Pickstone, J. V. (2000). *Ways of Knowing: A New History of Science, Technology and Medicine*. Manchester: Manchester University Press, and (2007). "Working Knowledges Before and After circa 1800"; Secord, J. A. (2004). "Knowledge in Transit". *Isis* 4 (95): 654-72; Rudwick, M. J. S. (2005). *Bursting the Limits of Time: The Reconstruction of Geohistory in the Age of Revolution*. Chicago: University of Chicago Press, pp. 3-4; Fox. "Fashioning the Discipline".

[12] But, paradoxically, only a small amount of López Piñero's extensive work was published in other languages than his mother tongue (Spanish). This has been – and still is – a characteristic phenomenon in general history in many European countries. Besides the question of language skills (see n. 38), the availability of considerably large national readerships for history is one of the reasons of this phenomenon. In addition, this disciplinary field offers a more diverse publishing scene than history of science in terms of language, with specialized publishers and journals whose major focus is almost exclusively national. For journals see European Science Foundation (2007). *European Reference Index for the Humanities (ERIH)*; *ERIH Initial List: History (2007)*, and *ERIH Initial List: History and Philosophy of Science (2007)*. Strasbourg: European Science Foundation [www.esf.org/erih].

[13] López Piñero. "La tradición de la historiografía de la ciencia", pp. 27-41. See also Pickstone, J. V. (2005). "Review Article: Medical History as a Way of Life". *Social History of Medicine* 18 (2): 307-23, and Homburg, E. (2008). "Boundaries and audiences of national histories of science: insights from the history of science and technology of the Netherlands". *Nuncius* [forthcoming]. We would like to thank Ernst Homburg for sending us a copy of his paper before publication.

[14] See Trenn, T. J. (1979). "Preface". In Fleck, L. *Genesis and Development of a Scientific Fact*. Chicago: The University of Chicago Press, pp. xiii-xxii, on pp. xiii and xviii.

[15] López Piñero. "La tradición de la historiografía de la ciencia", pp. 48-9.

[16] Gutmann Rosenkrantz, B. (1981). "[Review of *Genesis and Development of a Scientific Fact* by Ludwik Fleck]". *Isis* 72 (1): 96-9; Lenoir, T. (1984). "[Review of *Ludwik Fleck: Erfahrung und Tatsache* by L. Shäfer and T. Schnelle, eds.]". *Isis* 75 (4): 724-5; Harwood, J. (1986). "Review: Ludwik Fleck and the Sociology of Knowledge". *Social Studies of Science* 16 (1): 173-87.

[17] Although it is still most often integrated within the mainstream of socio-constructivism. See Golinski, J. (1998). *Making Natural Knowledge. Constructivism and the History of Science*. Cambridge: Cambridge University Press, on pp. 27-46.

[18] See for instance Bonah, C. (2002). "'Experimental Rage': The Development of Medical Ethics and the Genesis of Scientific Facts". Ludwik Fleck: An Answer to the Crisis of Modern Medicine in Interwar Germany? Society for the Social History of Medicine Millennium Prize Essay 2000. *Social History of Medicine* 15 (2): 187-207.

[19] See for example Olesko, K. (2006). "Science Pedagogy as a Category of Historical Analysis: Past, Present, and Future". *Science and Education* 15: 863-80; Topham, J. R. (2008). "Rethinking the History of Science Popularization/Popular Science". In Papanelopoulou, F.; Nieto-Galan, A., and Perdiguero, E., eds. *Popularising Science and Technology in the European Periphery, 1800-2000*. Aldershot: Ashgate.

[20] This problem is currently perceived as a major obstruction to the creation of "big pictures" in history of science. See Kohler, R. (2005). "A Generalist's Vision". *Isis* 96 (2): 224-9, on p. 224; Kaiser, D. (2005). "Training and the Generalist's Vision in the History of Science". *Isis* 96 (2): 244-51.

[21] If the "Scientific Revolution" constituted the developing core of the discipline, and the history of physics was certainly the most actively researched area for most of the twentieth century, in the last few decades, biology has been attracting the largest number of new scholars. In addition, as pinpointed by David Kaiser (see n. 19), the modern period has attracted the highest number of students in the last two decades with an important growth of studies on twentieth century science. Studies on periods previous to the nineteenth century have a stable number of students but are in the minority. Heilbron. "Science as a subject of history".

[22] Fox. "Fashioning the Discipline", p. 412.

[23] Kaiser. "Training and the Generalist's Vision", p. 249. Kaiser's numbers show however that, at the moment, North-American history of science scholarship is arguably the most international in the Western world. The increase of theses on American topics has gone together with the rise of the history of twentieth-century science in America, and has come to balance a research output which was in previous decades predominantly oriented towards European topics.

[24] To produce these statistics, we have analysed the list of theses provided by the British Society for the History of Science. Although the list is not complete and our analysis is a rough approximation (alas, mainly based on thesis titles), it is still useful for qualitative purposes. [www.bshs.org.uk/hstm/list_of_theses/] (accessed 26 September 2007). See also the *International Bibliography of History of Science, Technology, and Medicine* (HistSciTechMed).

[25] As shown for instance by an analysis of the lists of history of science theses produced in the last decades in France and in Spain. See http://www.sudoc.abes.fr, and http://teseo.mec.es/teseo.

[26] Kaiser. "Training and the Generalist's Vision", p. 246.

[27] In the meantime online resources are clearly changing our way of working and writing, particularly through their powerful search engines. Despite their benefits, they have introduced new problems that, if we do not make the effort to conceptualize them, will grow bigger in the near future. Some of these problems are, for example, the loss of contextuality, the loss of the material culture of the book, the definition of online source authority criteria, "copy and paste" and plagiarism, and linguistic determinism based on the illusion of lexicographic omniscience. Despite several conferences having been organized in the last few years around this topic, unfortunately many of them are too often characterized by a lack of reflection and are instead used as platforms to present different history of science web projects to the public.

[28] See Jeanneney, J.-N. (2007). *Google and the Myth of Universal Knowledge. A View from Europe*. Chicago: The University of Chicago Press.

[29] López Piñero. "La tradición de la historiografía de la ciencia", pp. 27-49.

[30] For instance, Francesc Bujosa observed in 1990 the profound localism afflicting the practice of the history of science and medicine in Spain. Although there is currently an important group of Spanish scholars working in this field at international level, they are a minority. The major reasons suggested by Bujosa (the endogamous and non-democratic nature of the university and research system, the lack of peer review, and parochialism in publication) are still pertinent. This perspective is in agreement with a recent report on the state of teaching and research in the humanities published by the Spanish Foundation for Science and Technology. Localism is also a characteristic of most of the history of science and medicine journals published outside the Anglo-American context, although there are some exceptions, and some of them have large national readerships (see n. 31). Bujosa. "Histoire de la médecine en Espagne", pp. 12-5, 21-3; Spanish Foundation for Science and Technology (2006). *White paper on research in the Humanities.* Madrid: FECYT, pp. 106-66 [http://www. fecyt.es], and FECYT (2005). "Informe sobre la investigación en Humanidades. Panel de Historia. [versión 25 de enero de 2005]", pp. 13, 20, 25; European Science Foundation. *ERIH Initial List: History and Philosophy of Science (2007), and ERIH Initial List: History (2007).*

[31] A recent quantitative and qualitative assessment of history and philosophy of science journals undertaken by the European Science Foundation shows that 53% of the journals in this area are published in English, 36% of which are produced by Anglo-American publishers. Furthermore, all the history of science, medicine and technology journals with the highest international standard (rated "A") are Anglo-American and have British or American editors (with only one or two exceptions). Anglo-American dominance is also the largest in the second group of journals (rated "B") listed by the report (in a global scale of 3 categories: A, B, C). The composition of the editorial boards of these journals is more nationally diverse though. As previously mentioned, in contrast, the linguistic and national distribution of general history journals and publishing media is more diverse. European Science Foundation. *European Reference Index for the Humanities (ERIH); ERIH Initial List: History and Philosophy of Science (2007).*

[32] We use here Golinski's broad definition of socio-constructivism. See Golinski. *Making Natural Knowledge.*

[33] Jacob. "Science Studies after Social Construction", pp. 96-7, 106-7; Secord. "Knowledge in Transit", pp. 659-60; Harris. "Thinking Locally, Acting Globally", pp. 135-6.

[34] See Jacob. "Science Studies after Social Construction", and Kaiser. "Training and the Generalist's Vision".

[35] See Kohler. "A Generalist's Vision".

[36] Findlen, P. (2005). "The Two Cultures of Scholarship". *Isis* 96 (2): 230-7; Jacob. "Science Studies after Social Construction", pp. 107, 111-6.

[37] This could be analogous to writing the modern history of British politics ignoring its Foreign Office and foreign policy, because its archives talked about foreign cultures and included documents in foreign languages.

[38] Followed by Italy (59%), Hungary (58%), Portugal (58%), Spain (56%), and Romania (53%). TNS Opinion & Social (2006). "Europeans and their Languages". *Eurobarometer* 243/Wave 64.3 (February), pp. 8-13; Wastiau-Schlüter, P., ed. (2005). "Foreign Language Learning: A European Priority". *Eurydyce* 2 (March), p. 3. See http://ec.europa.eu/public_opinion/archives/ebs/ebs_243_sum_en.pdf, and http://www.eurydice.org/.

[39] Rudwick. *Bursting the Limits of Time*, p. 4.

[40] Dearing, R.; King, L. (2006). *The Languages Review. Consultation Report: December 2006.* [http://www.dfes.gov.uk/consultations/downloadableDocs/6869-DfES-Language%20Review.pdf]; The British Academy. *Response to the Announcement of the Languages Review.* [http://www.britac.ac.uk/reports/dearing-2006/response-02-07.html] (accessed 10 September 2007).

[41] Löwy, I. (2007). "The Social History of Medicine: Beyond the local". *Social History of Medicine* 20 (3): 465-81, p. 478.

[42] Ibid.

[43] Holmes. "Justus Liebig"; Hakfoort. "The Missing Syntheses"; Christie. "Aurora, Nemesis, and Clio"; Köhler. "Focus: The Generalist Vision in the History of Science"; Kaiser. "Training and the Generalist's Vision".

[44] Secord, "Knowledge in Transit".

[45] Secord, J. A. (2000). *Victorian Sensation. The Extraordinary Publication, Reception, and Secret Authorship of Vestiges of the Natural History of Creation.* Chicago: The University of Chicago Press.

[46] Secord, J. (2007). "How Scientific Conversation Became Shop Talk". In Fyfe, A., and Lightman, B., eds. *Science in the Marketplace.* Chicago: Chicago University Press, pp. 23-59. We would like to thank Jim Secord for providing us with a copy of his paper before publication.

[47] The move from the context of knowledge production to its context of use has already happened in a wide range of disciplinary fields. The history of popular culture, the theory of education, the history of reading, the history of technology, and science communication studies have all contributed, through their focus on the non-elite members of society, the student, the reader and the user. And by giving communication processes a central place through their conceptualization as interactive and multidirectional phenomena taking into account both ends of the communicative process line. See for example Chartier, R. (1984). "Culture as Appropriation: Popular Cultural Uses in Early Modern France". In Kaplan, S. L., ed. *Understanding Popular Culture.* Berlin: Mouton Publishers, pp. 229-53; Gil-Pérez, et al. (2002). "Defending constructivism in science education". *Science & Education* 11: 557-71; Darnton, R. (1990). "First Steps Towards a History of Reading". In *The kiss of Lamourette: Reflections in Cultural History.* London: Faber & Faber, pp. 154-87; Oudshoorn, N., and Pinch, T. (2003). "Introduction: How Users and Non-Users Matter". In Oudshoorn and Pinch, eds. *How Users Matter: The Co-Construction of Users and Technologies.* Cambridge, Mass.: The MIT Press, pp. 1-25; Hiltgarner, S. (1990). "The Dominant View of

Popularization: Conceptual Problems, Political Uses". *Social Studies of Science* 20: 519-39.
[48] Secord. "Knowledge in Transit", pp. 660-7.
[49] Topham. "Rethinking the History of Science Popularization/Popular Science".
[50] Secord. "Knowledge in Transit", p. 656.
[51] See Gavroglu, K., et al. (2008). "Science and Technology in the European Periphery. Some Historiographical Reflections". *History of Science* 46: 1-23; Dolby, R. G. A. (1977). "The Transmission of Science". *History of Science* 15: 1-43. We would like to thank José R. Bertomeu and Agustí Nieto-Galan for allowing us reading a copy of the STEP paper before its publication.
[52] Chartier, R. "Culture as Appropriation", p. 233, and (1995). "Popular Appropriation: The Readers and Their Books". In *Forms and Meanings: Texts, Performances, and Audiences from Codex to Computer*. Philadelphia: University of Pennsylvania Press, pp. 83-97, on pp. 88-9.
[53] Sabra, A. I. (1987). "The Appropriation and Subsequent Naturalization of Greek Science in Medieval Islam: A Preliminary Statement". *History of Science* 25: 223-43, on pp. 227-8.
[54] Gavroglu. "Science and Technology in the European Periphery".
[55] See Palladino P., and Worboys, M. (1993). "Science and Imperialism". *Isis* 84 (1): 91-102
[56] Espagne, M., and Werner, M. (1987). "La construction d'une référence culturelle allemande en France, genèse et histoire". *Annales ESC* 42 (4): 969-92; Espagne, M. (1999). *Les transferts culturels franco-allemands*. Paris: PUF.
[57] Secord. "Knowledge in Transit", p. 669.
[58] Jacob, M. "Science Studies after Social Construction", pp. 96, 103-16. See also Lloyd, G. E. R. (1997). "The Comparative History of Pre-Modern Science: The Pitfalls and the Prizes". *Studies in History and Philosophy of Science* 28 (2): 363–68.
[59] Löwy. "The Social History of Medicine", pp. 466-7.
[60] Pyenson, L. (2002). "Comparative History of Science". *History of Science* 40: 1-33, on pp. 9-12.
[61] Morrell, J. B. (1972). "The Chemist Breeders: The Research Schools of Liebig and Thomas Thomson". *Ambix* 19: 1-46; Geison, G. L., and Holmes, F. L., eds. (1993). "Research Schools. Historical Reappraisals". *Osiris* 8; Hughes, T. P. (1983). *Networks of Power: Electrification in Western Society*, 1880-1930. Baltimore: The Johns Hopkins Press; Needham, J. (1979). *The Grand Titration: Science and Society in East and West*. Buffalo, NY: University of Toronto Press.
[62] Bloch, M. (1954). *The Historian's Craft*. Manchester: Manchester University Press.
[63] Sewell, W. H. (1967). "Marc Bloch and the Logic of Comparative History". *History and Theory* 6 (2): 208-18.
[64] See Detienne, M. (2002). "L'art de construire des comparables. Entre historiens et anthropologues". *Critique Internationale* 14 (1): 68-78; Saussy, H. (2006).

"Exquisite Cadavers Stitched from Fresh Nightmares: Of Memes, Hives, and Selfish Genes". In Saussy, ed. *Comparative Literature in an Age of Globalization.* Baltimore: Johns Hopkins University Press, pp. 3-42; Thomas, R. M. (1990). "The Nature of Comparative Education: How and Why are Education Systems Compared". In Thomas, ed. *International Comparative Education: Practices, Issues & Prospects.* Oxford: Pergamon Press, pp. 1-21.

[65] See for instance Fox. "Fashioning the Discipline", Pickstone. "Medical History as a Way of Life", and Rudwick, M.; Coleman, W.; Sylla, E., and Daston, L. (1981). "Review: Critical Problems in the History of Science". *Isis* 72 (2): 267-83, pp. 268-71. Excellent recent examples of this are Secord. *Victorian Sensation;* Topham, J. R. (2000). "Scientific Publishing and the Reading of Science in Nineteenth-Century Britain: A Historiographical Survey and Guide to Sources". *Studies in History and Philosophy of Science* 31 (4): 559-612; Rudolph, J. L. (2002). *Scientists in the Classroom: the Cold War Reconstruction of American Science Education.* New York: Palgave, and Chang, H. (2004). *Inventing Temperature: Measurement and Scientific Progress.* New York: Oxford University Press.

[66] Werner, M. and Zimmermann, B. (2006). "Beyond Comparison: Histoire Croisée and the Challenge of Reflexivity". *History and Theory* 45 (February): 30-50, pp. 35-6.

[67] See for example Glick, T. F. and Pi-Sunyer, O. (1969). "Acculturation as an Explanatory Concept in Spanih History". *Comparative Studies in Society and History* 11 (2): 136-54; Raj, K. (2007). *Relocating Modern Science: Circulation and the Construction of Scientific Knowledge in South Asia and Europe,1650-1900.* Basingstoke: Palgrave Macmillan, pp. 10-4; Buklijas, T. and Lafferton, E. (2007). "Science, medicine and nationalism in the Habsburg Empire from the 1840s to 1918". *Studies in History and Philosophy of Biological and Biomedical Sciences* 38: 679-86, pp. 679-81.

[68] Gross, R. A. (2000). "The Transnational Turn: Rediscovering American Studies in a Wider World". *Journal of American Studies* 34 (3): 373-93; Iriye, A. (2004). "Transnational History". *Contemporary European History* 13 (2): 211-22; Seigel, M. (2005). "Beyond Compare: Comparative Method after the Transnational Turn". *Radical History Review* 91 (Winter): 62-90; Curthoys, A. and Lake, M. (2005). *Connected Worlds: History in Transnational Perspective.* Canberra: Australian National University E Press; Werner and Zimmermann. "Beyond Comparison"; Boettcher, S. et al., eds. (2006). "Forum: Transnationalism". *H-German* January. [www.h-net.org/~german/discuss/Trans/ forum_trans_index]; Espagne, M. and Middell, M. eds. *Geschichte.Transnational.* [http://geschichte-transnational.clio-online.net]; Vleuten, E. van d. and Kaijser, A. (2006). "Prologue and Introduction: Transnational Networks and the Shaping of Contemporary Europe". In Vleuten and Kaijser, eds. *Networking Europe: Transnational Infrastructures and the Shaping of Europe, 1850-2000.* Science History Publications: Sagamore Beach, pp. 1-22.

[69] The proceedings of this conference have also been published. See Herran, N.; Simon, J.; Guillem-Llobat, X.; Lanuza Navarro, T.; Ruiz Castell, P., and Navarro, J. (2008). *Synergia: Jóvenes Investigadores en Historia de la Cienca*. Madrid: CSIC.

[70] See Kaiser. "Training and the Generalist's Vision".

[71] Harris. "Thinking Locally, Acting Globally", p. 134.

PART I

GEOGRAPHIES OF KNOWLEDGE
IN EARLY MODERN EUROPE

GEOGRAPHIES OF KNOWLEDGE IN EARLY MODERN EUROPE

TAYRA LANUZA-NAVARRO

The study of early modern science has experienced a deep transformation since the disciplinary consolidation of the history of science. The changes have not only affected the objects under scrutiny, but also the methods on which historians rely. The object was once absolutely clear "science", defined as universal rationality, independent of any social, cultural and geographical contingency.[1] According to this tradition, science had been conceptually and methodologically established by European "scientists" during the period of the Scientific Revolution. The method was therefore clear, for the activities and knowledge of seventeenth-century natural philosophers had to be studied as the origins of modern science. The Scientific Revolution thus became a central piece in the early development of history of science as a discipline, and for decades was the core element articulating and integrating history of science through a synthetic narrative, and in parallel, helping in the construction of Western cultural identity.

These assumptions have undergone a systematic reassessment in the last two decades. There has been an acknowledgement of the necessity of considering science in its geographical and socio-cultural context, taking into account elements such as patronage and other social, cultural, economic and political factors. Works such as Steven Shapin and Simon Schaffer's *Leviathan and the Air Pump* and Mario Biagioli's *Galileo Courtier* have decisively contributed to illuminate the social construction of science and its practice. Historians have also changed their attitudes towards the definition of science, emphasizing the importance of understanding early modern knowledge according to early modern criteria, questioning the applicability of modern definitions of science to this historical context, and further beyond that, abandoning the concept of science itself as a myth.[2] Thus, the term Scientific Revolution is debated, "revolution" being denounced as inadequate due to the emphasis on intellectual continuity (although a continuity free of the charge of progressive knowledge that it had in the past), and "scientific" being a label under dispute.

A still more recent concern is the obscuring of the macroscopic favoured by some aspects of social constructivism, and the failure of certain perspectives of early modern science to integrate historical accounts in larger frames.[3] While contributing to deconstruct traditional synthetic approaches, these debates have fostered historical reflection on the role that microscopic and macroscopic analysis should play in our narratives. A major challenge in this sense consists of defining the map of scientific knowledge and scientific practice in Europe, taking into account both their local and regional contingencies, and their integration in larger geographical scales through circulation of knowledge and its human agents.[4] As highlighted in the introduction to this volume, the study of communication and comparative history are two complementary ways of tackling this problem. This chapter aims to contribute to this debate through its focus on the structure of scientific knowledge and its practice in early modern Europe. In particular, we will investigate the place that mixed mathematics[5] and within these, astrology, had in the changing map of knowledge configured through their practice in France, the Iberian Peninsula and the German lands.

The case of astrology is particularly relevant. For decades it was relegated by historians of science to obscurity and considered foreign to a supposed early modern movement towards enlightened modernity, or deemed irrelevant in comparison to other disciplines such as mathematics and natural philosophy; it is still a historical object in need of further research.[6] As shown in the papers of this part, astrology was part of the mathematical disciplines that contributed to challenge the dominance of Aristotelian physics in the early modern map of knowledge. And, as argued by Patrick Boner, it played, for instance, a major role in the archetypical work of Kepler's fusion of mathematics and natural philosophy. Furthermore, as shown by Tayra Lanuza, astrology was not only part of the life of the Renaissance court but was also a subject taught in institutions designed for practical matters in relation to the control of the Spanish empire.

In his general survey of the key debates in the philosophy of science, Peter Godfrey-Smith described how people, events and theories from the period of the Scientific Revolution had carried a special weight in the discussions on the nature of science.[7] But related discussions on what is and what is not included in science after the nineteenth-century rebuilding of the term are distinct from studying what was considered as legitimate knowledge in the sixteenth and seventeenth centuries.

The evolution of several historiographies towards the study of knowledge in its own context, including debates on methodology, led historians of science to acknowledge that what was "science" in the early modern period had to be understood according to the models of rationality in force at that time.[8] Thus historians not only reflect on the definition of science and on what should be the object of its history,[9] but also study the structure of knowledge as understood by historical actors.[10] The study of the map of knowledge implies going beyond the contents of knowledge to study its context, organizational principles and, in particular, the historical processes shaping the boundaries between disciplines.[11] In this context, astrology plays an important role in historiographical debates about the map of knowledge in the early modern period. All this is now so clear that Margaret Jacob has joked that "none of the anxiety about what constituted true science in early modern Europe – so commonplace in the 1960s and directly correlated with internalist concerns – finds a place in the Cambridge History of Science volume devoted to early modern science".[12]

Leaving aside these old concerns, contemporary historians of science are more interested in acquiring a better understanding of how early modern people gained knowledge of the natural world, and what kind of map of knowledge they constructed. During the sixteenth and the seventeenth centuries, geometry and arithmetic were designated as "pure mathematics", while "mixed mathematics" were the disciplines characterized by the use of the tools provided by geometry and arithmetic in the study of nature. The term "mixed mathematics" came into use during this period,[13] and these changing denominations were part of a great and ongoing transformation in the definition of mathematics and the boundaries of its constituent disciplines. Change was due to several factors, including the famous dispute on the certainty of mathematics, the discussions on its value as acceptable knowledge, and the need, perceived by early modern scholars, for modification of the medieval Aristotelian classification of knowledge.[14] According to Nicholas Jardine, the challenge that mathematicians of this period posed to the previous structure of knowledge is illustrated by Galileo's *Dialogue* transgression of the accepted boundaries between natural philosophy and mixed mathematics.[15]

Mixed mathematics were considered as "subordinated" to the higher theoretical sciences (metaphysics, physics and pure mathematics), until a new polemic on the structure of knowledge and on the place of mathematics emerged between the late fifteenth and the sixteenth century.[16] In the Aristotelian tradition, resulting from previous scholastic disputes, these disciplines were referred to as the "more physical of the mathematical sciences", that is, mathematics concerned with physical problems, including optics, harmonics and astronomy.[17] In spite of the traditional distinction between astronomy and astrology proceeding from ancient works, these terms were for a long period interchangeable.[18] As emphasized by Eugenio Garin there was no clear separation in the Renaissance, between astronomy and astrology; [19] Astronomy and astrology were complementary, and the studies of the movements of the stars and of their effects were just the theoretical and practical parts of the same science.[20] As an example, in the Spanish kingdoms the term astronomy was not often used, and the term astrology lacked a univocal meaning.[21] Thus, in the different classifications of knowledge developed by mathematicians and natural philosophers during the sixteenth and seventeenth centuries, both terms appeared to refer to both aspects. As already mentioned, astrology was part of "mixed mathematics", a set of subjects whose critical emergence in this period is clear in the maps of knowledge devised in different parts of Europe by thinkers of the status of Christoph Clavius, Francis Bacon and Marin Mersenne.

Christoph Clavius (1538-1612), professor of mathematics at the *Collegio Romano* – the Jesuits' central college in Rome – was a reference for those arguing against the scholastic tradition that asserted that mixed mathematics did not produce genuine scientific knowledge.[22] In his description of how to promote mathematical disciplines in the studies of the Jesuit centres of learning, he stated that:

> Physics cannot be understood correctly without [the mathematical disciplines], especially what pertains to that part concerning the number and motion of the celestial orbs, of the multitude of intelligences, of the effect of the stars, which depend on the various conjunctions, oppositions, and other distances between them, of the division of continuous quantities to infinity, of the tides, of the winds, of comets, the rainbow, halos, and other meteorological matters...[23]

Clavius thus insisted that mathematics were indispensable for understanding physical phenomena, putting particular emphasis on the importance of astronomy, astrology and meteorology. His intellectual and political influence in the Jesuit order secured their plan of studies of 1585, which included a eulogy of mathematics, referring to its connections with various disciplines and professions, including poetry, history, politics, metaphysics, theology, law, farming and medicine.[24]

In his tree of knowledge, Francis Bacon (1561-1626) divided mathematics into pure and mixed. For him mixed mathematics included Aristotelian optics, harmonics and astronomy, but also cosmography, architecture and engineering.[25] On the other hand, Marin Mersenne (1588-

1648) explained in his *Traité de l'Harmonie Universelle* that the study of mixed mathematics included music, cosmography and astronomy, optics, mechanics and astrology.[26]

Polemics on the epistemological status, the wide scope of the disciplines included under its label and their critical role in the shaping of the Renaissance social, cultural and political world placed mixed mathematics in a privileged place in the cultural life in this period. The three articles constituting this part introduce each of them a different political and cultural context in which the early modern map of knowledge and the place of mixed mathematics were shaped. This part is introduced by a general overview in Daniele Cozzoli's paper to the role played by mathematics in the critical changes affecting the early modern map of knowledge between the sixteenth and the seventeenth centuries. Accordingly, Cozzoli provides a general account of how these changes resulted in a redefinition of the relations and relative status of mathematics and physics, and the subdisciplines composing them, and how the work of different types of practitioners contributed to reshape them into areas of knowledge and practice with new meanings.

The three papers in this part analyze particular case studies tackling the reshaping of the map of knowledge in this period by focusing on the study of the work of different authors. Cozzoli analyzes the different factors that shaped the translation of Descartes' *Meditations* from Latin into French with the intention of replacing Aristotelian philosophy and its associated map of knowledge. Patrick Boner shows how Kepler's defence of astrology was embedded in changes in the map of knowledge that contributed to the rise of mathematics as a discipline, in the context of the German universities and courts in which he was educated and worked. Tayra Lanuza devotes her paper to the study of the place of astrology and its practice in three Spanish institutions of learning beyond the universities, where this discipline appeared within the teaching of mixed mathematics.

All three papers show how the pedagogical context was fundamental for the successful communication and establishment of a new map of knowledge. The European network of Jesuit colleges, that of the German universities and courts, and the context of Spanish institutions for the training of mathematicians directed towards solving practical problems, were critical spaces where disciplines and their hierarchical relations were redefined. In all three cases the agency of this communication was not only intellectual (Kepler's project of creating a purely mathematical astrology or Descartes' use of the terms "a priori" and "a posteriori"), but also political (for example the Spanish empire in its geographical and political expansion), human (the circulation of mathematicians through the Jesuit network and the German universities network of universities and courts) and material (through books). The translation of works, such as that of Descartes, was just a particular case of processes of appropriation, in which knowledge and its taxonomical divisions were transformed in relation to intellectual, philosophical, political, pedagogical, linguistic and practical needs, all of which contributed to the rise of mathematics as a method and a system of knowledge allowing to explain life and nature.

Notes

[1] Lindberg, D. C. (1990). "Conceptions of the Scientific Revolution from Bacon to Butterfield: A Preliminary Sketch". In Lindberg, D. C., and Westman, R. S., eds. *Reappraisals of the Scientific Revolution*. Cambridge: Cambridge University Press, pp. 1-26; Shapin, S. (1996). *The Scientific Revolution*. Chicago: University of Chicago Press, and (1994). *A Social History of Truth: Civility and Science in Seventeenth-Century England*. Chicago: University of Chicago Press; Jacob, M. C. (1999). "Science Studies after Social Construction: The Turn toward the Comparative and the Global". In Bonnell, V. E., and Hunt, L., eds. *Beyond the Cultural Turn: New Directions in the Study of Society and Culture*. Berkeley: University of California Press, pp. 95-120, on pp. 101-9; Osler, M. J. (2000). *Rethinking the Scientific Revolution*. Cambridge: Cambridge University Press, pp. 3-4; Hellyer, M. (2003). *The Scientific Revolution. The Essential Readings*. Malden, Mass.: Blackwell Publishing; Livingstone, D. N. (2003). *Putting Science in its Place: Geographies of Scientific Knowledge*. Chicago: The University of Chicago Press, p. 90; Secord, J. A. (2004). "Knowledge in Transit". *Isis* 95(4): 654-72, pp. 655-6; Findlen, P. (2005). "The Two Cultures of Scholarship". *Isis* 96: 230-7, pp. 231-2.
[2] Ibid., and Livingstone, D. N., and Withers, C. W. J., eds. (2005). *Geography and Revolution*. Chicago: University of Chicago Press; Dear, P. (2005). "What Is the History of Science the History Of? Early Modern Roots of the Ideology of Modern Science". *Isis* 96 (3): 390-406; Park, K. and Daston, L. (2006). "Introduction: The Age of the New". In Park, and Daston, eds. *The Cambridge History of Science. Volume 3: Early Modern Science*. Cambridge: Cambridge University Press, pp. 1-20, on pp. 15-6.
[3] Secord. "Knowledge in Transit".
[4] Jacob. "Science Studies after Social Construction", pp. 101-16.
[5] See Brown, G. (1991). "The Evolution of the Term 'mixed mathematics'". *Journal of the History of Ideas*, 52(1): 81-102.
[6] Jacob. "Science Studies after Social Construction", pp. 108-9, and (2007). "The Importance of Early Modern European Science and the State of the Field", *Isis*, 98: 361-5.

[7] Godfrey-Smith, P. (2003). *Theory and Reality: An Introduction to the Philosophy of Science.* Chicago: University Of Chicago Press, p. 13

[8] An example for the case of astrology is found in the work of Nunes Carolino, L. M. (2003) *Ciência, astrologia e sociedade: a teoria da influência celeste em Portugal, 1593-1755.* Lisboa: Fundaçao Calouste Gulbenkian.

[9] Dear. "What Is the History of Science the History Of?".

[10] See Kelley, D. R. ed. (1997). *History and the Disciplines: the Reclassification of Knowledge in Early Modern Europe.* Rochester: University of Rochester Press, and Kelley, D. R., and Popkin, R. (1991). *The Shapes of Knowledge from the Renaissance to the Enlightenment.* Dordrecht: Kluwer Academic Publishers.

[11] See Feldhay, R. (1993). [Review of *The Shapes of Knowledge* by D. R. Kelley and R. Popkin, eds.]. *Isis* 84 (4): 793-5.

[12] Jacob. "The Importance of Early Modern European Science".

[13] Brown. "The Evolution of the Term 'mixed mathematics'".

[14] See, for instance, Heikki, M. (2001). "The Aristotelian Classification of Knowledge in the Early Sixteenth Century". In Pade, M. ed. *Renaissance Readings of the Corpus Aristotelicum.* Copenhagen: Museum Tusculanum Press, pp. 103-27; Weisheipl, J. A. (1978). "The Nature, Scope, and Classification of the Sciences". In Lindberg, D. ed. *Science in the Middle Ages.* Chicago: University of Chicago Press, pp. 461-82; Dear, P. (1995). *Discipline and Experience: The Mathematical Way in the Scientific Revolution.* Chicago: University of Chicago Press, and (2006). "The Meanings of Experience". In Daston and Park. *The Cambridge History of Science.*

[15] Jardine, N. (1991). "Demostration, Dialectic and Rhetoric in Galileo's *Dialogue*". In Kelley and Popkin. *The Shapes of Knowledge*, pp. 101-22.

[16] On the polemics on mathematics, see, among others, Jardine, N. (1988). "Epistemology of the sciences". In Schmitt, C.B; Skinner, Q., and Kessler, E. eds. *The Cambridge History of Renaissance Philosophy.* Cambridge: Cambridge University Press, pp. 685-712.

[17] On the discontinuity between the fourteenth century and the sixteenth century doctrines about the relations between mathematics and physics see Ariew, R. (1990). "Christopher Clavius and the Classification of the Sciences". *Synthese* 83(2): 293-300.

[18] A distinction that can be found, for instance, in Ptolemy's *Tetrabiblos,* in Isidore of Seville's *Etimologías*, in Roger Bacon's *Opus maius* and in Albertus Magnus' works.

[19] Garin, E. (1983). *Astrology in the Renaissance. The Zodiac of Life.* London: Routledge & Kegan Paul, p. 25.

[20] Capp, B. (1979). *Astrology and the Popular Press. English Almanacs, 1500-1800.* London: Brill, p. 16.

[21] López Piñero, J. M., dir., (2002). *Historia de la ciencia y de la técnica en la corona de Castilla. Vol. III: siglos XVI y XVII.* Salamanca: Junta de Castilla y León, p. 222.

[22] See Romano, A. (1999). *La contre-réforme mathématique: Constitution et diffusion d'une culture mathématique jésuite à la Renaissance, 1540-1640*. Rome: École française de Rome.

[23] Clavius, C. (1901). "Modus quo disciplinae mathematicae in scholis Societate possent promovere", in *Monumenta paedagogica Societatis Jesu quae primam rationem studiorum anno 1586 praecessere*. Matriti: A. Avrial, quoted and translated by Ariew, R. (1990). "Christopher Clavius", p. 298.

[24] Smolarski, D.C. (2002). "The Jesuit *Ratio Studiorum*, Christopher Clavius, and the Study of Mathematical Sciences in Universities". *Science in Context* 15 (2): 447-57.

[25] Bacon, Francis (1623). *De Dignitate et Augmentis Scientiarum*, Book 1, ch. VI. In Spedding, J.; Ellis, R.L., and Heath, D.D. eds. (1963-1989). *The Works of Francis Bacon*. Stuttgart: Holzboog, p. 578.

[26] M. Mersenne (2003). *Traité de l'harmonie universelle*. Paris: Fayard, [originally published in 1627], pp. 38-9.

BEYOND MIXED MATHEMATICS:
HOW A TRANSLATION CHANGED THE STORY
OF DESCARTES' PHILOSOPHY OF MATHEMATICS

DANIELE COZZOLI

This paper uses the study of the translation of René Descartes' *Meditationes de Prima Philosophia* from Latin into French as a way to map the rapidly changing relationships between disciplines in the mid-seventeenth century. Descartes' work is particularly relevant in unveiling these changes, since it led the rise of mathematics against Aristotelian physics, and therefore has proportional importance in defining the main characteristics of seventeenth-century scientific investigation.

As we will see, the study of such a minutious question as Descartes' use of the expressions *a priori* and *a posteriori*, and their translation into French, provides an overview of the Renaissance map of knowledge and its hierarchical structure. This study reveals a fundamental linguistic tension between Latin and French that echoes the tension between tradition and the new outlook introduced by the Scientific Revolution. Furthermore, it assigns a leading role to translation and translators. The French translation of Descartes had a fundamental role in changing the *Wirkunggeschichte* of Descartes' theory of proof in subsequent centuries.

However, in Descartes' time this translation had a clear intention that put together the interests of the author, the translator and the publisher of this work: reaching a wider public by replacing Aristotelian philosophy in schools and universities. This aim, together with the rapid changes affecting the seventeenth-century map of knowledge, and the concomitant development of French as a literary language, led the translator, the publisher and the author of the *Meditations* to appropriate the language of Aristotelian philosophy for their own purposes. Paradoxically, their use of the Aristotelian language and conceptual framework was only a strategy, aimed at favouring the communication of Descartes' philosophy to a readership familiar with Aristotelian philosophy, in order to eventually replace it with Descartes' new philosophical and scientific frame.

This paper thus links the study of discipline configuration and knowledge taxonomies with the study of translation and appropriation. In

this sense it problematizes the famous statement commonly attributed to
Giordano Bruno, and widely communicated by early-modern translators in
order to strengthen the social support of their craft: "from translation all
science has its offspring".[1] Furthermore, it shows the complexity of the
configuration and transmission of knowledge through the use of the
concept of appropriation, as proposed by Roger Chartier and Abdelhamid
Sabra.[2]

In the first part of this paper I will provide an overview of the
seventeenth-century map of knowledge and the rapid changes affecting it.
I will argue that Descartes' work had a fundamental role in this process
and thus, that its study offers the opportunity to grasp the complexities of
the changing relationships between disciplines in this period. Accordingly,
in the second part of this paper I will introduce Descartes' outlook through
a study of his work and in particular, of his *Metaphysical Meditations*.
Subsequently, I will contend that translation, translators, publishers and
the educational context had an active role in redefining the seventeenth-
century map of knowledge. Furthermore, I will analyse the consequences
that the special characteristics of the French translation of Descartes'
Meditations had in the cultural context of his time and in subsequent
scholarship.

The seventeenth-century map of knowledge
and its changes

One of the main features of the outlook produced by the seventeenth-
century Scientific Revolution was the new role played by mathematics
within the investigation of natural phenomena. In the sixteenth century, the
revival of Platonism had placed mathematics at the centre of a web
consisting of philosophy, cosmology, astronomy and optics, but also of
wizardry, astrology and music. The influence of Platonism on the making
of Copernicus' and Kepler's astronomy has been admitted for a long time.[3]
Lindberg has emphasized the influence of authors such as Cardano,
Scaliger, Patrizi and Della Porta, on the development of Johannes Kepler's
optics.[4] But the golden century of mathematics was the seventeenth
century, which witnessed a radical change in the cultural place of
mathematics within the system of knowledge. In this period, the classical
distinction between pure mathematics (geometry and arithmetic) and
mixed mathematics (astronomy, astrology, mechanics, music, optics,
hydrostatics, pneumatics) was redefined. Astrology and astronomy became
two separate disciplines. Until the sixteenth century, in fact, authors
considered astrology and astronomy synonymous. For instance,

Bernardino Baldi, in his *Le Vite de' matematici,* introduced Copernicus' life in the following way:

> His birth was studied by Francesco Giuntino in Ptolemy's treatise which is entitled *On Stars Prognostics.* Giuntino claims that Copernicus had Mercury at the beginning of Aries, and the Moon placed in a trigonic position in the seventh degree of Sagittarius. These facts, he says, determined the excellence of Copernicus' intelligence. Giuntino also claims that this fact can be discovered from Copernicus' extant mathematical works, because he was a very talented man in astrology, and, as he calls himself, the German Ptolemy.[5]

The quarrel about astrology dated back to the fourteenth century. As pointed out by Eugenio Garin, the humanistic struggle against astrology was part of a more general quarrel with Renaissance natural philosophy and science.[6] Toward the end of the sixteenth century Kepler, whose mathematical studies were deeply rooted in a Platonic, Hermetic view, showed his perplexity on the value of judiciary astrology.[7] However, only at the end of the seventeenth century did the rise of the new mechanical philosophy force a gap between astronomy and astrology. Thanks to the recovery of Archimedes and the work of Renaissance engineers and mathematicians, mechanics ceased to be the art of practitioners and became part of physics. Galileo's *Mechanics* (1593) overcame the Greek distinction between what belonged to the natural ($\phi\upsilon\sigma\iota\varsigma$) and what belonged to the artificial ($\mu\eta\chi\alpha\nu\eta$).[8]

The mechanization of the world picture marked a distance between mathematics and the neo-Platonic tradition, which was permeated with magical, astrological and numerological elements. Kepler's theory of light was also a product of the same neo-Platonic Renaissance tradition, which had produced the work of Francesco Patrizi and Giovanni Battista Della Porta.[9] By contrast, Descartes' optics was devoid of these elements.

Relations between the disciplines, which aimed at explaining natural phenomena, changed dramatically. Natural philosophy abandoned the qualitative explanations of Aristotle's physics in favour of quantitative explanations of the new mechanical philosophy. As Galileo claimed, the book of nature was written in mathematical language. The seventeenth century witnessed the creation of a gap between logic and mathematics and a much closer relation between mathematics and physics.[10] Mathematics slowly replaced Aristotle's logic, which had been developed by the Aristotelians of the School of Padua[11] toward a method of scientific investigation, the so-called *regressus* theory, as the key for understanding nature. By contrast, geometry became the paradigm of certainty for any discipline, including philosophy.[12]

The seventeenth century was thus a crucial period for the redefinition of the relations between mathematics, and physics (as defined by Aristotle's work). The work of René Descartes is particularly illustrative of these rapid changes. In 1641, he published a compendium of his new physics, whose aim was to replace scholastic text-books based on Aristotle's physics. Descartes' work was entitled *Principia philosophiae,* where *philosophia* meant natural philosophy. However, he himself referred to the content of his book as "my physics",[13] referring either to his natural philosophy or his metaphysics (or both).[14]

Renaissance Aristotelians [15] usually followed Thomas Aquinas' hierarchy of knowledge. Accordingly, metaphysics was first, followed by physics and mathematics, in that order. Mathematics encompassed pure mathematics, arithmetic and geometry, and mixed mathematics was composed of astronomy (which included astrology), mechanics, optics, music (which included acoustics), pneumatics, hydrostatics and measuring and weighing techniques. Authors discussed the place of mathematics within the system of knowledge. In particular, they focused on relations between mathematics and physics. A number of Aristotelians, who were influenced by Averroes, were inclined to think that physics was more important than mathematics, for the latter derived its principles from the former. In 1542, Ludovico Boccadiferro, celebrated lecturer of natural philosophy at Bologna, [16] stated the superiority of physics over mathematics, stressing Aristotle's idea that geometry derived the idea of the continuum from physics.[17]

In the fifteenth and sixteenth century, mathematicians were mainly concerned with the recovery of Greek mathematics. As explained by Mario Biagioli,[18] the work of recovery of ancient Greek mathematics was also a product of the "courtier gestalt" of its members. By contrast, late sixteenth and seventeenth century mathematicians not only worked on reconstructing Greek mathematics, but also introduced new mathematical theories. Viète and Descartes redefined the relationship between algebra and geometry, Fermat gave important contributions to number theory, Leibniz and Newton introduced calculus, Pascal and Fermat established the modern understanding of probability, Desargues invented projective geometry. In one of his letters, Descartes recalled that, for the first time in history, he overcame the Greeks, for he solved a question they were not able to solve, the so-called Pappus' problem.[19]

Descartes' map of knowledge and his use of *a priori* and *a posteriori* in the *Meditations*

Descartes was a major actor in the events leading to the changing relationship between mathematics and physics. In the *Discourse on Method*, he claimed that only mathematicians could find proofs, which were defined by him as certain and unquestionable reasons.[20] Descartes thought that any investigation needed a method, a set of rules, which might aid in solving any given problem. However, the *regressus* of the Aristotelians turned out to be useless, for, according to the French philosopher, it was an instrument apt only for explaining what had already been discovered. Logic did not possess any heuristic power, but had a merely explanatory function.[21] Mathematics, by contrast, played a crucial role in the quest for method. Indeed, both in the *Rules* and in the *Discourse,* Descartes explained that his method of discovery had been inspired by ancient geometrical analysis and by modern algebra.[22] In the *Geometry,* he described the method of solving geometrical problems as a form of analysis combining modern algebra with ancient analysis.[23]

Mathematics influenced not only Descartes' view concerning scientific investigation but also his philosophical enquiry. On July 31st 1640, in a letter to Costantin Huygens, he compared his reasoning in the *Metaphysical Meditations* to Apollonius' long proofs.[24] Furthermore in his *Meditations*, he compared his reasoning to ancient analysis. In fact, in replying to one of the objections collected by Father Mersenne, in the second set of objections to the *Metaphysical Meditations,* he claimed to have used only ancient analysis in order to prove the existence of God and the immortality of the soul. According to Descartes, the *via analytica* of the ancient geometers was a better way to teach philosophy, for it showed the path of discovery.[25] The author of the objection, who was very likely Mersenne himself,[26] argued that Descartes should have followed the *more geometrico* instead of the unusual mode he had used. In order to persuade the reader of the superiority of analysis, Descartes described in some detail the modes of proof of analysis and synthesis, and stated that analysis was *a priori,* namely "from what is prior" and that synthesis was *a posteriori,* namely "from what is posterior".[27]

In 1647, Claude Clerselier and the young Duke of Luynes translated Descartes' *Meditations* into French. Six years earlier, Descartes had chosen to compose the *Meditations* in Latin because he wanted only a small number of theologians to be able to read it.[28] On the one hand, Descartes remembered the criticisms he received to his 1637 *Essais,* which had been written in French. On the other hand, he aimed to convince theologians that his metaphysics could be a better support for theology than Aristotle's philosophy. For this reason, the *Meditations* were printed in Latin, jointly with a set of objections which had been composed by a selected number of theologians and philosophers and with Descartes' own replies to these objections. When he agreed to the translation of his *Meditations* into French, he probably realized that Latin had not prevented critics from attacking him. Most of the theologians, in fact, were not convinced by Descartes. Furthermore, a French translation would make his work available to a broader public.

It is worth noting that in the second half of the seventeenth century, Aristotelianism did not lose out to Descartes' philosophy in academic teaching.[29] By contrast, Cartesianism enjoyed great success among nobles and courtiers. In 1690, Father Gabriel Daniel, a Jesuit who was the author of the *Voyage autour du monde de Descartes,*[30] complained that most of the books which had been recently published followed Descartes' philosophy.[31] Apparently above all ladies appreciated Descartes' philosophy.[32]

The French translation differed from Descartes' original Latin text in various respects. Many passages were more clear and elegant. Yet a number of concepts were translated in such a way that their meaning appeared to have changed. The two expressions of *a priori* and *a posteriori,* for instance, had been replaced by an account in terms of causes and effects. Indeed, we read that analysis "shows how effects depend upon their causes"[33] and that synthesis proceeds "as if examining causes by means of their effects".[34]

A present day reader will find the French translation's account in causal terms even more disconcerting than that found in the original Latin text. Today, mathematical logic textbooks provide us with Alfred Tarski's model-theoretic definition of the relation of logical consequence. This is the relation between the premises and the conclusion of an inference. This definition is generally along the lines of: a statement A is a logical consequence of a set of statements G, if every interpretation making G true, also makes A true. In *On the Concept of Logical Consequence* (1936), Tarski held that the relation of logical consequence possessed the property of being *a priori,* where by *a priori* he meant "not depending on experience".[35]

A seventeenth-century reader, however, would not have been puzzled by the French translation. The translation explained the matter by means of the Aristotelian notion of logical consequence. According to Aristotle, the relation of logical consequence is a causal relation between the premises and the conclusion. Clerselier, the French translator of the *Obiectiones,* identified Descartes' analysis with the Aristotelians' *demonstratio quia* and Descartes' synthesis with their *demonstratio propter quid*.

Historians have not questioned this identification. However, they have found it difficult to explain the meaning of the passage, because usually Aristotelians described the *demonstratio quia* as a "proof going from the effect to the cause" and the *demonstratio propter quid* as a "proof going from the cause to the effect". This is the opposite of what we read in Descartes' text.

Ferdinand Alquié has argued that the French translation makes the text incomprehensible and that one should understand *a priori* and *a posteriori* in a broad sense.[36] Stephen Gaukroger explains that in the order of knowing analysis can be *a priori* and synthesis *a posteriori*, because one can know effects prior to the causes. Gaukroger stresses Descartes' use of *tanquam* (as if). Descartes in fact wrote that analysis was *tanquam a priori*, and synthesis was *tanquam a posteriori*.[37] However, Gaukroger is not entirely satisfied with this explanation and argues that it can be revised.[38] Benoît Timmermans believes that for Descartes analysis was *a priori*, because "instead of following (i.e. already knowing) order from principles to consequences, it worked out *a priori* an unknown order".[39] This interpretation is not very satisfactory, for it considers *a priori* and *a posteriori* as asymmetrical. *A posteriori* is in fact understood as "from what is posterior", but *a priori* is not understood as "from what is prior". The definition of *a priori* is not clear.

In this paper, I show that Descartes' notion of analysis and synthesis cannot be identified with Aristotle's *demonstratio quia* and *demonstratio propter quid*. I will argue that Descartes' use of terms, which are familiar to Aristotelians, did not entail that he endorsed Aristotle's view on proof. Descartes' use of *a priori* and *a posteriori* may be understood only by referring to his attempt of renovating philosophical language by means of mathematics. First, I will draw a historical and conceptual map of Renaissance Aristotelian's theory of proof and of the mathematical notion of analysis and synthesis in order to show their differences. I will argue that Descartes derived his view on what a proof should be only from the mathematical tradition. Secondly, I will analyse the French translation. Finally I will explain what Descartes possibly meant by *a priori* and *a posteriori*.

The Renaissance theory of *regressus*

Clerselier interpreted Descartes' notion of *a priori* in the light of the theory of proof of Renaissance Aristotelians. In this section, I will briefly reconstruct Renaissance theory of *regressus* in order to explain Clerselier's choice.

In the Renaissance, there was a well-established view of what a proof should be, derived from Aristotle's *Posterior Analytics*. Seventeenth-century Aristotelians made their own contribution to Aristotle's theory by developing the notion of *regressus*. In chapter thirteen of the second book of his *Posterior Analytics,* Aristotle described two different kinds of syllogism. According to Aristotle, the *syllogism of the reason why* is a form of logical reasoning which shows the cause of a stated fact, while the *syllogism of the stated fact* is a form of logical reasoning that shows only the existence of a stated fact.[40] Aristotle explained his idea through the following example. If we want to deduce that planets are near the earth from the fact that they do not twinkle, we can set forth the following syllogistic reasoning: planets do not twinkle and what does not twinkle is near the earth; therefore, planets are near the earth. For Aristotle, the former syllogism proves the existence of the stated fact, but it does not allow us to know its cause; in fact, it is not because planets do not twinkle that they are near, but rather because they are near that they do not twinkle.[41]

Let us consider the following syllogism: what is near the earth does not twinkle, and planets are near the earth; therefore, the planets do not twinkle.[42] According to Aristotle, this is a better form of reasoning because it shows the cause of the stated fact.[43] Aristotle thought that the most perfect form of reasoning was a causal chain of syllogisms, where the middle term of every syllogism was the proximate cause of the conclusion, and where premises were stated prior to the conclusion and were known prior to the conclusion.[44] For instance, Aristotle pointed out that Anacarsis' explanation that there were no flutists among the Scythians because there were no vines in their country, was flawed because it skipped an intermediate cause. He claimed that the right explanation was that since there was no wine, Scythians could not get drunk and they were, therefore, never in the mood to sing.[45]

As explained by Mario Mignucci,[46] Themistius and Averroes provided the two most important interpretations of Aristotle's theory of proof. The theory of *regressus* was derived from Themistius' interpretation of this Aristotle's two-fold notion of proof. In his paraphrase of Aristotle's *Posterior Analytics,* Themistius referred to Aristotle's *syllogism of the reason why* as a proof *by cause* and to Aristotle's *syllogism of the stated fact* as a proof *by sign*.[47] According to Themistius, a proof *by cause* was a proof going from causes to effects and a proof *by sign* was a proof going from effects to the causes.[48] Themistius' interpretation suggested a development of Aristotle's two-fold theory of proof towards a methodology of investigation. If we look, in fact, at this topic, keeping in mind Themistius' interpretation, we can say that collecting observational data is a process from effects to causes, and arranging collected data and hypotheses into an axiom-like structure is a process going from causes to effects.

Around 1472, Ermolao Barbaro, a diplomat who had learned Greek at Verona and studied arts at Padua, translated Themistius' commentary to Aristotle's *Posterior Analytics,* which was published in 1481. Barbaro promoted a philological methodology of textual emendation based upon an exacting collection of ancient manuscripts, which reflected the interest of partisans of the Venetian cultural revival.[49] In Barbaro's translation, Themistius' proof *by sign* was translated as *demonstratio a signo* and Themistius' proof *by cause* was translated as *demonstratio ex causa.*[50]

Averroes provided a different reading of Aristotle's theory of proof which influenced the Renaissance view on the topic. In the Latin translations of Averroes' works (1126-98), Aristotle's *syllogism of the reason why* is called either demonstratio *propter quid,* or *per causas* and *demonstratio quia.* Aristotle's *syllogism of the stated fact* called *demonstratio quod est, demonstratio signi* or *demonstratio inventionis.* Averroes interpreted the first as the reasoning showing the proximate cause and the second as the reasoning showing the remote cause.[51] Aristotle's characterization of the *syllogism of the reason why* as the reasoning showing the proximate cause suggested that the *syllogism of the stated fact* should show the remote cause.[52] Averroes also described the *demonstratio propter quid* as a proof from what is *natura notiora* (better known by nature) but *nobis latentiora* (farther from us) or *de posterioribus ad priora* (from what is posterior to what is prior) and the *demonstratio quod est* as a proof from what is *nobis notiora* (better known by us) to what is *natura latentiora* (farther from nature).

Although Themistius' and Averroes' interpretations could have some textual basis,[53] they were not compatible. This was due to the fact that the middle term expressed either the remote cause or the effect. In the Renaissance, in fact, there were Aristotelians who attempted to make the two understandings compatible. Pedro da Fonseca,[54] one of the Coimbran fathers[55] explained that a *demonstratio quod est* proceeded either from the effect or from the remote cause. Lunar phases could be explained either by the fact that the moon does not have a shadow or by the opposing positions of the sun and the moon on the ecliptic plane. The foundation of the former explanation was the effect. The latter explanation was based on the remote cause. This same explanation can be found in Eustachius a Santo Paulo.[56]

According to Themistius' understanding, the two proofs were sometimes combined in a methodology of investigation called *regressus*. This methodology was a two-stage procedure. In the first stage, causes are discovered from their effects. In the second stage, effects were deduced from the causes discovered in the first stage.[57] As Stephen Gaukroger points out, the circular nature of the *regressus* was only apparent, because the *regressus* was intended in a broad sense as a methodology. An inference from the effect to the cause was a sort of inference to the best explanation. The inference from the cause to the effect is an attempt to prove that the cause is appropriate to explain the particular phenomenon.[58]

In the sixteenth century, several Italian Aristotelians, such as Agostino Nifo, Vincenzo Maggi, Antonio Trombetta and Pietro Pomponazzi, endorsed the methodology of the *regressus*. The *regressus* was of paramount importance in the work of Giacomo Zabarella (1533-89). In his *De methodis*, Zabarella described the *resolutio* as a proof *a posteriori* and "from the effects to the causes" and the *compositio* as a proof *a priori* or "from the causes to the effects".[59]

In the sixteenth and seventeenth centuries, the *demonstratio propter quid* was also referred to as *compositio* and was defined variously as *a priori*, *ex causa*, and *ex causa proxima*. The *demonstratio quia* was also called *resolutio* and described as *a posteriori*, *ex effectu* and *ex causa remota*.[60] Eustachius a Santo Paulo considered *proximate causes* the synonym of *ex causa* and of *a priori* and *remote causes* the synonym of *ex effectu* and of *a posteriori*. For instance, Eustachius argued that the proof that the soul was the act of the natural organic body was *a posteriori* or *ex effectu*.[61] The interpretation of the Renaissance Aristotelians also affected non-Aristotelian authors. Pierre Gassendi, in the *Exercitationes Paradoxicae adversus Aristoteleos* (written in 1624, but published in 1658), identified proofs *a priori* with proofs *ex causa* and proofs *a posteriori* with proofs *ex effectu*.[62]

In summary, Renaissance Aristotelians called the *demonstratio propter quid* as *compositio* and as *a priori*, *ex causa*, and *ex causa proxima;* the *demonstratio quia* as *resolutio*, *a posteriori*, *ex effectu*, and *ex causa remota*. Slightly different notions were considered the same. Furthermore, Clerselier's translation of Descartes followed the general trend of interpreting *a priori* as "by the cause" and *a posteriori* as "by the effect". Moreover, the two Aristotelian proofs were also called *resolutio* and *compositio*. This fact would contribute to interpret Descartes' ideas in the light of Aristotle's thought.

Descartes certainly knew the Renaissance Aristotelians' theory of proof, because he had studied in the Jesuit college of La Flèche. At la Flèche, the teaching of logic consisted of reading Aristotle's *Organon* and Porfirius' *Isagoge*. The teacher usually read Aristotle through one of his Jesuit commentaries. To study Aristotle's logic, Fathers at La Flèche were advised to use either the textbook by Pedro da Fonseca or that by Francisco de Toledo. [63] In September 1640, Descartes recalled the Coimbrans, Francisco de Toledo and Antonio Rubio as the philosophers who had the greatest following among the Jesuits in the 1620s.[64] Descartes also demonstrated then knowledge of Eustachius' *Summa philosophica* and of Charles François d'Abra de Raconis' *Summa totius philosophiae*.[65]

Nevertheless, I will argue that by *a priori* and *a posteriori* Descartes did not mean what the Aristotelians meant. In order to prove these statements, we first need to look more closely at the notions of analysis and synthesis. In the next section, I will outline a conceptual map of the mathematical notions of analysis and synthesis in order to clarify the difference between them and the Aristotelians' theory of proof.

The synthesis and analysis of the ancient geometers

The *mos geometricus* or the synthesis was the mode of proving statements in Euclid's *Elements*. Given a number of preliminary definitions, the synthesis assumed some primitive statements, logical axioms and specific axioms for the given theory, whose truth was considered self-evident. A proof was then a finite succession of statements, any of which was either a primitive statement (i.e. an axiom or a postulate) or had been deduced from axioms and from statements deduced from axioms.

Seventeenth-century scholars found the concept of synthesis straightforward, as they could easily refer to Euclid's *Elements*. It followed that request of those who objected to Descartes' work was very clear: to expound the arguments within an axiomatic framework. By contrast, analysis was something quite obscure to a seventeenth-century scholar. This can be explained by the fact that, with the exception of very few passages in Aristotle and Proclus, analysis had been described extensively only in Pappus of Alexandria's *Mathematical Collection*, which was translated into Latin only in 1588.

A few examples of ancient analysis could be found in Euclid and Apollonius. Moreover, Simplicius quoted Eudemus of Rhodes' lost history of geometry, and attributed a number of analytical proofs to Hyppocrates of Chios (c. 430 B. C.). Hyppocrates' proofs regarded the squaring of the lunulae (i.e. the space between two circumferences having different radius but the same extremes).[66] His method was likely to have consisted in assuming what was to be proved as if it had already been proved, and then trying to go backward to the principles on which the statement depended. Using this procedure, some intermediate new theorems were established.[67] For instance, Hyppocrates proved that the area of the lunula which could be inscribed in a semi-circumference was equivalent to that of the triangle whose base was the diameter of the same circle and whose height was the radium itself. In order to prove this theorem, he assumed the problem, and then tried to reduce it to principles which were already known. The proof of this theorem also established the intermediate lemma stating the proportionality between areas of circles and squares whose basis is the diameter of the circle.

A full description of the method of analysis was contained in Pappus' *Mathematical Collection,* which described the so-called method of analysis-synthesis. As we shall see, the interpretation of Pappusian method is still debated, as is whether his proofs really follow it.[68] Because of this reason, I will not illustrate Pappus' method by means of an example, but will just report his own description. Pappus explained that his method was a technique that helped students solve problems once they had finished studying Euclid's *Elements.*[69] His method consisted of two stages. The first stage was a heuristic stage called analysis. The second was a demonstrative stage called synthesis. In the first place, one assumed that the problem had already been solved. Then one tried to go backward to the principles on which the problem depended. Michael Mahoney has emphasized that Pappus' analysis was conclusive only in negative cases. Indeed, in the case of a theorem, if the conclusion of the analysis was false, then the analysis was a valid *reductio ad absurdum.* If the conclusion of analysis was true, analysis played a heuristic role in order to find a synthetic proof.[70] By contrast, Wilbur Knorr has pointed out that even the first stage of Pappus' method, analysis, was a demonstrative stage. It followed that synthesis was superfluous. According to Knorr, Pappus added a second superfluous stage because of Aristotle's influence.[71]

Descartes between Pappus and the *regressus*

There was a certain similarity between Pappus' method and the *regressus*; they were both a two-stage procedure. However, even if one accepts Knorr's thesis, it should be underlined that neither Pappus' method nor Euclid's axiomatic mode of proof necessitated Aristotle's causal notion of logical consequence.

In 1588, Guidobaldo Del Monte published Federigo Commandino's Latin translation of Pappus' *Mathematical collection.* After Commandino's translation, mathematicians started working towards reconstructing ancient analysis. Marin Mersenne in *La verité des sciences* (1625) described the recovery of ancient analysis as one of the most important enterprises of his time.[72] Commandino, who was probably following Francesco Barozzi's translation of Proclus' *Commentary to the First Book of Euclid's Elements,* translated analysis and synthesis as *resolutio* and *compositio* respectively. His translation led him to identify Aristotle's *demonstratio quia* and *demonstratio propter quid* with Pappus' analysis and synthesis.[73] It must be stressed that the two mathematical ways of proving did not imply any commitment towards Aristotle's notion of logical consequence.

François Viète, who was amongst the mathematicians concerned with recovering ancient analysis, explained in the *In Artem Analyticem Isagoge* (1591) that he aimed to transform the ancient analysis into a general art of solving problems in mathematics. Viète mantained the Greek terms, probably hoping to avoid confusion between Aristotle's theory of proof and his own methodology for solving mathematical problems. Analogously, Descartes did not use the terms of *resolutio* and *compositio*, but those of analysis and synthesis, most likely, because he did not want to identify mathematical analysis and synthesis with Aristotle's two kinds of proofs.

Nevertheless, in the second half of the seventeenth century, the two mathematical modes of reasoning (i.e. analysis and synthesis) virtually became synonyms of Aristotle's two kinds of proof. This is at least true, for all the scholars who accepted Aristotle's notion of logical consequence in terms of a causal relation such as Hobbes, who, in *De Corpore* (1655), identified the geometrical mode of reasoning of analysis and synthesis with Aristotle's two kinds of proof. Another scholar who contributed to the interpretion of the mathematical tradition in the light of Aristotelianism was Gassendi, who, in his *Institutio Logica* (1658), identified analysis and synthesis with the *inventio* and *judicium.*[74] Even scholars close to Descartes, such as the authors of the *Port Royal's Logic* and Adrien Baillet – Descartes' official biographer – used the terms *resolution* and *composition* to indicate analysis and synthesis. Although they did not identify ancient analysis with the *regressus,*[75] terminological use probably contributed to the misunderstanding.

We can thus claim that Clerselier's translation followed the general trend of explaining the mathematical notions of analysis and synthesis in the light of the theory of *regressus.* Descartes, however, had rejected

Aristotle's logic and its theory of logical consequence in causal terms. Indeed, he had always criticized Aristotle's logic for its sterility. In the *Rules*, he held that the notion of logical consequence was grounded on intuition. According to Descartes, deductive reasoning was a chain of inseparable acts of intuition.[76] Moreover, in the *Discourse on Method* he criticized syllogism for its sterility. Interpreting *a priori* and *a posteriori* in causal terms entailed that one had to accept Aristotle's causal notion of proof. Had Descartes accepted Aristotle's account of proof in the *Meditations*, this would have been an implicit revision of his judgment on Aristotle's logic. Apart from Clerselier's translation, there was no textual evidence that Descartes had ever changed his judgment concerning Aristotle's logic.[77]

Did Descartes approve his translator's work?

So far I have shown that Clerselier's translation misrepresented Descartes' view. But was Descartes aware of this fact? It could then be argued that Descartes did not revise the French translation of his book. Clerselier might have misunderstood Descartes because he was not aware of the *Rules*. This is a plausible explanation, since Clerselier only read the *Rules* when he inherited Descartes' papers, after his death.

However, we should take into consideration a number of facts. Firstly, in the introduction addressed by the publisher to the reader of the French translation, we read that the translators tried to clarify and render more familiar a number of notions that were quite novel. We also read that the author himself asked for these revisions. Moreover, the publisher informed the reader that the author had revised and approved the translator's work.[78] Adrien Baillet, Descartes' biographer, claimed that the translators found it difficult to render a number of Latin technical words into French, and that it was Descartes himself who modified these passages.[79] Furthermore, a number of passages lead us to think that Descartes used the terms of *a priori/a posteriori* as synonyms of cause/effect terminology. On May 10 1632, he commented on the importance of knowing the natural order of all the fixed stars:

> Knowing this order is the key and the ground of the highest and the most perfect science which may be known by men, concerning material objects, for by means of it we can know *a priori* all forms and essences of terrestrial bodies, whereas without it we can but guess *a posteriori* and by means of their effects.[80]

Furthermore in *Le Monde,* he explained that,

Thus, those who are able to examine sufficiently the consequences of these truths and of our rules will be able to recognize effects by their causes. To express myself in scholastic terms, they will be able to have *a priori* demonstrations of everything that can be produced in this new world.[81]

In my opinion Descartes probably accepted the French translation because it made the reasoning clearer to the reader. As Roger Ariew has convincingly argued, from 1640, Descartes tried to make his thoughts more compatible with Scholasticism.[82] As I have already noted, he wanted his own philosophy taught in schools and universities. Although he knew that Aristotelianism would not be discharged soon, he hoped that if the Jesuits took his philosophy into their teaching jointly with Aristotle's, he would have won the battle against the latter. It should be recalled that the reader of the *Meditations* was not supposed to be talented in mathematics. Indeed, as Descartes had explained to Constantin Huygens[83] – who was complaining that his French friend did not send him the *Meditations* – he wanted the book to be read by theologians and philosophers only, who were not supposed to know Pappus or Viète. The teaching of mathematical disciplines in fact did not include such arguments. I think that Descartes realized that explaining mathematical modes of proving in terms of cause and effect would have made his reasoning more easily comprehensible to his readers. In November 1643, Descartes reminded Princess Elisabeth that experience had shown him that people who knew metaphysics hardly understood algebraic reasoning and vice versa.[84] By *experience,* Descartes was clearly referring to the criticisms to his *Meditations.*

How, then, should Descartes' use of *a priori* and *a posteriori* be understood?

I have shown that Descartes' notion of analysis and synthesis cannot be identified with the *demonstratio quia* and with the *demonstratio propter quid.* Moreover, textual evidence let us incline to think that Descartes accepted the French translation of his work. Why, then, did Descartes claim that analysis is *a priori* and synthesis is *a posteriori?* And how should his notions of *a priori* and *a posteriori* be understood?

So far I have shown that Descartes' concepts of analysis and synthesis derived only from the mathematical tradition of ancient analysis and modern algebra. In this section we will take a closer look at Descartes' use of *a priori* and *a posteriori* in mathematical contexts. From the reading of a number of letters concerning mathematical themes, we can see a number of features that an *a priori* proof should possess.

On January 1638, Descartes compared the mode of proving of his *Geometry* to that of Fermat. Descartes claimed that Fermat's proof was by *reductio ad absurdum*, while his own proof was obtained by means of a new theory of equations and followed the noblest mode of proving, which was usually called the *a priori* proof.[85] On March 1638, he stated that the proof about refraction, which was contained in his *Optics,* was *a priori.*[86] Descartes probably meant that his proof of the refraction law was very certain.

From these texts, we can infer that for Descartes an *a priori* proof was preferable than an *a posteriori* proof, and that it was a direct proof. We can also deduce that not all analytical proofs were *a priori.* On February 22 1638, Descartes explained to Father Vatier that he had proved the statements of his *Metereology* "a posteriori"*,* and that in order to prove them *a priori* he should have explained all the principles of his physics.[87] In the *Principles of Philosophy,* he claimed to have deduced his physics from a number of more general and certain principles. Therefore, we can infer that an *a priori* proof was deduced from principles that were general and certain. On August 1641, Descartes claimed that, concerning human behaviour, one cannot get the same certainty as science. He pointed out that one can prove this fact *a priori*, for the human mind was not corruptible, while the mind-body compound was corruptible. Moreover, one could prove it more easily *a posteriori* from the consequences which would derive from it.[88] We can then claim that, according to Descartes, an *a posteriori* proof might be more perspicuous, and that it might be a *reductio ad absurdum,* which was (for Descartes) less certain than a direct proof.[89] On February 1639, commenting on the tangent lines inverse problem, Descartes reminded Florimond Debaune that:

> I do not believe that in general it is possible to find the converse of my rule for tangent lines, nor that used by Monsieur de Fermat. However, one can deduce *a posteriori* a number of theorems about all curve lines, which are expressible by means of one equation [...]. There is also another way, which is more general and *a priori,* say by means of the intersection of two tangent lines [...].[90]

In summary, we can ascribe to an *a priori* proof the following features: it is a direct proof, it is more general than an *a posteriori* proof, and it is rooted in one (or more) general principles. By contrast, an *a posteriori* proof may be *ad absurdum*, it provides *ad hoc* solutions and it may be more perspicuous. One should prove *a priori,* but in some cases only *a posteriori* proofs are possible.

The picture I have drawn above explains why Descartes claimed that analysis was in some sense *a priori,* and synthesis *a posteriori.* Most of the features of an *a priori* proof were also the features of an analytical proof, and most of the features of an *a posteriori* proof were the features of a synthetic proof. Indeed according to Descartes, an analytical proof showed the discovery path, it was *a priori,* and its propositions followed a natural order, while a synthetic proof was *a posteriori,* it did not show the discovery path and its statements were not conveniently ordered, but it was easier to understand.[91]

So far I have described the features of *a priori* and *a posteriori* proofs according to Descartes' use of the two terms in his scientific correspondence. However, I have not provided a clear definition of how Descartes understood *a priori* and *a posteriori.* I believe that the notions of *a priori* and *a posteriori* belonged to those concepts Descartes used, without being able to provide a clear definition of them. In my view, Descartes' notions of *a priori* should be understood in the same way as the concept of "determination to move", which Descartes used in the *Optics* and in the *Principles of philosophy* without being able to explain it clearly. Descartes in fact was only able either to describe its use or to explain it by means of analogies.[92] Although Renaissance Aristotelians used the concept of *determinatio,* none of Descartes' correspondents was able to understand his unfamiliar use of it. As Abdelhamid Sabra has pointed out, the notion of *determinatio* should grasp the idea of vector.[93] The notion of vector could not be clear at Descartes' time. Concerning *a priori/a posteriori,* Descartes probably tried to grasp the idea of a proof, which was more general and it was rooted in more general principles. He could not avoid using scholastic terminology, but he did not want to accept Aristotle's theory of proof.

Conclusion

In the *Traité de l'Harmonie universelle,* which was published in 1627, Marin Mersenne described the structure of the map of mathematical knowledge in the following terms:

> Geometry looks at continuous quantity which is pure and deprived from matter and from everything which falls upon the senses; arithmetic contemplates discrete quantities, i.e. numbers; music concerns harmonic numbers, i.e. those numbers which are useful to the sound; cosmography contemplates the continuous quantity of the whole world; optics looks at it jointly with light rays; chronology talks about successive continuous quantity, i.e. past time; and mechanics concerns that quantity which is

useful to machines, to the making of instruments and to anything that belongs to our works. Some also add judiciary astrology. However, proofs of this discipline are borrowed either from astronomy (that I have comprised under cosmology) or from other sciences.[94]

However this account would soon belong to the past, for Mersenne's close friend René Descartes was redefining the whole context of mathematics and physics. Mechanics, optics and astronomy were becoming part of physics. In 1637, Descartes in fact published the *Optics,* which, as explained in the *Discourse on Method*, was nothing but that part of physics he could explain without referring to Copernicus' dangerous idea.[95] In the Scholastic system of knowledge, geometrical optics and Aristotle's theory of light were compatible. Both used the notion of *intentional species* in order to describe visual perception and the propagation of light.[96] In his *Optics,* Descartes, following Kepler's work, attacked the very notion of *intentional species,* and proposed a new explanation of refraction which was based upon a new physics. In 1637, Descartes also redefined the distinction between arithmetic and geometry. In his *Geometry,* he explained that numbers might be represented as right lines by means of the proportion theory. Once arithmetic operations were translated into geometrical terms, a geometrical problem might hopefully be treated algebraically.[97] Mersenne's distinction between discrete and continuous quantity was overcome.[98] The rest was made by Beeckmann, Galileo and the other proponents of the new mechanics. The new science of motion needed new physical principles.

In this paper, I have shown that in his attempt to establish a new physics and in his attempt of redefining boundaries between disciplines, Descartes, had to build new concepts, a new language and new readerships. In some cases, however, he only had vague notions which helped him to grasp the concepts; therefore *a fortiori* he could not explain them to his readers. The terms of *a priori* and *a posteriori* were instances of this kind of notion.

Clerselier's translation allowed Descartes to explain notions which he did not seem to be able to communicate otherwise, and he aimed to engage new readerships and to replace their Aristotelian conceptual frame by his new philosophical frame. The French translation managed to communicate because the reader had in mind the causal theory of proof of Renaissance Aristotelians.

Descartes perhaps did not imagine that, since many authors identified analysis and synthesis with Aristotle's *demonstratio quia* and *demonstratio propter quid*, Clerselier's translation could have been the source of a major misunderstanding, even among his followers. In 1670, the Cartesian Nicolas Poisson – who subsequently edited Descartes' *Opuscula Postuma* – published *Les rémarques sur la méthode de M. Descartes*. Poisson, who aimed at defending Descartes' ideas and at making Cartesianism more compatible with Aristotelianism, explained Descartes' notion of analysis and synthesis in terms of Aristotle's theory of proof.[99]

The cultural context of publishing, translation, authorship and readership did not only have a fundamental role in reshaping the map of knowledge in the seventeenth century, but it has also had an important – and not always acknowledged – influence on the development of present scholarship.

Notes

[1] Bruno, G. (1994). *Un'autobiografia*. Napoli: Procaccini, pp. 40-1; Pellegrini, A. M. (1943). "Giordano Bruno on Translations". *ELH* 10(3): 193-207.
[2] Chartier, R. (1984). "Culture as Appropriation: Popular Cultural Uses in Early Modern France". In Kaplan, S. L., ed. *Understanding Popular Culture*. Berlin: Mouton Publishers, pp. 229-53; Sabra, A. I. (1987). "The Appropriation and Subsequent Naturalization of Greek Science in Medieval Islam: A Preliminary Statement". *History of Science* 25: 223-43.
[3] See Kuhn (1957). *The Copernican Revolution*. Cambridge, Mass.: Harvard University Press.
[4] See Lindberg C. D. (1986). "The Genesis of Kepler's Theory of Light: Light, Metaphysics from Plotinus to Kepler". *Osiris* 2: 5-42, and Grafton, A. (1997). "From Apotheosis to Analysis: Some Late Renaissance Histories of Classical Astronomy". In Kelley, D. R., ed. *History and the Disciplines. The Reclassification of Knowledge in Early Modern Europe*. Rochester: University of Rochester Press, pp. 261-76.
[5] See Baldi, B. (1998). *Le Vite de' Matematici. Edizione annotata e commentata a cura di Elio Nenci*. Milano: Franco Angeli, p. 404. Unless stated otherwise all translations from Latin, French and Italian are by the author of this paper.
[6] Garin, E. (1994). *Lo zodiaco della vita. La polemica sull'astrologia dal Trecento al Cinquecento*. Roma-Bari: Laterza, n. 5, p. 28.
[7] See Grafton. "From Apotheosis to Analysis". For a more detailed view on the place of astrology in Kepler's thought see Patrick Boner's essay in this volume.
[8] See Laird, W. R. (1986). "The Scope of Renaissance Mechanics". *Osiris*, 2: 48-68; Micheli, G. (1995). *Le origini del concetto di macchina*. Firenze: Olschcki; Rossi, P. (2002). *I filosofi e le macchine. 1400-1700*. Milano: Feltrinelli, and Albert Presas' introduction to Aristòtil (2006). *Qüestions mecàniques*. Barcelona: Fundació Bernat Metge.

[9] See Lindberg. "The Genesis of Kepler's Theory of Light". On Patrizi's theory of light see Doetz, L. (1999). "Space, Light and Soul in Francesco Patrizi's *Nova de Universis philosophia (1591)*". In Grafton, A., and Siraisi, N. *Natural Particulars. Nature and the Disciplines in Renaissance Europe*. Cambridge, Mass.: MIT Press, pp. 139- 70.

[10] See Cellucci, C. (1998). *Le ragioni della logica*. Roma-Bari: Laterza.

[11] See Randall, J. (1961). *The School of Padua and the Emergence of Modern Science*. Padova: Antenore, and Gilbert, N. W. (1961). *Renaissance Concepts of Method*. New York: Columbia University Press.

[12] See Arndt, H. W. (1971). *Methodo scientifica pertractatum, Mos geometricus und Kalkülbegriff in der philosophischen Theorienbildung des 17. und 18. Jahrhunderts*. Berlin: de Gruyter, and Engfer, H. J. (1982). *Philosophie als Analysis : Studien zur Entwicklung philosophischer Analysiskonzeptionen unter dem Einfluss mathematischer Methodenmodelle im 17. und fruhen 18. Jahr*. Stuttgart-Bad Cannstatt: Frommann.

[13] See for instance AT I 144, AT I 194, AT III 298. All quotations made in the text are from Descartes, R. (1996). *Œuvres de Descartes, edited by Paul Adam and Charles Tannery*. 12 vols, Paris: Presse Universitaire de France, with AT followed by volume and pages.

[14] Concerning the metaphysical roots of Descartes' physics see Garber, D. (1992). *Descartes' Metaphysical Physics*. Chicago: Chicago University Press.

[15] For a clear discussion of what was Renaissance Aristotelianism see Schmitt, C. B. (1973). "Towards a Reassesment of Renaissance Aristotelianism". *History of Science* 11: 159-93.

[16] Baldi. *Le Vite de' Matematici*, p. 528.

[17] See Boccadiferro, L. (1558). *Explanatio libri 1. Physicorum Aristotelis. Ex Ludouici Buccaferreae, ... lectionibus except*. Venezia: in Academia Veneta, f. 2. See also Gilbert. *Renaissance Concepts of Method*, pp. 165-7.

[18] See Biagioli, M. (2003). "The Social Status of Italian Mathematicians, 1450-1600". *History of Science* 27: 41-95, p. 60.

[19] See AT I 478.

[20] See AT VI 19.

[21] See AT VI 19.

[22] See AT X 377 and AT VI 18.

[23] See AT VI 372-76. A reconstruction of Descartes' method of solving geometrical problems may be found in Jullien, V. (1996). *Descartes. La Géométrie de 1637*. Paris: Presse Universitaire de France, p. 80. See also Bos, H. J. M. (1981). "On The Representation of Curves in Descartes' Geometry". *Archive for the History of Exact Sciences* XXX, and (1998). "La structure de la Géométrie de Descartes". *Revue d'Histoire des Sciences* 51 (2/3): 98. Giorgio Israel has shown that Descartes' method of solving geometrical problems reflects quite trustfully his method of the *Regulae*. Israel, G. "From the *Regulae* to the *Géométrie*". *Revue d'Histoire des Sciences* 51 (2/3): 183-236.

[24] See AT III 751.

[25] See AT VII 155-6.

[26] Claudio Buccolini and Sergio Landucci argue that Mersenne was the author of the objections, whereas Daniel Garber thinks that Morin may be behind this objection. Buccolini C. (1998). "Dalle Objections di P. Petit contro il Discours alle Secundae objectiones di Marin Mersenne". *Nouvelles de la Republique des lettres* 1: 2-28; Landucci, S. (2001). "Contributi di filologia cartesiana". *Rivista di Storia della filosofia* 1: 5-24; Garber, D. (2001). "J.-B. Morin and the Second Objections". In Garber. *Descartes embodied*. Cambridge: Cambridge University Press, pp. 64-85.

[27] See AT VII

[28] See AM IV 127. All quotations made in the text are from Descartes, R. (1936). *Correspondance. Publiée avec une introduction et des notes de Ch. Adam et G. Milhaud*. Paris, with AM followed by volume and pages.

[29] Only in The Netherlands Descartes' philosophy replaced Aristotelianism at the very end of the Seventeenth century. See Schmitt. "Towards a Reassesment of Renaissance Aristotelianism", p. 163; Verbeeck, T. (1992). *Descartes and the Dutch: Early Reactions to Cartesian Philosophy, 1637-1650*. Carbondale and Edwardsville: Southern Illinois University Press.

[30] "Travel around The World of Descartes".

[31] See Daniel, G. (1690). *Voyage autour du Monde de Descartes*. Paris : Vve de S. Bénard, and (1693). *Suite du voyage du Monde de Descartes, ou nouvelles difficultées proposées à l'auteur du Voyage du Monde de Descartes. Avec la réfutation de deux défenses du Système général du Monde de Descartes*. Paris.

[32] See Daniel *Voyage*, p. 267.

[33] "L'analyse fait voir comme les effets dependent des causes". AT VII 155.

[34] "Comme en examinant les causes par leurs effets". AT VII 156.

[35] See Tarski, A. (1969). "On the Concept of Logical Consequence". In Tarski. *Logic, Semantics, Metamathematics*. Oxford: Clarendon Press .

[36] See Descartes, R. (1997). *Œuvres philosophiques, Edition de F. Alquié*. Paris: PUF.

[37] See AT VII 155-6.

[38] See Gaukroger, S. (1989). *Cartesian Logic*. Oxford: Oxford University Press, pp. 99-102.

[39] Timmermans, B. "The Originality of Descartes' Conception of Analysis as Discovery". *Journal of the History of Ideas* 60 (3): 433-47, p. 445.

[40] See *Aristotle An. Post.* 78 a.

[41] See *Aristotle An. Post.* 78 b .

[42] See *Aristotle An. Post.* 78 a-78 b.

[43] See *Aristotle An. Post.* 71 b.

[44] See *Aristotle An. Post.* 71 b.

[45] See *Aristotle An. Post.* 78 b.

[46] Mignucci, M. (1975). *L'argomentazione dimostrativa in Aristotele. Commento agli Analitici secondi*. Padova : Antenore.

[47] See Themistius (1900). *Commentaria in Aristotelem Graeca*. Berolini: typis et impensis Georgii Reimeri, vol. V, p. 28.

[48] See Themistius. *Commentaria*, p. 27.

[49] See Branca, V. (1963). "Ermolao Barbaro and Late Quattrocento Venetian Humanism". In Hale, J. R., ed. (1973). *Renaissance Venice*. Totowa: Rowman and Littlefield, pp. 218-43; Delph, R. K. (1994). "From Venetian Visitor to Curial Humanist: The Development of Agostino Steuco's 'Counter'Reformation Thought". *Renaissance Quarterly* 47 (1): 102-39, p. 107.

[50] See Barbaro, E. (1542). *Paraphrasis in Aristotelis Posteriora... Hermolao Barbaro,...interprete...* Venetiis: Girolamo Scoto, pp. 8-9.

[51] See Averroes. (1562-74). *Aristotelis opera cum commentariis Averrois*. Venetiis: apud Iunta, vol. I, 2a, pp. 208ff.

[52] See Mignucci. *L'argomentazione*, p. 294.

[53] Ibid., p. 298.

[54] See Pedro da Fonseca (1964). *Institutionum dialectcarum libri octo*. Coimbra: Universidade de Coimbra, vol. 2, p. 456.

[55] The Coimbran fathers were a group of Jesuits mainly based at the University of Coimbra who at the end of the sixteenth century wrote a number of commentaries to Aristotle's works which were adopted in many universities. See Schmitt. "Towards a Reassesment of Renaissance Aristotelianism".

[56] See Eustachius a Santo Paolo (1609). *Summa philosophica quadripartita*. Parisiis: Chastelain, 2 vols, p. 222.

[57] During the Middle Age, Robert Grosseteste and Roger Bacon combined the two proofs in a methodology of investigation. Grosseteste also uses Themistius' commentary to Aristotle's *Posterior Analytics*. See Crombie, A. C. (1970). *Robert Grosseteste and the Origin of Experimental Science 1100-1700*. Oxford: Oxford University Press, p. 47.

[58] See Gaukroger. *Logic*, p. 75.

[59] See Zabarella, J. (1985). *Jacopi Zabarellae. De Methodis libri quatuor. Liber de regressu. Edited by C. Vasoli*. Bologna: Clueb, p. 82.

[60] See, for instance, Arriaga, R. (1639). *Cursus philosophicus*. Parisiis: Bechet Denis, p. 197; Eustachius a Santo Paolo (1609). *Summa*, vol. I, p. 223; Fonseca, P. da *Institutionum*. vol. 2, p. 456, and Rubio, A. (1605). *Logica Mexicana sive Commentarii in universam Aristotelis logicam...* Coloniae: sumptibus Arnoldi Mylii Birckmannip, p.520.

[61] See Eustachius a Santo Paolo. *Summa*, vol. I, p. 259.

[62] Gassendi, P. (1727). *Exercitationes Paradoxicae adversus Aristoteleos*. In *Opera omnia in sex tomos diuisa curante Nicolao Aueranio aduocato florentino*. Florentiae: Tartini, Giovanni Gaetano & Franchi, Santi Stamperia Reale, vol. III, p. 177.

[63] For further details see Sirven, J. (1930). *Les années d'apprentissage de Descartes.* Paris: Vrin.

[64] See A T III 184-5.

[65] See Ariew, R. (1999). *Descartes and the Last Scholastic.* Ithaca: Cornell University Press, pp. 26-8.

[66] See Heath, T. L. (1981). *A History of Greek Mathematics.* New York: Dover, vol. I, pp. 183-201.

[67] Interpretation of Hippocrates' analysis is controversial. Hintikka and Remes has argued that it had a mere didactical role, while Cellucci thinks it was a methodology of discovery. Hintikka, J., and Remes, U. (1976). *The Method of Analysis.* Dordrecht: Reidel; Cellucci. *Le ragioni della logica.*

[68] Leibniz tried to reconstruct how really Pappusian procedure worked. See Leibniz, G. W. (1950). *Sämtliche Schriften und Briefe.* Berlin: Akademie, vol. II, 456-8.

[69] Marco Panza has shown the evidence for the application of Pappus' method for solving problems in the works of Heron, Apollonius and Archimedes. See Panza, M. (1997). "Classical Sources for the Concepts of Analysis and Synthesis". In Otte, M., and Panza, M. eds. *Analysis and Synthesis in Mathematics.* Dordrecht: Reidel, pp. 385-95.

[70] Mahoney, M. (1968-9). "Another Look at Greek Geometrical Analysis". *Archive for History of Exact Sciences* 5: 319-48; Szabó, A. (2000). *L'aube des mathématiques grecques.* Paris: J. Vrin, pp. 261-2; Mäempää, P. "From Backward Analysis to Configuration Analysis". In Otte and Panza. *Analysis*, p. 202.

[71] See Knorr, W. R. (1986). *The Ancient Tradition of Geometric Problems.* Boston: Birkhauser, p. 359.

[72] See Mersenne, M. (1969). *La vérité des sciences. Contre les sceptiques ou Phyrrhoniens.* Faksimile-Neudruck der Ausgabe. Stuttgard-Bad Canstatt: Frommann, p. 751.

[73] In his *Admirandum illud geometricum problema* (1586). Barozzi was probably the first to explain analysis and synthesis in Aristotelian terms. See Sergio, E. (2007). *Verità matematiche e forme della natura da Galileo a Newton.* Roma: Aracne, pp. 27-8.

[74] See Hobbes, T. (1839-45). *The English Works Of Th. Hobbes.* London: John Brown, vol. I, pp. 58-80; Gassendi, P. (1981). *Pierre Gassendi's Institutio Logica: A Critical Edition with Translation and Introduction.* Assen: Van Gorcum, pp. 70ff.

[75] See Arnauld, A., and Nicole, P. (1993). *La logique ou l'art de penser éd. critique par P. Clair et R. Girbal.* Paris: Vrin, p. 291; Baillet, A. (1970). *La vie de M. Descartes.* Genève: Slatkine, vol II, p. 138.

[76] See AT X 369.

[77] Descartes' severe judgment explains the lack of logic in his writings. Nevertheless, there was some debate after Descartes' death concerning "Cartesian logic". See Borghero, C. (1988). "Méthode et Géométrie. Interpretazioni

seicentesche della logica cartesiana". *Rivista di filosofia* LXXIX: 25-57. Jacobus
Revius, in his *Methodi cartesianae consideratio theologica* (1648), and, Cyriacus
Lentulus, in his *Nova Renatii Descartes sapientia* (1651), criticized Descartes for
the lack of logic in his writings. A number of Descartes' followers and thinkers
close to him tried to develop some "Cartesian logic". See Savini, M. (2004). *Le
développement de la méthode cartésienne dans les provincies-Unies.* unpublished
PhD Dissertation. Paris : EPHE, Univ. Paris IV, Sorbonne, pp. 171-267; Borghero,
C. (1990). "La méthode senza la geometria. Poisson e la diffusione del metodo
cartesiano". In Belgioioso, G.; Cimino, G.; Costabel, P., and Papuli, G. eds.
(1990). *Descartes. Il Metodo i Saggi.* Roma: Istituto dell'Enciclopedia Italiana,
vol. II, pp. 587-95.

[78] See AT IX 2-3..

[79] Baillet. *La vie de M. Descartes*, vol. II, pp. 171-73

[80] AT I 250-1. Constructing astronomy by means of general principles was
considered by Copernicus one of the advantages of his astronomy. In fact in
Ptolemy's system, epycicles and deferents of planets can be modified
independently, while, in Copernicus' systems, they are direct consequences of
observation. See Kuhn. *Copernican Revolution*, pp. 175-7.

[81] AT vol. VI 47. Translated in Cottingham, J.; Stoothoff, R., and Murdoch, D. *The
Philosophical Writings of René Descartes*, vol. I, p. 97.

[82] See Ariew, R. (1999). "Descartes among the Scholastic". In Ariew. *Descartes*,
pp. 7-35, on p. 29, and "Descartes and the Jesuits of la Flèche". In ibid., pp. 140-
54.

[83] See AM IV 128.

[84] See AT IV 46.

[85] See AT I 490

[86] See AT II 31.

[87] See AT I 563.

[88] See AT III 422.

[89] See AT I 490.

[90] See AT II 514.

[91] Descartes, R. (1984). "Meditations on First Philosophy". In Descartes. *Writings*,
vol II, pp. 110-1.

[92] See Gabbey, A. (1980). "Force and Inertia in the Seventeenth Century:
Descartes and Newton". In Gaukroger, S. ed. *Descartes. Philosophy, Mathematics
and Physics*. Brighton: Harvester Press, pp. 230-320, on p. 258; Eastwood, B. S.
(1984). "Descartes on Refraction. Scientific versus Rhetorical Method". In
Eastwood. (1989). *Astronomy and Optics from Pliny to Descartes,* London:
Variorum Reprints, pp. 481-502, on p. 502; Sabra, A. I. (1967). *Theories of Light
from Descartes to Newton.* London: Oldbourne, pp. 110-1, Garber, D. (1992).
Descartes Metaphysical Physics. Chicago: Chicago University Press, pp. 188-93;
Damerow, P.; Freudenthal, G.; McLaughlin, P., and Renn, J. (2004). *Exploring the
Limits of Preclassical Mechanics. A Study of Conceptual Development in Early*

Modern Science: Free Fall and Compounded Motion in the Work of Descartes, Galileo and Beeckman. New York: Springer, pp. 130-2.
[93] Sabra. *Theories*, pp. 110-1.
[94] Mersenne, M. (2003). *Traité de l'harmonie universelle.* Paris: Fayard, p. 38-9.
[95] See A M IV199-200.
[96] Lindberg. "The Genesis of Kepler's Theory of Light".
[97] See AT VI. See also Jullien. *Descartes.*
[98] Mersenne himself was, in the 1630s, more cautious about the classification of the disciplines. In the *Questions physiques et mathematiques* (1634), in fact, he remarked that: "These subalternations contribute nothing to the subalternated sciences, which must possess their own particular principles, which have to be as sure and evident as those that one calls subalternating. And it may be that this subalternation has not been understood, and that it would be more expedient to say that geometry and arithmetics are the general rules, which serve for constructing the demonstrations, and for deducing all the conclusions of the other sciences, if one gives their true principles". Mersenne, M. (1985). *Les Questions theologiques, physiques, morales, et mathematiques.* Paris, pp. 179-180.
[99] See Poisson, N. (1670). *Remarques sur la méthode de M. Descartes, où on établit plusieurs principes généraux, necessaires pour entendre toutes ses œuvres.* Vandosme: chez Sebastien Hip, Imprimeur de son Altesse, pp. 69-70. On Poisson's reading of Descartes see Borghero. "La méthode senza la geometria".

KEPLER'S DEFENCE OF ASTROLOGY AND THE EARLY MODERN EMERGENCE OF MATHEMATICS

PATRICK J. BONER

Johannes Kepler (1571-1630) is generally acknowledged as having pioneered the establishment of modern astronomy through his efforts to merge Aristotelian physics with mathematical astronomy. Although "flawed at the foundations" on account of its false notion of inertia,[1] Aristotelian physics featured fundamentally in Kepler's novel notion of planetary motion as a physical problem. Yet Kepler's endeavours in astronomy are also characterised by the cultivation of an ancient Pythagorean-Platonic tradition, in which mathematical regularities accounted entirely for the movements of the heavens. Identified amongst the "spiritual children of Claudius Ptolemy",[2] Kepler is seen according to this second interpretation as a mathematician who "essentially solved the problems of ancient astronomy" by applying mathematical models to the explication of celestial motions with unprecedented precision.[3] Such a two-current conception, in which Kepler's emphasis on mathematical principles is identified alongside his unflinching commitment to physical causes, elucidates the emergence of a double-faced Janus in Keplerian scholarship, according to which historians stress the "striking polarity" of Kepler's "intellectual giftedness".[4] It is clear, however, that these two currents flowed within a single stream, and that Kepler considered them collectively in his revolutionary conception of the role of the mathematical astronomer. Comparable to the idea of the "mathematical architect" elaborated by Daniele Barbaro (1513-1570) in his commentary on Vitruvius' *De architectura*, Kepler's revolutionary conception of the astronomer entailed the professional pursuit of a "complete, mathematically detailed and physically grounded representation of the cosmos".[5]

Kepler's reappraisal of the role of the astronomer can be considered as a radical example of a more general rise to prominence of mathematics in the sixteenth and seventeenth centuries. Collectively known as the mathematical *quadrivium*, arithmetic, astronomy, geometry and music experienced a "renaissance" in early modern Europe, attracting the attention of such intellectual and theological authorities as Philip Melanchthon (1497-1560) and Christoph Clavius (1538-1612). Melanchthon, author of the Augustinian Confession and creator of a Lutheran educational curriculum at the University of Wittenberg, was rivalled in his promotion of a more rigorous mathematics curriculum by Clavius, whose efforts as the first teacher of mathematics at the Collegio Romano significantly contributed to the constitution of a growing group of Jesuit mathematics instructors. Praised by Melanchthon as a "necessary preparation for knowing God",[6] mathematics was similarly advocated by Clavius as "first amongst all the sciences" in natural philosophy.[7] While a significant increase in the number of mathematical chairs in institutions of learning across Lutheran Germany occurred between 1530 and 1560, knowledgeable mathematics instructors were incorporated into Jesuit educational institutions across Europe in the late sixteenth and early seventeenth centuries. Such a sudden swelling in intellectual and sociocultural recognition is regarded as "the emergence of an autonomous disciplinary field of mathematics vis-à-vis that of philosophy",[8] whose principal protagonists ushered in a new intellectual and institutional order.

As an essential part of astronomy studying "the language of the stars", astrology played a prominent role in the early modern emergence of mathematics. Melanchthon attributed to astrology a place of importance in the educational curriculum of the University of Wittenberg, providing a "rational basis" for astrology by including it in his definition of Aristotelian physics.[9] Involving the observation and prediction of sublunar sympathies with the stars, astrology was also seen by Melanchthon as exemplifying certain "ordinances of God".[10] One of the institutions most heavily influenced by the curriculum of the University of Wittenberg was the University of Tübingen, where Kepler studied theology. At Tübingen, Kepler was taught astrology by his mathematics instructor, Michael Mästlin (1550-1631). Despite being depicted as "decidedly modern in an age obsessed with celestial forecasting",[11] Mästlin's critical attitude towards astrology did not undermine his acceptance of it as an essential part of the mathematical *quadrivium*. If anything, Mästlin's apparent reluctance to provide "only the barest of signification" in his published works was typical of his time,[12] when astrology took on as many forms of acceptance and exposition as its practitioners. From his earliest astrological formulations following his theological studies, Kepler similarly exhibited a critical attitude towards astrology,[13] yet his conviction in a core set of astrological principles never wavered.

In what follows, Kepler's emphasis on the mathematical foundations of astrology is presented as an extreme example of the rise of mathematics from the auxiliary ranks of early modern natural philosophy. Echoing the claims of Clavius, Kepler commended such classical scholars as Proclus for attributing to mathematics a place of pre-eminence in acquiring knowledge about nature.[14] Kepler's critical attitude towards astrology resulted in what he regarded as a return to its mathematical roots, an astrology cleansed of the cultural conventions obscuring its mathematical core. His cleansing efforts as a mathematics instructor in Graz (1594-1600) form the focus of the following narrative. Particular emphasis is placed on Kepler's early correspondence with one of his principal patrons, Bavarian Chancellor Hans Georg Herwart von Hohenburg (1522-1611), who openly criticised Kepler's astrology as "slippery and uncertain".[15] Herwart remained unconvinced by Kepler's efforts to reform astrology, applying the same mathematical principles employed by Kepler to effect their undoing. For his own part, Kepler remained firmly committed to his cause, countering the criticisms of his correspondent by returning again and again to the mathematical archetypes which he believed lay at the core of astrology and cosmology more generally. Realised momentarily by certain configurations of the planets, Kepler's archetypes exemplified the metaphysical blueprint according to which he conceived of the material cosmos. They underlay not only his astrology, but all other areas of observable phenomena, unifying nature through knowable, mathematical foundations.

Conserving the kernel: Kepler's archetypal conception of astrology

On occasion, Kepler took great pains to differentiate between astrology and astronomy. In his *Tertius interveniens* (1610), for example, he referred to popular astrology as "a curiosity", "a foolish daughter" whose "most reasonable mother, astronomy", relied on predictive principles rather than "individual circumstances".[16] In a letter of December 1599, he briefly outlined what eventually would become the five books of his *Harmonice mundi* (1619), corresponding respectively with geometry, arithmetic, music, astrology and astronomy. Describing the fifth, astronomical book as the examination of "the causes of the periodic motions", Kepler set aside astrology, the focus of the fourth book, as the study of "the causes of the planetary configurations".[17]

Despite the certainty he so often attributed to mathematical knowledge, he appears to have allowed for doubt and ambiguity in astrology.[18] It may well be argued that Kepler's astronomy, burdened by a potentially boundless number of archetypal principles, was similarly plagued by the spectre of incertitude,[19] but there can be no mistaking the marked contrast between what Kepler claimed to be two very different degrees of demonstrability. He often compared astrology to the imprecise art of the physician, associating the two by their shared dependence on *a posteriori* observations.[20] And, whether or not he held out for astrology the possibility of becoming more precise, he anticipated no such prospect occurring within the foreseeable future.

However, despite the many differences Kepler identified between astrology and astronomy, he suggested that the two shared in geometry the same metaphysical foundation. As a result, geometrical principles could be similarly applied to the two areas by way of *analogia*, or points of parallel, which he also considered pertinent to the study of music.[21] In fact, according to Kepler, all phenomena of the material world, whether the motions of the planets, the measurements of musical notes or the effects of the heavens on earthly weather conditions, derived from the same exclusive set of geometrical principles. Astrology, astronomy and music, in other words, shared the same archetypal origins. This ubiquitous "presence" of geometrical principles accounted for the underlying consistency of Kepler's cosmos, in which the idea of "harmony" encompassed far more meaning than modern standards.[22]

> . . . The *analogia* [of astrology] with music and astronomy is absolute. I show that the analogy must necessarily be considered in such a way, since the origins of all things are derived from geometry. Nature affirms these principles . . . and employs the very same ones in all things which are suitable to them.[23]

Kepler's association of astrology with astronomy did not rely on ontological commonalities alone. In addition, he claimed that humankind had been intellectually equipped with the very principles underlying both mathematical branches. He conceded that such principles appeared to be the "inventions of men", but that their conception had been no mere coincidence. Man, created in "the image of the Creator", had arrived at these notions as a consequence of his own inherent essence.[24] Kepler stressed that knowledge of the archetypal principles had not come about by choice, since he did not regard "the election of man" to be "the measure of nature".[25] Rather, archetypal knowledge originated from the actualisation of man's inherent essence. In other words, the material world exemplified certain principles previously present to the human intellect. Astrology and astronomy, then, stemmed from the same general system, in which the tools for acquiring knowledge and the practitioners making use of them relied upon a single epistemological scaffolding. As Gérard Simon suggests, in Kepler, "the reformer in astrology" cannot be readily distinguished from "the renovator of astronomy".[26]

It is clear that Kepler's scepticism in astrology reflected the foundational principles of his natural philosophy. To lend measure to the magnitude of his sceptical attitude towards astrology, he compared his appraisal of it in a letter of May 1599 to the incredulous stance of Nicolaus Raimarus Ursus (1551-1600) towards astronomy. In his letter, Kepler not only associated himself with Ursus, but he considered his disbelief in astrology even greater than Ursus' suspicion of astronomy. He therefore did "the same in astrological matters as Ursus [did] in astronomy", except with an attitude which he described as "more mistrustful".[27]

To appreciate the significance of Kepler's surprising comparison,[28] let us recall the sceptical onslaught waged by Ursus against astronomy from the time of his tenure as Imperial Mathematician to his death in 1600. Personally appointed by Holy Roman Emperor Rudolf II in August 1591, Ursus published his *De hypothesibus astronomicis tractatus* six years later, a violent assault against "critics and detractors" including Christoph Rothmann (1550-ca.1600), Helisaeus Röslin (1544-1616) and Tycho Brahe (1546-1601). Amidst malicious insults and often uncompelling commentary, Ursus presented in the *Tractatus* what can be considered four serious arguments against the veracity of astronomy: (1) he interpreted astronomical hypotheses as "illegitimate or false postulates" which possessed connotations of "spuriousness"; (2) the accurate prediction and retrodiction of apparent celestial coordinates did not, according to Ursus, constitute the truth of an astronomical hypothesis, thus rendering irresolvable "the most fundamental issues of astronomy"; (3) similar to other areas of mathematics such as algebra, astronomy was seen as capable of arriving at true conclusions from false postulates; (4) all previous astronomical hypotheses, Ursus argued, had incorporated "crass absurdities", and there was no reason to believe that astronomers were "ever to come up with true hypotheses".[29]

Through his involvement in the extended controversy between Ursus and Tycho, the product of which became an unpublished defence of Tycho now known as the *Apologia pro Tychone contra Ursum* (1600), Kepler was made well aware of the extent of Ursus' critical, if unoriginal, campaign against astronomy. Contracted in 1600 to work for Tycho on the understanding that he compose a defence of his prospective patron, Kepler set out to undermine Ursus' misrepresentation of astronomical hypotheses point by point. In response to Ursus' ascription of "spurious" astronomical postulates, Kepler professed in the *Apologia* that "false hypotheses, which together yield the truth once by chance, do not in the course of a demonstration in which they have been combined with many others retain this habit of yielding the truth, but betray themselves".[30] Further, in the face of the observational equivalence of competing predictive systems, Kepler cited the differences brought about by physical considerations. "Thoughtless" astronomers such as Ursus were therefore condemned for paying attention solely to numbers and believing that "the same result follows from different hypotheses".[31] Alternatively, emphasis upon the physical implications of astronomical hypotheses allowed Kepler to distinguish between competing systems and to curtail the criticisms of sceptical and relativistic adversaries. In his unconventional account of qualitative considerations, Kepler attributed particular importance to the physical principles of simplicity, coherence and precise archetypal patterns.

Given his ardent opposition to Ursus' astronomical scepticism, Kepler's "mistrustfulness" of astrology may in fact appear to have been extremely severe. Further, his persistence in the publication of astrological calendars, nativities and learned expositions on astrology may even seem insincere. However, despite the fact that several scholars have taken this to be the case,[32] it should be stressed that Kepler questioned the predictive powers of astrology and not its fundamental truthfulness. Kepler described himself as "more mistrustful" than Ursus not because he disputed the metaphysical foundations of astrology, but because he found little efficacy in forecasting the actions of earthly inhabitants, whose fickle paths of flight proved far less foreseeable than the perennial motions of the planets. Kepler considered testimony to the existence and astrological influence of certain metaphysical archetypes all but undeniable: from its very creation, the cosmos had embodied the same metaphysical archetypes by which the celestial and sublunary realms continued to interact. Kepler claimed that such archetypes were momentarily exemplified by certain configurations of the planets, which, as a source of stimulus for the sublunary realm, were variously responded to by the inhabitants of the earthly ambit. How the earthly inhabitants responded to these celestial stimuli, however, was a matter altogether uncertain, a source of "mistrustfulness" for which Kepler held out little immediate hope.

Yet what did Kepler mean by the metaphysical archetypes he saw as so essential to the understanding of astrology? In April 1599, Kepler offered an answer to this question in a long letter to Herwart von Hohenburg, Bavarian Chancellor and Kepler's scholarly patron in Munich. Commended by Mästlin for his "extraordinary industry" as a chronologist and credited with laying the foundations for the formulation of logarithms,[33] Herwart regularly wrote to Kepler through a correspondence network involving several highly stationed readers. The letters exchanged between Herwart and Kepler were first dispatched to the imperial court at Prague, where dignitaries such as Peter Casal, Capuchin priest and secretary to Archduke Ferdinand, were requested to read them. An ardent Catholic and an outspoken statesman, Herwart ensured that his correspondence with Kepler passed before the eyes of numerous non-Protestants. Such an arrangement has been said to have "raised Kepler from the mass of his [non-Catholic] colleagues", whereby "he found special consideration with the ruling Catholic party".[34]

In a letter of March 1599, Herwart expressed his desire for "some sure [astrological] method, together with a directory of the causes of the conditions of the weather and an indication of when each of the conditions would more or less proliferate".[35] For such a purpose, Herwart predicted that it would be pertinent "to examine accurately first the conditions of the weather, and then the observations and ensuing traditions of the ancients".[36] In his response of April 1599, Kepler began without recourse to the ancients, preferring to affirm first the interaction of the celestial and sublunary realms through ordinary observations. As he expressed it in a marginal note to Herwart's letter, the interaction involved a "connection between the heavens and earth and the effects thence issuing forth".[37] In considering the connection, Kepler described Aristotle's *Meteorologica* and *De generatione et corruptione* as "fairly pertinent", whereas Ptolemy, "led astray by the fables of fools", was described as "having left nature behind" in his negligence of observations.[38] Kepler cited recurrent phenomena, such as the associations of "[earthly] humours with the light of the Moon" and the sequence of the tides "in accordance with the motion of the Sun and the Moon",[39] as compelling experiential evidence of influences originating from beyond the lunar orbit. As his most persuasive form of evidence, however, he pointed to the aspects, the configurations of the planets whose effects upon the sublunary realm derived directly from the metaphysical archetypes.

In his long letter to Herwart, Kepler outlined eight influential configurations of the planets. Also known as "aspects" or "planetary radiations", the configurations conveyed influences "by means of reason".[40] Seen as the centre of a circle whose circumference corresponded with the zodiac, the Earth served as the principal reference point for the planets. Ostensibly motionless, the Earth remained at the centre of the circle while the planets continually changed positions on the surrounding circumference. Kepler considered all latitudinal motion negligible, as he conceptualised the circle, and consequently the positions of the Earth and the planets, in terms of two dimensions. When two planets separated themselves by a certain distance or angular section with respect to the central reference point of the Earth, an aspect was momentarily formed. Since the planets proceeded almost continually in motion, an influential arrangement swiftly passed into obscurity as soon as it was formed. Kepler listed the aspects in order of ascending angular separation: conjunction (0°), sextile (60°), quintile (72°), quadrature (90°), trigon (120°), sesquiquadrature (135°), biquintile (144°) and opposition (180°).[41]

Of the eight aspects Kepler enumerated, five were originally espoused by Ptolemy in the *Tetrabiblos*,[42] the leading astrological treatise of the early modern era.[43] Kepler therefore considered the aspects of conjunction, sextile, quadrature, trigon and opposition the cardinal configurations, whose potential influence surpassed that of the three "new" aspects he introduced, quintile, sesquiquadrature and biquintile.[44] In response to questions concerning the three new aspects, Kepler claimed that they were consistent with "the very same reasoning which embraces those time-honoured and familiar aspects".[45] He called upon the "talents of natural philosophers" to inquire into the possibility of additional aspects by using as examples the unmistakeable effects of the original five. Conditions of the weather which conspicuously occurred at the onset of certain configurations were immediately seized upon as signs of evidence.

> Behold on this day, when two planets measure by their separation the difference of 89°, nothing new occurs in the conditions of the weather [*meteoris*]. The day after, when they measure by their separation a full 90° difference, that is, [when they are] in quadrature, a storm suddenly springs forth.[46]

"A rational angle", such as that produced by the separation of two planets by "the sum of 90°", stimulated striking meteorological occurrences in the earthly atmosphere.[47] Kepler employed similar observational evidence in support of the three new aspects, yet he considered another source of evidence even more compelling.

For each of the eight ratios which Kepler held as harmonic — unison (1:1), minor third (6:5), major third (5:4), fourth (4:3), fifth (3:2), minor sixth (8:5), major sixth (5:3) and octave (2:1) — he identified a corresponding aspect.[48] He did this by comparing the amount of the circumference cut off by a particular planetary configuration with the circumference of the circle as a whole. The larger part of the proportion measured the whole of the circumference, while the smaller part measured the segment separated by the circumferential span of the two planets. Sextile, for example, cut off 60° or 5/6 of the circumference, and therefore corresponded with the harmonic ratio of 6:5, the minor third; opposition, cutting off 180° or 1/2 of the circumference, corresponded with the harmonic ratio of 2:1, the octave. The three new aspects, quintile (72°), sesquiquadrature (135°) and biquintile (144°), rounded out the group of eight, in accordance with the ratios 5:4, 8:5 and 5:3, respectively. In a later letter, Kepler illustrated to Herwart how the aspects achieved such an agreement. He straightened into a line the circumferences of each of the eight aspects and equated the angle of individual planetary configurations with the ratios formed by corresponding harmonic proportions.

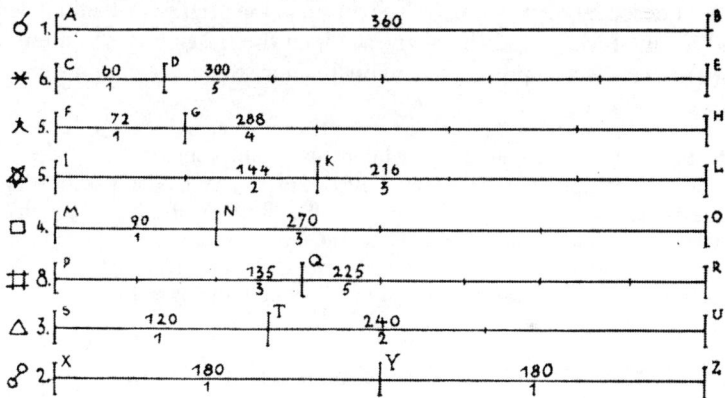

Figure 1. Correspondence of the aspects with the harmonic consonances
(*KGW*, vol. 13, p. 349)

In a letter of May 1599, Herwart expressed concern over Kepler's account of the aspects. Diplomatic in tone, Herwart delicately intimated his doubts about his correspondent's wide-ranging rejection of "all things astrological except for the aspects of the planets".[49] Since "even the smallest foundation [in astrology] may appear to derive from sheer superstition",[50] Herwart questioned Kepler's motivation to maintain what amounted to yet another problematic part of astrology. Accordingly, although Herwart acknowledged his attempt to prevent the amount of influential configurations from reaching an immeasurable "multitude", he still did not see why there could not be included "very many, if not an infinitude of aspects", which derived from the same principles as those of the original eight.[51] He saw no limiting principle, no arguable means of minimising the aspects to an exclusive number. In the end, rather than raise questions against what Kepler considered "the final refuge" of astrology, he chose "to remain quiet".[52] His careful tact did little to conceal his doubt from the discerning eyes of his correspondent, who chose to elaborate on his idea of the aspects in another letter.

Admittedly, the prospect of "an infinitude" of aspects had previously occurred to Kepler. However, he claimed to have resolved "the entire matter" in the twelfth chapter of his *Mysterium*,[53] where he discussed the twelve signs of the zodiac, the products of "human invention, for which of course nothing natural of any sort serves as a basis", and the aspects, the configurations which, "in the [two-dimensional] plane of a straight line, possess something in common with the harmonic consonances".[54] Yet despite his reference to the *Mysterium*, Kepler felt compelled to elaborate on why he accepted only eight aspects as potentially influential. In a second letter of May 1599, he acknowledged the possibility of an endless number of aspects as a problem previously addressed by several prominent scholars.

> For Ptolemy gave a cause for why men should choose chiefly these [aspects] and omit all the rest, but he did not prove that nature herself does not possess any more. On the contrary, the election of man is not the measure of nature. Reinhold, in his commentary concerning the matter of music in Peurbach's *Theoricae novae planetarum*, offers just such an objection, as does Cardano in his commentary on Ptolemy's *Almagest*. Yet neither suffices for the attentive reader . . .[55]

Why, in Kepler's opinion, were scholars unable to resolve the number of the aspects? He recognised the existence of "not more than seven or eight harmonic divisions of a string",[56] yet it was not their correspondence with the harmonic ratios which established the eight aspects as potentially influential. More significantly, the aspects shared with the consonances the same metaphysical foundations, the same mathematical archetypes constituted by clearly identifiable constraints. Kepler claimed that such archetypes were limited to a select group of regular polygons constructed solely by a ruler and compass, through inscription in a circle. He therefore considered acceptable the aspect of quadrature, since the separation of two planets by 90° on the circumference of the circle formed the side of a square, whose vertices touched the interior of the circle at four equidistant points. Trigon was similarly admitted, as the angular displacement of 120° amounted to the side of an inscribed triangle, whose three vertices split the circle into three equidistant parts. The numbers of angles and sides of such figures were not as important to Kepler as their ability to be constructed "rationally", that is, by means of geometrical constraints which prevented such polygons as the heptagon, whose angle of separation between each of its sides was neither rational nor divisible by 360, from being accepted as archetypal.

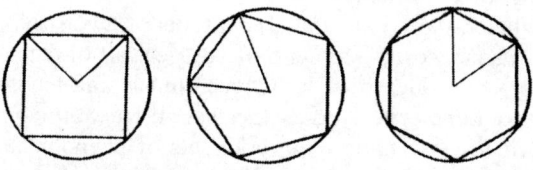

Figure 2. Agreement of quadrature (90°), quintile (72°) and sextile (60°) with the sides of certain polygons (*KGW*, vol. 1, p. 42)

Kepler thus saw the inscription of regular polygons in a circle as a means of accessing pre-established archetypes, geometrical patterns momentarily exemplified by the configurations of the planets. Defined as figures whose "sides and outward-facing angles" were equal and whose inscription in a circle was determined by "the proportion of the side of the figure to the diameter of a circle", regular polygons were rationally comprehended and constructed. [57] In fact, their knowability and constructability were considered two sides of the same cosmological coin. "To know in geometrical matters", Kepler suggested, "is to measure by a known measure, which for our present concerns is the diameter of a circle". [58] And, just as he did not believe that irrational figures could be constructed in a rational way, he considered it inconceivable that irrational figures, in the form of celestial configurations, could exert any influence on earthly affairs. Accordingly, he accepted eight astrological configurations whose constitution consisted of the same archetypal principles underlying the harmonic consonances. In his letter to Herwart, Kepler compared the exclusive place of the aspects to a residential compound, in which only eight inhabitants were allowed admission to the house inside.

> It may be, though, that nevertheless the door by which the [aspects] are admitted is within a dwelling site: so, when this is opened, the house still remains closed. The multitude of infinite aspects remains outside, shut out by the front door. [59]

Innumerable aspects could be considered for candidacy, Kepler contended, but only a small number could be admitted according to certain geometrical principles. The suggestion that the aspects were "futile", [60] then, was proven wrong not by recourse to the harmonic proportions, but rather by reference to more fundamental mathematical archetypes. The limits of rationality, represented by the inability of the intellect to realise certain figures, prevented the vast majority of configurations from being considered potentially influential.

Kepler further explored the archetypal roots of the astrological aspects and musical consonances in another letter to Herwart of August 1599. The "*analogia*", as Kepler identified it, between music and the configurations of the planets could be conceived as "perfect" if Herwart considered more closely the formal cause underlying both sets of phenomena.[61] Variously realised in the form of rational, regular polygonal archetypes, geometry underlay astrology, music and all other areas of natural phenomena. Nature, the material manifestation of a divine metaphysical blueprint, was seen as unified through the geometrical archetypes. That man was created in "the image of the Creator", Kepler claimed, allowed him to understand the archetypes and, as a consequence, the formal causes of natural occurrences. As he communicated to Herwart,

> . . . I therefore conclude that up to now I may adhere to that opinion which subsequently introduces three new aspects into astrology, such that the *analogia* between music and astronomy may be perfect. I show that this *analogia* must necessarily be considered in such a way, since the origins of all things are derived from geometry. Nature affirms these principles in the creation of a single kind, and employs these same [principles] altogether in all things which are capable of them. As a result, [they are] in music, in the motions of the planets, in the effects of the planets, in the measurement of musical notes, in the cause of the weather, in the dancing of men and in the structure of songs. For although these [principles] are the inventions of men, nevertheless man is the image of the Creator.[62]

Despite Kepler's best efforts to convince his correspondent of the theological roots of the aspects, Herwart remained incredulous. As he confessed in a letter of August 1599, Kepler did not persuade him to see the aspects as anything but the products of "opinions and superstitions".[63] In the face of such a matter which seemed to him "slippery and uncertain",[64] Herwart preferred to continue playing the role of devil's advocate by rejecting the aspects along with the rest of astrology. Without offering any sound explanation in his favour, Herwart affirmed the all-out rejection of astrology as a consequence of the questionability of the aspects.

> However, up to now I would not in fact see any firm foundation by which such a power and effect, which is customarily attributed to the aspects of

the luminaries, may be confirmed, so that also all of astrology . . . indeed seems to me to depend upon mere opinions and superstitions.[65]

Herwart interpreted any agreement between the aspects and the musical consonances as coincidental, a curious means by which "some numbers of the consonants coincide most remarkably with the numbers of angles which subtend the sides of figures inscribed in a circle".[66] And, if he did not consider the aspects more compelling than any other component of astrology, why would he believe in anything at all?

Conclusion

Apparently, Herwart's incredulity had little effect on Kepler's enduring conviction in the aspects as "the final refuge" of astrology. Kepler identified the mathematical archetypes underlying the aspects as the original tools of creation, whose existence in the mind of man made knowledge possible. To question the archetypes, whose construction and comprehension constituted for Kepler the core of knowledge, was to question far more than Kepler's astrology. Kepler would later expand the number of aspects to thirteen, accepting five additional configurations in his *Harmonice mundi.* [67] Yet his departure from a one-to-one correspondence with the eight harmonic consonances did not imply a deviation from his earlier line of reasoning. Kepler's cosmos remained the material realisation of mathematical archetypes, the building blocks of a divinely inspired blueprint. Fundamentally mathematical, Kepler's defence of astrology is a radical example of the emergence of mathematics at the eve of its formalisation as an autonomous discipline.

Notes

[1] Stephenson, B. (1987). *Kepler's Physical Astronomy*. New York: Springer-Verlag, p. 2.
[2] Field, J. V. (1988). *Kepler's Geometrical Cosmology*. London: Athlone Press, p. 190.
[3] Stephenson. *Kepler's Physical Astronomy*, p. 1.
[4] See, for example, Koyré, A. (1973). *The Astronomical Revolution: Copernicus-Kepler-Borelli*, trans. by R. E. W. Maddison. Ithaca: Cornell University Press, p. 284; Caspar, M. (1993). *Kepler*, trans. and ed. by C. Doris Hellman. New York: Dover, p. 382.
[5] On the influence of Barbaro's commentary on Kepler's early conception of harmony, proportion and the role of the astronomer as a "mathematical architect",

see Jardine, N. (1998). "The Places of Astronomy in Early-Modern Culture". *Journal for the History of Astronomy* 29: 49-62, pp. 53-5, 58.

[6] Kusukawa, S. (1995). *The Transformation of Natural Philosophy: The Case of Philip Melanchthon*. Cambridge: Cambridge University Press, pp. 139-40. On the creation of a "critical mass" of mathematical astronomers in Lutheran Germany equipped to understand the technicalities of Nicolaus Copernicus' *De revolutionibus orbium coelestium* (1543), see Westman, R. S. (1980). "The Astronomer's Role in the Sixteenth Century: A Preliminary Study". *History of Science* 18: 105-147, p. 121.

[7] Romano, A. (2004). "Réflexions sur la construction d'un champ disciplinaire: les mathématiques dans l'institution jésuite à la Renaissance". *Paedagogica historica* 40: 245-59, pp. 258-9.

[8] Ibid., p. 255.

[9] Barker, P., and Goldstein, B. R. (2001). "Theological Foundations of Kepler's Astronomy". *Osiris* 16: 88-113, pp. 88-90.

[10] Field, J. V. (1984). "A Lutheran Astrologer: Johannes Kepler". *Archive for History of Exact Sciences*, 31: 189-271, p. 208.

[11] Jarrell, R. A. (1975). "Mästlin's Place in Astronomy". *Physis* 17: 5-20, p. 19.

[12] Ibid., p. 20.

[13] For a more thorough consideration of Kepler's critical attitude towards astrology in his early career, see Boner, P. J. (2006). *Kepler's Living Cosmos: Bridging the Celestial and Terrestrial Realms*. unpublished PhD Dissertation. Cambridge: University of Cambridge, pp. 15-38.

[14] On Clavius' positive appraisal of Proclus, see Romano. "Réflexions", pp. 258-9. On Proclus' importance as a "mediator between [Plato's] *Timaeus* and [Euclid's] *Elements*" in Kepler's conception of natural knowledge, see Field. *Kepler's Geometrical Cosmology*, pp. 167-8.

[15] Dyck, W. von; Caspar, M., et al., eds. (1937-). *Johannes Kepler Gesammelte Werke*. Munich: C. H. Beck [hereafter *KGW*], vol. 14, p. 60. Unless stated otherwise, all translations from Latin and German are by the author of this paper.

[16] *KGW*. vol. 4, pp. 161, 181.

[17] Ibid., vol. 14, p. 100.

[18] See Barker and Goldstein. "Theological Foundations"; Barker, P. (2004). "Astronomy, Providence, and the Lutheran Contribution to Science". In Menuge, A. J. L., ed. *Reading God's World: The Scientific Vocation*. St Louis: Concordia Publishing House, pp. 157-87; Rabin, S. J. (1987). *Two Renaissance Views of Astrology: Pico and Kepler*. unpublished PhD Dissertation. New York: City University of New York, p. 157.

[19] Martens, R. S. (2000). *Kepler's Philosophy and the New Astronomy*. Princeton: Princeton University Press, pp. 97-8, 146-68. On the ambiguity of the identity of Kepler's archetypes, see Donahue, W. H. (2002). "Philosophy in Kepler's Astronomy". *Journal for the History of Astronomy* 33: 296-297, p. 297.

[20] See *KGW*. vol. 4, pp. 164, 177. On the use of *a posteriori* observations in Renaissance logic, its scholastic origin and its degree of certainty, see Daniele Cozzoli's paper in this volume.

[21] Originally intended by Plato to refer to a mathematical proportion (ἀναλογία), the word *analogia* came to signify in classical and early modern Latin "a method of reasoning from parallel cases"; see Clarke, P. G. W. (1985). *Oxford Latin Dictionary*. Oxford: Clarendon Press, p. 126.

[22] On Kepler's comprehensive conception of harmony, see Stephenson, B. (1994). *The Music of the Heavens: Kepler's Harmonic Astronomy*. Princeton: Princeton University Press. A general account of geometry and certainty in Renaissance natural philosophy can be found in Daniele Cozzoli's paper in this volume.

[23] *KGW*. vol. 14, pp. 38-9.

[24] Ibid., p. 39.

[25] *KGW*. vol. 13, p. 348.

[26] Simon, G. (1975). "Kepler's Astrology: The Direction of a Reform". *Vistas in Astronomy* 18: 399-426, p. 447; Kusukawa. *Melanchthon*, p. 188.

[27] *KGW*. vol. 13, p. 354.

[28] On the scale of Ursus' scathing rejection of astronomical hypotheses, see Westman. "The Astronomer's Role", p. 126; Rosen, E. (1986). *Three Imperial Mathematicians: Kepler Trapped between Tycho Brahe and Ursus*. New York: Abaris Books, p. 199. On Ursus' composition of the *De hypothesibus astronomicis tractatus* (1597) and his ensuing conflict with Tycho, see Launert, D., and Müller, W. (1993). *Nicolaus Reymers: ein berühmter Mathematiker und Naturwissenschaftler aus Dithmarschen und seine* Arithmetica analytica. Meldorf: Meldorfer Gelehrtenschule, pp. 3-8. For the most thorough and discerning discussion of the controversy between Ursus and Tycho, see Jardine, N. (1984). *The Birth of History and Philosophy of Science: Kepler's* A Defence of Tycho against Ursus *with Essays on its Provenance and Significance*. Cambridge: Cambridge University Press.

[29] Ibid., pp. 211-14.

[30] Ibid., p. 140; cf. p. 89.

[31] Ibid., p. 141; cf. p. 90.

[32] See, for example, Tondorf, E. A. (1904). "Kepler's Attitude toward Astrology". *Popular Astronomy* 12: 303.

[33] *KGW*. vol. 17, p. 67.

[34] Caspar, *Kepler*, pp. 69-70.

[35] *KGW*. vol. 13, pp. 297-298.

[36] Ibid., p. 298.

[37] Ibid., p. 300.

[38] Ibid.

[39] *KGW*. vol. 13, p. 310.

[40] Ibid.

[41] Ibid.

[42] See Ptolemy. *Tetrabiblos*, 34c.

[43] On the importance of Ptolemy's *Almagest* and *Tetrabiblos* in Melanchthon's transformation of the educational curricula of institutions of learning in Lutheran Germany, see Methuen, C. (1998). *Kepler's Tübingen: Stimulus to a Theological Mathematics*. Aldershot: Ashgate, pp. 97, 219.

[44] *KGW*. vol. 13, p. 310.

[45] Ibid.

[46] Ibid. "*Meteora*" has been taken here to refer to "the conditions of the weather". The term appears to have stemmed from the original Greek μετέωρα, "rising up, levitating", which signified all those phenomena interpreted by Aristotle as atmospheric occurrences produced by mixtures of moist and dry exhalations from the Earth. In addition to those phenomena we continue to consider atmospheric, such as clouds, hail, hurricanes, lightning, rain, rainbows, snow, thunder, tornadoes and wind, Aristotle classified as *meteora* comets, meteor showers and shooting stars; see Aristotle. *Meteorologica*, 340b, 341b, 359b-360b, 370a. Although Kepler did not consider comets atmospheric phenomena, condemning such a claim as "doubtful and untrue", he accepted all other *meteora* Aristotle originally identified; see, for example, *KGW*. vol. 8, p. 229. On the multiple meanings of *meteora* in ancient Greece and Rome, see Taub, L. C. (2003). *Ancient Meteorology*. London: Routledge, pp. 1-2.

[47] *KGW*. vol. 13, p. 310.

[48] On Kepler's early coupling of the aspects and the harmonic consonances and his eventual rejection of their correspondence in the *Harmonice mundi*, see Simon, G. (1979). *Kepler astronome astrologue*. Paris: Gallimard, pp. 44-6.

[49] *KGW*. vol. 13, p. 333.

[50] Ibid.

[51] Ibid.

[52] Ibid.

[53] *KGW*. vol. 13, p. 348.

[54] *KGW*. vol. 1, pp. 39, 43.

[55] *KGW*. vol. 13, p. 348.

[56] Ibid., p. 349.

[57] *KGW*. vol. 6, pp. 20-1.

[58] *Ibid.*, p. 21.

[59] *KGW*. vol. 13, p. 349.

[60] Ibid., p. 348.

[61] *KGW*. vol. 14, pp. 38-9.

[62] Ibid.

[63] *KGW*. vol. 14, p. 60.

[64] Ibid.

[65] Ibid.

[66] Ibid.

[67] For a brief summary of Kepler's acceptance of thirteen aspects in the fourth book of his *Harmonice mundi*, see Boner, P. J. (2005). "Soul-searching with Kepler: An Analysis of *Anima* in His Astrology". *Journal for the History of Astronomy* 36: 7-20, pp. 10-1.

ASTROLOGY IN SPANISH EARLY MODERN INSTITUTIONS OF LEARNING

TAYRA LANUZA-NAVARRO[*]

According to Margaret Jacob's recent assessment of the state of the field of studies on early modern European science, the third volume of the Cambridge History of Science shows that historians have liberated early modern science from straitjackets such as the dichotomies science/religion and science/magic.[1] This would mean, as far as astrology is concerned, that the criticisms expressed years ago by Paolo Rossi, Francis Yates and Simon Schaffer,[2] among others, concerning the dismissal of studies on the "occult sciences" within the history of science, have already been addressed and amended. Brian Copenhaver's sentiment in presenting "a few of the larger questions that suggest how relevant occultism is to the history of early modern science" in the volume *Reappraisals of the Scientific Revolution* accord with this idea.[3] However, the actual occurrence of astrology in the scientific institutions of the early modern period – if mentioned – is still usually considered a secondary matter that can be overlooked without misrepresenting mainstream science and/or the activities of early modern scientists. Even if historians of science admit that astrological ideas were not only common in the popular context, but that the influence of the stars was a generally accepted idea in the early modern intellectual world,[4] studies defining what kind of astrology was a part of the training of astronomers and physicians, and how it was taught at the universities and other centres of learning are scarce.

As a consequence, the position of astrology in the map of knowledge during the early modern period has not been well defined, and the general view of the structure of this map is for this reason incomplete and distorted. Indeed, studies on early modern astronomy and medicine, as well as studies on the universities, acknowledge that astrology was present in the institutions of learning.[5] Several historians have referred to the statutes or "constitutions" of the universities and the rules on the teaching of astrology that they expressed and to its general presence.[6] But there are only a few studies that have tried to examine in depth the content of the lectures on astrology, its actual teaching and its role as a discipline in what was considered legitimate knowledge, that is, in modern terms, science.[7] Outstanding among these studies is the recent work by Darrel Rutkin on the teaching of astrology in Italian universities.[8] Monica Azzolini, Ana Ávalos and Borek Neškudla have also considered its occurrence in early modern institutions of learning in certain locations: Bologna, the New Spain and Prague.[9] In the case of Spain, scholars have not failed to stress that astrology played a role in the universities and other institutions where science was taught.[10] However, there has never been an in-depth study of the actual presence of this discipline, its contents in the teaching context, and the relationship of astronomers, cosmographers and physicians with astrology while they taught in universities and other learning centres. This paper aims at stressing the relevance of the teaching of astrology in Spanish institutions of learning, between the end of the sixteenth century and the first decades of the seventeenth century, and the critical value of defining its place in the history of early modern science.

The Spanish case is particularly interesting because it is necessary to consider as institutions of knowledge not only universities but also other centres that were created as a response of the monarchical state to the necessities of the Empire. Mathematics, astronomy, medicine and cosmography, along with astrology, were taught at the universities of Salamanca, Alcalá de Henares, Valencia and Seville.[11] However, a study of the presence of astrology as a discipline related to other areas of knowledge in the Spanish early modern academic world would not be complete without including the lectures at the Casa de Contratación, in Seville, and the Academia de Matemáticas and the Colegio Imperial, in Madrid. Both groups of centres should be studied with the objective of understanding in what way astrology was a part of the study of nature. This article will focus on the second group of institutions, which, although constituting an heterogeneous set, had a fundamental role in the running of the Spanish kingdom and its Empire.

Firstly, I will analyse printed and manuscript sources related to the Casa de Contratación, the Academia de Matemáticas and the Colegio Imperial in order to find evidence of the presence of astrology in these institutions. Secondly, I will describe the content of these sources in order to fully comprehend the astrological theories and ideas contained in them. Using these descriptions and the presence of astrology in the institutions as a basis, my purpose is to unveil the importance of astrology in the seventeenth-century map of knowledge. My final objective is to contest the usual assumption that the distinction between natural and judiciary astrology was a clear and unambiguous one, which situated the former as some kind of "scientific astrology"[12] and the latter as superstition, whose presence was thus restricted to popular culture, and therefore never present in institutions of learning.

The distinction between these two aspects of astrology – natural and judiciary – was characteristic of the medieval and early modern period.[13] Natural astrology was the part of the discipline that studied the influences of the planets and signs on the natural world, being thus important for agriculture, meteorological predictions, navigation and medicine. Natural astrology was intended to make only general predictions. By contrast, the objective of judiciary astrology was to interpret stellar influences and make predictions related to human affairs: individual, political and religious prognostications. Predictions were usually called "astrological judgements".[14] The astrologers of the sixteenth and seventeenth century divided their discipline in several subject matters: genethlialogy (predictions referred to an individual made according to the position of the planets and stars at birth), elections (predictions referring to the result of a concrete action or enterprise made according to the position of the planets and stars at the moment of consultation of the astrologer), and interrogations (predictions on any question, also based on the sky at the moment of the consultation). This early modern distinction of natural versus judiciary astrology has led historians to assume that the difference between both aspects was clear and unambiguous. However, I will argue that the boundaries between what was natural and what was judiciary was not so obvious, and that, in fact, professors of mathematical disciplines included in their works non-natural astrology, in many cases using genethlialogy, elections and interrogations.

Astrology and the division of knowledge

Early modern historians of science hardly need to be reminded that the early modern map of knowledge was structured by disciplines and disciplinary boundaries different to those handled by historians of other periods. Astrology had an important role in the study of the natural world because early modern approaches to nature included astrological ideas that represented the general understanding of the heavens. Even if the nineteenth-century definition of "science" as a category is obviously not relevant for early modern naturalists, *scientia* was a term used to include many aspects of knowledge and practices which our contemporary conception of science has discarded.[15] This included not only the "occult sciences", such as astrology, but also any discipline using logic according to the scholastic tradition, as well as the humanities.[16] The Spanish cosmographer Juan Cedillo Díaz defined astrology as the "science of the stars, because it is about knowledge, order, bigness and aspects of them, and about their passions and accidents with respect to us".[17]

During the early modern period, astrology had strong relationships with other disciplines of a physico-mathematical nature, especially with the group of disciplines that were known during the seventeenth century as 'mixed mathematics'. Astronomy and the geometric calculations required to locate the planets were often seen as knowledge aimed at practical applications, such as the making of calendars and of tools for navigation, meteorology and agriculture, and the making horoscopes and astrological predictions. Geography and cartography were also related to astrology: the traditional division of the Earth into several regions was usually linked with a description of the nature of places and the peoples living in them according to the planets and signs of the Zodiac, that were supposed to have an influence over them. Medicine was one of the disciplines more strongly related to astrology. Astrological knowledge provided physicians with tools for diagnosing diseases, and for establishing the correct days for taking medicines or for purging, according to the influences of the planets and signs on the human body and its humours.[18]

Even if the combination of interests in mathematics, astronomy, medicine, cosmography and philosophy with an interest in astrology seems unlikely today, it was perfectly consistent in the seventeenth century, in the same way that the combination of interests in the Hebrew language, chemistry and philosophy was perfectly reasonable.[19] Astrology was taught at the universities from the medieval period as part of the *quadrivium*, in the context of teaching mathematics, natural philosophy and medicine. As Darrel Rutkin has stressed, by retracing the development of these disciplinary patterns, it is possible to more accurately characterize the pre-modern scientific map of knowledge and its educational institutionalization.[20]

The aim of this article is to contribute to this task, by analysing with the status of astrology in the work of the teachers of mathematics, astronomers and cosmographers at the Casa de Contratación, the Academia de Matemáticas and the Colegio Imperial.

The "Casa de Contratación"

The Casa de Contratación[21] was the Spanish royal institution created to control exploration, colonization and exchange with the territories of the new world. Besides controlling every ship and every trip to America, this institution was also a centre of learning for sailors, and has recently been defined as "a chamber of knowledge".[22] It was the only place where captains could obtain a licence to travel to the West Indies, and where navigators were taught how to sail. Among the people who worked at the Casa de Contratación, there were cosmographers and mapmakers with teaching obligations.[23] Here I discuss the content of the works published by the cosmographers who worked in this institution. As I will show they were indeed interested in astrology in its relations to the geographical description of the Earth and to medicine.

One of the first cosmographers of the Casa de Contratación was Alonso de Santa Cruz (1505-1567), whose services to the king Charles I of Spain (Charles V of the Holy Roman Empire) during more than twenty years included activities related to mathematics as well as to astrology.[24] During the early years of the Casa de Contratación, founded in 1503, astrology was part of the training in cosmography and navigation. The first evidence on astrology being taught at this institution are the words of Alonso de Chaves (appointed "Cosmographer and Master chart-maker" in its early years). They were written in 1545, in the context of a discussion on how to find a solution to the problems caused by the lack of skill and inexperience of pilots:

> There is also need of a man learned in astrology, cosmography and navigation chart, who would read every day a public lesson in this house, so that pilots and other people are taught in the aforementioned arts.[25]

This notable cosmographer clearly considered that astrology needed to be included in the lessons at the Casa de Contratación. The Latin term *astrologia* was commonly used in this period to designate a discipline including both astronomy and astrology in all their aspects, but Chaves could have been talking only about the "theoretical part" of *astrologia,* that is, positional astronomy, obviously necessary for navigation. However, the boundaries between the study of that theoretical part and the practice of prognostication were not clearly defined; and it is not unlikely that astrological teaching was included, along with teaching of the use of sailing instruments and astronomical knowledge. This idea relies also on the importance given to astrology for meteorological prediction during the early modern period. It was a part of the practice of the so-called natural astrology, an aspect of the discipline considered as legitimate knowledge, as valid and unquestionable as natural philosophy. Natural astrology was also considered licit by the Inquisition. In this sense astrology was usually referred to as useful for navigation.

The works published by the cosmographers of the Casa de Contratación show that they were interested in astrology and practised it, and therefore probably taught some aspects of astrology in their lessons.

Jerónimo de Chaves, the son of Alonso de Chaves, was appointed as the professor for the newly created chair of cosmography in the Casa de Contratación in 1552. In 1550 he had published *Cronografía o repertorio de los tiempos,* a very popular work reprinted at least nine times during the sixteenth century. "Cronographies", or repertoires, were quite a popular literary genre in the period. They contained physical descriptions of lands, usually including natural astrology: meteorological predictions, medical advices and instructions for navigation. Several authors published works of this kind that were very popular, such as that of Andres de Li.[26]

The *Cronografía* by Jerónimo de Chaves had a wide and detailed astrological content. The book was divided in four parts, two of which were mainly astrological. In the second part of the work, dealing with the general description of the world,[27] the author included the positions and order of the planets and signs, as well as "the qualities and command that each of them has". The third part of the book dealt with calendar issues, and it included a lunar table with the eclipses until the year 1600.

In the first part of the work Chaves explained the division of time – that is, of history – into ages, referring to "things that have happened". In this section on historical issues he exposed the division of the world in parts according to physicians and astrologers, describing the qualities of each of these. These qualities were related to those of the planets and signs, establishing their influences over the parts of the day. The text also made reference to the dominion of a planet in each of the different ages of men. In the second part Chaves included a description of the elements and issues ruled by each planet: metals and animals, physical characteristics and personalities of men, diseases, colours, disciplines or aspects of knowledge, and the parts of the human body. Most of these belonged to natural astrology, but the inclusion of an individual's personality was often considered as a judiciary practice. Chaves considered, however, that these predictions, even performed on individuals, were natural knowledge.

From the point of view of astrological content, the most interesting part of his book is its fourth chapter. This part included medical questions, including the lists of critical days that were important for medical therapeutics, and the astrological elections for the proper moments to purge and bleed patients. It also included astrological-meteorological matters, described as: "natural prognostications, also called rustic astrology, on the changes of weather, that is, on calm, rains, winds, storms, colds, earthquakes, pestilences and scarcity"[28]. These aspects were obviously also of interest for sailors. Thus, basic astrological ideas were seamlessly included in this kind of knowledge, displayed in the work and teaching of the mathematical practitioners appointed at the Casa de

Contratación. Cosmographers considered astrological qualities as an ordinary part of their works on territorial description. Chaves' work shows how this was perfectly combined: astrology was mingled with geographical description, with meteorological predictions and with medical advice.

Another cosmographer at the Casa de Contratación was Rodrigo Zamorano, the author of the first Spanish translation of the *Elements* of Euclid. His chronology and repertoire, *Cronología y Reportorio de la razón de los tiempos,* had also great popular success, with fifteen editions in fifty years.[29] His book was both rich and broad in astrological contents. In its title page, Zamorano contended that his book would certainly be "a very useful work for physicians, astrologers, farmers, sailors and historians and in general for the entire republic".[30] Actually, the link between those various professional practices was provided by astrology. The work included expositions of the significance of the Zodiacal signs and the character traits of the people influenced by them. Zamorano also referred to the places on earth ruled by each sign, as well as "the means to understand the effects that eclipses will cause"[31]. Several chapters were devoted to the signs (meteorological and astrological) that should be taken into account to understand the changes in the weather and "the abundance or scarcity of the years and those that will be [years] of sickness or pestilence"[32]. This cosmographer included also a guide to judge the effects of comets. Zamorano's work shows the relationship of astrology with the practices of physicians, astronomers, sailors and historians, thus proving that this discipline played a fundamental role in the articulation of the early modern map of knowledge.

Zamorano's successor as cosmographer was Antonio Moreno, who was professor of cosmography from 1612 to 1634 at the Casa de Contratación.[33] Zamorano considered that he fulfilled all the requirements for the job, "because he knows well the mathematical sciences, astrology and cosmography".[34]

Hence, astrology was regarded as important knowledge by those in charge of teaching at the Casa de Contratación. The cosmographers Chaves and Zamorano considered natural astrology as an ordinary part of their interests, and they dealt with judiciary astrology as well, in referring to predictions on men.

The Academy of Mathematics of Madrid

It is also possible to find traces of astrological activity in the Academia de Matemáticas (Academy of Mathematics), established in the court of

Phillip II in the 1580s with the aim of training mathematicians and technicians, experts in arithmetic, geometry, astronomy, and cosmography as well as in other professions.[35] The regular members of the Academy were considered from the beginning as servants of the king, a position that included several privileges and rights, such as dependence on royal jurisdiction and residence at court. According to the *Instituciones*, the Academy's rules written by Juan de Herrera, the Academia had the aim of giving a scientific and technical training to young courtiers.[36] The *Instituciones* established that training in cosmography also meant training in astrology as a secondary discipline;[37] however, this *astrologia* most probably referred only to astronomy, because, when fully developed, the knowledge of an astrologer was supposed to be in this context, geometry, astronomical tables and movements, as well as astronomical instruments.[38]

In 1607, the Chief Cosmographer of Indies, Andrés García de Céspedes, obtained the position of professor at the Academia, and the course of mathematics established there coincided with that of the University of Salamanca,[39] making no reference to astrology. Despite the absence of astrology in the rules of the Academia de Matemáticas, Juan Cedillo Díaz – Céspedes' successor from 1611 – wrote a treatise on astrology, which remained in manuscript form, most probably for being lecture notes.[40] This manuscript – which had previously not been studied – is an important source to understand the place of astrology in early modern Spain, as Cedillo was one of its most important mathematicians and cosmographers.

Cedillo began his treatise with a clear explanation of what should be understood by astrology, establishing a distinction that nowadays seems clear, but that was not so in the seventeenth century. After defining astrology as "science of the stars", as already explained in the introduction to this article, he argued that this "science" had two parts, a practical and a speculative one. The theoretical or speculative part studied the movements and positions of the stars from a geometrical and arithmetical point of view, "and this is the one the moderns call Astronomy". Astrology, on the other hand, had another object:

> The practical one is called judiciary, prognosticatrix and divinatory, because by using several physical and experimental reasons, the strengths and operations of the stars in the bodies and elemental elements can be divined, from which judgements are made on the qualities, inclinations and temperaments of men and on the changes of weather, and the rest of things that are effects of celestial causes, and this is the one the moderns call Astrology.[41]

According to Cedillo's definition, judiciary astrology actually included natural astrology. As I have already argued this opposes the traditional views of historians referring to this discipline. In fact Cedillo's definition follows logical reasoning: if what astrology does is "to judge" according to certain moments of figures in the sky, every kind of astrology is then judiciary. This is another proof that the distinction between both aspects of astrology was not so clearly established.

Judiciary astrology was divided in two parts, the first of them called by Cedillo introductory and the second one, "executive". [42] Introductory judiciary astrology was on the principles used in the judgements, the qualities of signs and planets and their combinations. Therefore, this introductory part was astrological theory. "Executive" judiciary astrology was the part involving actual and concrete predictions, as defined by the cosmographer: "The executive is the one that predicts following those principles the events and occurrences of things". [43] Cedillo divided this practical astrology into four parts, the first of them being catholic astrology, that is, universal predictions regarding cities, empires and monarchies, as well as about wars, plagues, sects, earthquakes and rains. This kind of astrology was difficult, according to Cedillo, because the effects depended on the influence of the fixed stars and of planetary conjunctions. [44] This was quite a popular theory in Spain during that period, even if it was of Arabic origin. The second aspect of judiciary astrology referred to weather predictions, the third to predictions related to the lives of men and their inclinations, and the last one on "natural elections. This is an interesting classification of astrological knowledge, especially because it is followed by the assertion that "the other things that the Arabs judge using astrology are forbidden, such as interrogations, robberies, fortunes and elections that depend on free will". [45]

It has usually been assumed by historians that natural astrology was accepted as scientific astrology, while judiciary astrology was considered divinatory and therefore was forbidden and persecuted by the Inquisition. However, Cedillo's classification implies that parts of what is generally regarded by historians as judiciary astrology could be considered natural astrology by early modern practitioners. Furthermore, the term judiciary was not necessarily identified with forbidden practices. Thus, Cedillo's opinion was that there was something akin to a "natural judiciary astrology", involving particular predictions on the lives of men and their inclinations, as well as some elections. This interesting aspect of the astrological treatise of this cosmographer is another proof that the division between natural and judiciary astrology was a thin line frequently redefined according to the opinion of individual practitioners.

The astrological content of Cedillo's treatise began with descriptions of the planets, classified as beneficial and malefic, as fecund and infertile, and a few others categories. At this point Cedillo considered necessary to explain, following the medieval scholastics, that it should be understood that the planets "did not have those qualities in their own nature, but that they had effective virtues of those qualities, and that they affected, through their movement and radiation, the aforementioned qualities in [terrestrial] elements and elemental bodies".[46] The planets and the stars, made of the fifth element of ether according to the Aristotelian theory, could not have themselves the qualities of the four inferior elements. They only transmitted it. This was what Thomas Aquinas explained as having the qualities not "essentially" but "potentially".

Cedillo went on to describe the nature of each planet, and then described the signs of the Zodiac. He detailed the astrological "essential dignities" of planets in the signs, such as exaltations,[47] and also referred to "the division of the aspects".[48] Te treatise was apparently not concluded, because the aspects were not detailed. Another manuscript by Juan de Cedillo described, instead, those aspects at length, including also some predictions.[49]

Thus, the cosmographer Juan Cedillo accepted catholic astrology in its relation to meteorology ("the changes of air"), as well as to universal predictions affecting regions and cities. But he also considered some judgements on human inclinations to be acceptable, showing again that this aspect of genethlialogy was not completely excluded from the practice of the discipline in institutions of learning.

The Colegio Imperial of Madrid

After the death of Cedillo in 1625, the teaching of mathematics at the Academia was carried on by the Jesuits from the Colegio Imperial of Madrid. This institution had been founded by King Philip IV with two positions for the teaching of mathematics. According to the institution's original plan, written in 1625, studies were intended to include lessons "on mathematics, where a teacher will read in the morning the *sphaera*, astrology, astronomy, the astrolabe, perspective and prognostications".[50] There was therefore a distinction between astronomy and astrology, and the rules specified that the teachers should give lessons on prognostications.

For the Colegio Imperial the case of the Scottish Jesuit Hugonis Sempilius (Hugh Sempill) is very relevant. Sempilius (born in Scotland around 1590, and dead in Madrid in 1654), was professor of mathematics at the Colegio. His work *De Mathematicis disciplinis Libri duodecim* (printed in Ambers in 1635) was dedicated to King Philip IV, and circulated in Europe through the wide network of Jesuit colleges.[51] Among the twelve books on the mathematical disciplines constituting his treatise, book eleventh was devoted to astrology. In addition, its tenth book, on astronomy, included a chapter on the properties of the planets. There Sempilius introduced basic elements necessary for the subsequent exposition of astrology. These bases included the qualities of heat/cold and moist/dry attributed to the planets, and the distinction between beneficent and malefic, and feminine and masculine qualities as applied to the planets. The aforementioned four basic qualities were the simplest principles for astrological prognostication. The other distinctions were a part of the astrological tradition, also fundamental to the prediction. Sempilius explained also some other astrological elements, such as the houses of the planets, their dignities and detriments, the signs of exaltation, triplicities, and even the terms and decans, conventionally used for judiciary predictions. Although he only mentioned them briefly, directing the reader to specialized works for further details, it is however significant that this astrological knowledge was included in an encyclopaedic work on the mathematical disciplines.

Sempilius quoted in his book several astrological texts that were reference works for astrologers, including the *Quadripartito* by Ptolemy and other ancient sources, such as Julius Firmicus' *Mathesis*; Arabic authors such as Albumasar, Haly and Alcabitius, reference Renaissance authors such as Girolamo Cardano, Giovanni Antonio Magini and David Origanus, and even treatises of the Hermetic tradition.[52]

The lists provided of houses, signs, terms and other astrological elements can be seen as a simple extension of competence of astronomy within matters related to the geometrical division of the sky. However, the mention of qualitative attributions of properties, such as the fundamental qualities of the elements, or genre and humoural properties, as well as establishing places of exaltation and detriment for the planets, are purely astrological issues.

The book on astrology was centred on the issue of the compatibility between astrology and Christianity. Hence, Semipilius included in it the complete text of the Bull *Coeli et Terrae,* published by Pope Sixtus V to condemn divination and particularly judiciary astrology. Furthermore, the first chapter of his book on astrology focused on the distinction between the aspects of astrology that were allowed and those which were forbidden for Catholics, and it exposed the conventional arguments against the discipline.

The main astrological content of Semipilius' work is presented in the second chapter, which begins with a clear exposition of the parts of astrology accepted by Semipilius:

> I admit a triple astrology, whose greatness, usefulness and honest pleasure is licit to enjoy. The first concerns the changes of the air [and] the changes of weather in a general and in a particular sense. The second is on the genitures in the natural aspects. The third observes times, days and places, as necessary for physicians, farmers and sailors. I am suspicious of any other astrology.[53]

Thus, the distinction between natural and judicial astrology was flexible, and besides the traditional aspects belonging to natural astrology, agriculture, medicine and navigation, scholars considered that some aspects of nativities were also natural astrology. Therefore, Semipilius, like Cedillo, considered acceptable the predictions on "natural" inclinations of people indicated by observation of the stars.

Semipilius' description of astrological knowledge began with the first topic of natural astrology, that is "the judgements on the changes of the air and of weathers", which were "general [predictions], because they refer to many places and moments, such as floods of water or opposite dryness, and other similar things; which use to come from the great conjunctions of the planets and from the eclipses of the luminaries".[54] Meteorological effects were commonly ascribed to eclipses in the academic context; however, Jesuits like Semipilius also mentioned planetary conjunctions as influences for weather changes. This was idea was in fact part of the Arab theory of the great conjunctions, not seen favourably by the Inquisition. Astrological predictions related to meteorology were licit from the Catholic Church's point of view, but the theory of the conjunctions was forbidden. Semipilius did in fact reject this theory elsewhere. Even so, by mixing both aspects of astrology – astrometeorology and conjunctions – we can see that Semipilius assumed that, for meteorological predictions, the use of conjunctions could be considered licit as it referred to natural practices. Thus, again the boundaries between judiciary and natural astrology were flexible, even if theoretically restricted by Church censorship.

The following part of Semipilius' text dealt with eclipses and their effects. The social influence of solar and lunar eclipses in the early modern period was extensive, and its study was considered as "scientific astrology" in the academic context. Professors of mathematics such as Núñez de Zamora, of the University of Salamanca, also wrote prognostications on eclipses.[55] In his treatise, Semipilius considered that "men whose nativity figure coincides with the eclipse must be noted".[56] Therefore, again, he included predictions usually considered to be judiciary astrology, yet in the context of natural astrology. Nativities were forbidden, but, in the academic context, mathematicians used figures of the sky – that is, astrological charts and horoscopes – to make predictions that they classified as natural.

Sempilius continued with a list of the regions and cities governed by each sign of the Zodiac, composed following the work of the humanist theologian Pedro Sánchez Ciruelo, professor at the Universities of Alcalá and Paris during the first half of the sixteenth century.[57] He concluded this part by summarizing his perspective on the use of astrology, stating:

> "and this way I have said enough about the condition of the air, the predispositions of the inhabitants, diseases, fruitfulness and familiarity of regions and cities with the signs of heaven and planets". [58]

Furthermore, he took care in assuring the reader that all the effects he had referred to were inside the limits of natural astrology. However, as I have argued, the theory of conjunctions was mainly used not for predictions on medicine, agriculture and navigation, but for political and religious predictions.

The third chapter of Sempilius' book was devoted to astrological elections, beginning with a distinction between licit and non-licit practices. Sempilius did not reject every election, "but only those that have no ground in nature, or those fortuitous and voluntary, or those that inquire after causes in the stars that are impious".[59] He accepted natural elections, related to the moment for sowing, to the care of livestock and similar things. He also admitted elections for travelling, accepting predictions of robberies on the highways. This was always a very popular aspect of astrological elections, a question that people usually asked the astrologers for, but predictions of this kind were problematic. Prognostications on robberies were often considered superstitious practices and consequently it regularly featured in the early modern polemics on astrology.

Sempilius also considered that a certain part of nativities should be considered licit. In his opinion "the astrologer must choose the lord of the birth chart, show the station of planets, pay attention to the signifiers of nutrition, judge if the temperament will be sanguine, phlegmatic, choleric or melancholic; indicate the form, height and inclination of the body, as well as natural predispositions, diseases, and insinuate the natural death of the new-born".[60] In fact, some of these aspects, considered licit practices by Sempilius, were to the subject of polemics. Predictions related to death from a birth chart were for instance persecuted practices, forbidden by the Church.

Another aspect that should be highlighted in Sempilius' work is his defence of astrological medicine, based in the theory of humours. The basis of this understanding of the medical practice fitted perfectly in the Galenic theory, assuming that the humoural constitution of a body depended on the nature of the planets and Zodiacal signs that had influenced the person in the moment of his birth. Since it was founded in natural grounds, this aspect of astrology was always allowed. In spite of this, Sempilius' was clear in condemning – in a Catholic perspective – prognostications based in the houses of the planets, as each of the houses could be assigned to predictions on "the fortune or misery of the new-born, as well as if the future days will be of abundance for him…",[61] on sisters and brothers, on parents and legitimacy of the new-born or on friends and enemies.

Sempilius' work proves thus that it was difficult to avoid going beyond the theory of licit natural astrology, because there were always issues located on a flexible boundary between one type of astrology and the other. This was the case with the qualities of men, which if attributed to natural influences could be considered part of permitted practices, but if used for concrete prognostications on the life of a man based in those qualities were instead considered as judiciary astrology.

Conclusions

The content of the printed works and the manuscripts, written by some of the mathematicians, astronomers and cosmographers active between the late sixteenth century and the early seventeenth century at some of the major Spanish institutions of learning, show that astrology arguably had an important role in the map of knowledge of this period.

The repertoires written by the cosmographers of the Casa de Contratación are rich in astrological content, and are not restricted to what was defined as natural astrology. These publications do not imply that the cosmographers taught astrology in the Casa de Contratación; however, there is no doubt that at least some of them were also astrologers, or at least had a deep knowledge of how to practise this art. Jerónimo de Chaves and Rodrigo Zamorano focused on astrological medicine and meteorology, but they also accepted certain personal predictions based in the technique of nativities in aspects that they considered to be also result of natural causes.

The work of Juan Cedillo Díaz, cosmographer at the Academia de Matemáticas de Madrid, reveals an interest in astrology to the point of devoting a whole treatise to this subject. In addition to explaining the theory of astrology and its use in meteorology and medical astrology, Cedillo also referred to the natural inclinations of people according to the influences of the stars in their birth charts, and to the way to practise prognostications on other aspects that were usually considered judiciary astrology. These prognostications were considered superstitious and in fact, some of the astrologers who found themselves accused before the Inquisition for the practice of astrology were condemned because of prognostications related to robberies.

In the Jesuit Colegio Imperial of Madrid, Hugonis Sempilius showed greater reticence regarding the acceptance of astrology. His main concern was the distinction between licit and illicit astrology according to the Catholic Church. The largest part of his chapter on astrology only dealt with meteorology and medical astrology. Although he insisted that he only admitted natural astrology, Sempilius included a certain part of genethlialogy in that natural astrology. Thus, nativities, if used only for those kinds of natural prognostications on the inclinations of the person, could, in his opinion, be licit practices.

Two main general conclusions can be extracted from the cases exposed in this article. First, that between the late sixteenth century and early seventeenth century, astrology had a presence in the scientific institutions where the mathematical sciences were taught. Cosmographers and professors in the Casa de Contratación, the Academia de Matemáticas and the Colegio Imperial practised it, or had a least a wide knowledge of it. A certain part of astrology was rejected and not considered, but the influence of the stars – a conventional belief in that period – was never denied.

Second, the boundaries between disciplines should not be taken for granted, and they often offer a complex and difficult picture. The common assumption of the distinction of natural and judiciary astrology breaks down when we consider its practice within the context of mathematical teaching for practical purposes in early modern Spain. We have thus seen that, in fact, natural astrology could include aspects usually classified as judiciary astrology, such as genethlialogy or Arab theories. Always considering that they were only referring to natural causes, different authors crossed the boundaries of a strict natural astrology by including political and personal prognostications and referring to birth charts. The limits of natural astrology depended on the criteria of each author, or perhaps on his context of practice and social status: some aspects of astrology, in the texts of cosmographers and professors, could be considered natural and thus licit, while the same kind of predictions ended in an Inquisitorial trial for popular astrologers.

This suggests that the distinction to make was not strictly natural versus judiciary astrology, but a distinction between the practices of astrology in the popular realm and in the academic context. Popular astrology was a practice considered as superstitious and illicit. [62] In contrast, academic astrologers did cross the boundaries of licit astrology, by extending the realm of natural astrology to territories commonly attributed to judiciary astrology. In addition to the social status of these practitioners, another reason could perhaps explain this contrast: Academic astrology was considered a part of natural philosophy, or of mathematical knowledge within the discipline of 'mixed mathematics', and this protected it against the activities of the Inquisition. The disciplinary structure of the map of knowledge cannot be understood without considering social factors that contributed to determine its definition, and the case of early modern astrology is a relevant example. Further research on a wider range of practitioners will allow completing the picture, this paper being a first step in this direction.

Notes

[*] This article has been possible thanks to the fellowship "Beca Postdoctoral d'excel·lència per estades en centre estrangers" of the Generalitat Valenciana during 2006 in Bath Spa University, and in the European University Institute, Florence, during 2007.
[1] Jacob, M. (2007). "The Importance of Early Modern European Science and the State of the Field". *Isis* 98: 361-5. For instance, William Eamon's *Secrets of Nature* began explaining that "the conception of knowledge embodied in this account [in a letter of Hermetic studies] blurs the distinction between religious and scientific knowledge". Eamon, W. (1994). *Science and the Secrets of Nature. Books of Secrets in Medieval and Early Modern Culture*. Princeton: Princeton University Press, p. 20.
[2] Among other examples of this criticism, see Rossi, P. (1968). *Francis Bacon: From Magic to Science*. London: Routledge and Kegan Paul; Yates, F. (1972). *The Rosicrucian Enlightenment*. London: Routledge and Kegan Paul; Schaffer, S. (1985). "Occultism and Reason". In Holland, A. J. *Philosophy, its History and Historiography*. Dordrecht: Reidel, and (1987). "Godly Men and Mechanical Philosophers: Souls and Spirits in Restoration Philosophy". *Science in Context* 1: 55-85.
[3] Copenhaver, B. (1990). "Natural Magic, Hermetism and Occultism in Early Modern Science". In Lindberg, D. C., and Westman, R. S. eds. *Reappraisals of the Scientific Revolution*. New York: Cambridge University Press. The terms

'occultism' and 'occult sciences' have been abandoned in more recent publications for being anachronistic.

[4] As Lynn Thorndike showed long ago. Thorndike, L. (1923-58). *History of Magic and Experimental science.* London: Macmillan & Co.

[5] This can be seen in many early modern science scholarship. Some examples are a recent issue of *Science and Education* devoted to science teaching in Early Modern Europe (spring 2006), and Feingold, M., and Navarro, V. (2006). *Universities and Science in the Early Modern Period.* Dordrecht: Springer.

[6] Among others, Federici Vescovini, G. (2001). "Astronomia e Medicina all'Universitá di Bologna nel secolo XIV e agli inizi del XV". In Battistini, P. *Seventh Centenary of the Teaching of Astronomy in Bologna.* Bologna: Cooperativa Libraria Universitaria; pp. 123-49; Jardine, N. (1998). "The Places of Astronomy in Early-Modern Culture". *Journal for the History of Astronomy* 29: 49-63; Siriasi, N. (1987). *Avicenna in Renaissance Italy. The Cannon and Medical Teaching in Italian Universities after 1500.* Princeton: Princeton University Press.

[7] Discussions on the modern term 'science' and its use in referring to the practices of past societies are a constant in history of science, as well as in sociology and anthropology. On the use of the word 'science' by historians of the early modern period, see Dear, P. (2005). "What is the History of Science the History of?". *Isis* 96 (3): 390-406; from the point of view of education, see Smith, M. U. and Scharmann, L. C. (1999). "Defining versus Describing the Nature of Science: A Pragmatic Analysis for Classroom Teachers and Science Educators". *Science Education* 83 (4): 493-509; from a socio-cultural point of view, see Harding, S. G. (1998). *Is Science Multi-cultural?: Postcolonialisms, Feminisms, and Epistemologies.* Bloomington: Indiana University Press. I use the term in this article in its wider meaning to refer to scientific practices, considering that any interest in the physical, natural world was a form of science. With this anachronistic term I mean to be talking about what the seventeenth century scholars would call natural philosophy, but also natural history, mathematics (not included in the previous seventeenth century terms), and several other disciplines that were knowledge, in a broad sense.

[8] Rutkin, D. (2002). *Astrology, Natural Philosophy and the History of Science c. 1250-1700: Studies toward and Interpretation of Giovanni Pico della Mirandola's* Disputationes Adversus Astrologiam Divinatricem. unpublished PhD Dissertation. Indiana University, and "Changing Curricular Patterns: Astrology and the Italian Universities, ca. 1300-1800". Paper presented at the HSS Conference in Washington D.C., November 2007.

[9] Azzolini, M. (2005). "Reading Health in the Stars. Politics and Medical Astrology in Renaissance Milan". In Oestmann, G.; Rutkin, D., and Stuckrad, K. von eds. *Horoscopes and Public Spheres: Essays on the History of Astrology.* Berlin: Walter de Gruyter, pp. 183-206; Ávalos, A. (2007). *As Above, so Below. Astrology and the Inquisition in Seventeenth-Century New Spain.* unpublished PhD thesis. Florence: European University Institute; Neškudla, B. (2002). "Astronomy

and Astrology at Prague University before the Battle at White Mountain". *Acta Historica Astronomiae* 16: 388-92.

[10] See for instance, Navarro Brotons, V. (2006). "The Teaching of Mathematical Disciplines in Sixteenth-Century Spain". *Science & Education* 15 (2-4): 209-33; Esteban Piñeiro, M., and Vicente Maroto, M. I. (2002). "Los cosmógrafos y otros 'oficios matemáticos'". In López Piñero, J. M. dir. *Historia de la ciencia y de la técnica en la corona de Castilla.* Vol. III.: siglos XVI y XVII. Salamanca: Junta de Castilla y León.

[11] Some results of my research on astrology in the Castilian universities and in the University of Valencia were the subject of two previous articles. See Lanuza Navarro, T. (2006). "La Astrología en las Universidades Castellanas durante el siglo XVII". In Batlló Ortiz, J.; Ferran Boleda, J., and Piqueras Carrasco, M. *Actes de la VII Trobada d'Història de la Ciència i de la Tècnica.* Barcelona: Societat Catalana d'Història de la Ciència i la Tècnica, pp. 577-82, and (2005). *Astrología, ciencia y sociedad en la España de los Austrias.* unpublished PhD thesis. València: Universitat de València, pp 131-4.

[12] Vernet, J. (1973). *Astronomía y astrología en la España del Renacimiento.* Barcelona: El Acantilado.

[13] Furthermore, the distinction was sanctioned by the writings of the Fathers of the Church and its Doctors, particularly by Thomas Aquinas in his *Summa Theologica,* part II, 2, q . 95, art. 5.

[14] In the early modern period, the words 'judge' and ''judgement' were usually employed to refer to astrological prognostications. 'To judge' was a synonymous of drawing a celestial chart and making a prediction. The works containing astrological predictions, as well as the predictions themselves, were usually called 'judgements'.

[15] Findlen, P. (1996). *Possessing nature. Museums, Collecting, and Scientific Culture in Early Modern Italy.* University of California Press, p. 9.

[16] Kelley, D., and Popkin, R. eds. (1991). *The Shapes of Knowledge from the Renaissance to the Enlightenment.* Dordrecht: Kluwer Academic Publishers, p. 2.

[17] Cedillo, J. *Tratado de astrología.* Madrid: Biblioteca Nacional, Mss. 9091, p. 14.

[18] Humanism intensified the practice of astronomy and astrology among humanist physicians, who saw it as offering excellent support to the interpretation of the Hippocratic texts. Navarro Brotóns, V. (2002). "El Colegio Imperial de Madrid. El Colegio de San Telmo de Sevilla". In López Piñero, dir. *Historia de la ciencia.* On the relationship of astrology and medicine as expressed in Spanish works, see Lanuza Navarro, T. (2007). "Astrology and Medicine in Early Modern Spain". *Cronos* 9 (1): 49-63.

[19] Coudert, A. (1991). "Forgotten Ways of Knowing: the Kabbalah, Language and Science in the Seventeenth Century". In Kelley and Popkin. *The Shapes of Knowledge*, pp. 83-100.

[20] Rutkin. *Astrology*, p. 17.

[21] "The House of Trade".

[22] Barrera-Osorio, A. (2006). *Experiencing Nature: The Spanish American Empire and the Early Scientific Revolution.* Austin: University of Texas Press.

[23] On the teaching of mathematical disciplines at the Casa de Contratación, see Navarro. "Mathematical Disciplines".

[24] Esteban Piñeiro. "Cosmógrafos".

[25] Chaves, A. de (1545). *Parecer sobre lo errados que estaban los instrumentos construidos por Diego Gutiérrez y sus dos hijos,* quoted in Esteban Piñeiro. "Cosmógrafos". Unless stated otherwise all translations from Spanish and Latin are by the author of this paper.

[26] Andres de Li, Spanish author from the fifteenth century, published his *Reportorio de los tiempos* for the first time in Zaragoza in 1492. His work was very popular and reprinted several times in the sixteenth century. See Delbrugge, L. (1999). Andres de Li: *Reportorio de los tiempos.* New York: Rochester.

[27] Chaves, J. *Cronografia o repertorio de los tiempos.* Sevilla: Joan Gutiérrez. (Chapter II: *Descripción general del mundo*).

[28] Chaves. *Cronografia,* Fol. 8v.

[29] Esteban Piñeiro, M. (2002). "La casa de Contratación y la Academia Real Matemática". In López Piñero, dir. *Historia de la Ciencia.*

[30] Zamorano, R. (1585). *Cronologia y repertorio de la razón de los tiempos.* Sevilla: Andrea Pescioni y Iuan de Leon.

[31] Zamorano. *Cronología.*

[32] Ibid.

[33] Vicente Maroto, M.I., and Esteban Piñeiro, M. (1991). *Aspectos de la ciencia aplicada en la España del Siglo de Oro.* Salamanca: Junta de Castilla y León.

[34] See Vicente Maroto and Esteban Piñeiro. *Ciencia aplicada.*

[35] On the Academia de Matemáticas, see Esteban Piñeiro, and Vicente Maroto. "Casa de Contratación y Academia", and (1988). "Una reflexión sobre la existencia y el significado de la Academia de Matemáticas". In *Estudios sobre historia de la ciencia y de la técnica. IV Congreso de la Sociedad Española de Historia de la Ciencia y de las Técnicas.* Valladolid: Junta de Castilla y León; Navarro. "Mathematical Disciplines".

[36] Esteban Piñeiro and Vicente, Maroto. "Casa de Contratación y Academia".

[37] Herrera, J. de, *Instituciones,* fol. 13r.

[38] Herrera, J. de, *Instituciones,* fol. 11r y 11v.

[39] Vicente Maroto and Esteban Piñeiro. *Ciencia aplicada.*

[40] Cedillo Díaz, J. *Tratado de astrología.* MBN: Mss. 9093. Cedillo's manuscripts with reference numbers 9091 to 9093 in the Biblioteca Nacional of Madrid were identified by M. Esteban and I. Vicente. See Vicente Maroto and Esteban Piñeiro. *Ciencia aplicada,* pp. 152-62.

[41] . Cedillo. *Astrología,* Fol. 14r.

[42] "exercitatoria".

[43] Cedillo. *Astrología,* Fol. 14r.

[44] Cedillo. *Astrología,* Fol. 14r.

[45] Cedillo. *Astrología*, Fol. 14v.
[46] Cedillo. *Astrología*, Fol. 14v.
[47] Essential dignities were the places of the Zodiac where planets were supposed to have their greater strength.
[48] Cedillo. *Astrología*, Fol. 19r.
[49] Cedillo Díaz, J. (1620). *Dianoya de los aspectos de los planetas. Pensamiento nuevo*. MBN: Mss. 9092, Fol. 8r a19v.
[50] Navarro Brotóns. "Colegio Imperial y San Telmo".
[51] On Sempilius and the Colegio Imperial, see Navarro Brotons, V. (1992). "La actividad astronómica en la España del siglo XVI: perspectivas historiográficas". *Arbor* 117: 185-216, and (2003). "Tradition and Scientific Change in Early Modern Spain: The Role of the Jesuits". In Feingold, M. ed. *Jesuit Science and the Republic of Letters*. Cambridge, Mass.: The MIT Press, pp. 336-8.
[52] Sempilius, H. (1635). *De Mathematicis disciplinis Libri duodecim*. Ambers: Baltasare Moretti. MBN: 3/21647.
[53] Sempilius. *Mathematicis*, p. 237.
[54] Sempilius. *Mathematicis*, p. 237.
[55] Núñez de Zamora (1600). *Pronóstico del eclipse de Sol que se hizo el año 1600*. Salamanca: Andres Renaut. MBN: R-36982.
[56] Sempilius. *Mathematicis*, p. 237.
[57] The aim of Ciruelo's work, *Apotelesmata Astrologiae Christianae,* was the defence of licit Christian astrology.
[58] Sempilius. *Mathematicis*, p. 240.
[59] Sempilius. *Mathematicis*, p. 240.
[60] Sempilius. 1635. MBN: 3/21647, p. 241.
[61] Sempilius. *Mathematicis*, p. 241.
[62] On the Inquisitorial trials, see Lanuza-Navarro, T. and Ávalos-Flores, A. (2008) "Astrological prophecies and the Inquisition in the Iberian World". In *Proceedings of the Second International Conference of the European Society for the History of Science* (Cracow, 2006) [forthcoming].

PART II

COMMUNICATING SCIENCE AND PEDAGOGY

COMMUNICATING SCIENCE AND PEDAGOGY

JOSEP SIMON

The study of science pedagogy from a historical perspective has experienced an important revival in history of science in the last decade. This introduction intends to assess qualitatively the re-emergence of this topic, to analyse the structure and dynamics of this research field and to propose new avenues for its improvement. In this context, I will emphasize four related arguments, namely the need for a higher scholarly internationalism, and for promoting interdisciplinarity, and the interest in approaches focusing on communication, and on comparison, in an international perspective.

A major flagship of the new rise of interest in pedagogy in history of science is often considered to be represented by David Kaiser's study of the uses of Feynman diagrams and his editorship of the book *Pedagogy and the Practice of Science*, and by Andrew Warwick's path-breaking study of the Cambridge Mathematical Tripos.[1] Kaiser was also invited to contribute to a recent *Focus* section in *Isis* devoted to reflection on the need of a "generalist vision in history of science". In his paper, he argued that bringing pedagogy to the centre of historical analysis could be a way of avoiding the fragmentation afflicting our discipline. As a feature common to all scientific disciplines and central to science, culture and society in general, the study of training can connect history of science with other disciplines within the historical field, as well as in sociology and anthropology.[2]

However, as pinpointed by John Rudolph – a historian of education trained in history of science[3] – historical research on science education is fragmented in different academic compartments, such as history of science, history of education, science education and various subfields within general history, that rarely interact.[4] Rudolph's overview and a recent historiographical essay by Kathryn Olesko,[5] are perhaps the first general surveys on this topic in more than a decade.[6] From the late 1970s, historians of science such as William Brock and Roy MacLeod, and historians of education such as Edgar Jenkins, contributed to establish fruitful relations between these two academic fields in Britain.[7] However, in the last decade, history of education has suffered significant decay in British universities through the orientation of education departments towards more immediate targets of school education. By contrast, in Continental Europe, the discipline has a more vigorous existence, and indeed there are teams of researchers developing relevant projects on the history of science education.[8] However, as I argue in this introduction, their work is not always acknowledged in the history of science, due to the existence of barriers that present difficulties in communication across disciplinary and national boundaries.

This state of affairs is well represented by Rudolph's expression – in his review of Kaiser's edited volume – of the perplexity that readers of the journal *History of Education Quarterly* would feel in remarking the narrow and unorthodox use of the term "pedagogy" in the aforementioned book.[9] And it is also well illustrated by Kaiser's and Warwick's bold and inaccurate statement in the conclusion of this book: "...although there is an enormous literature on the history of education, virtually none of it is concerned with the relationship between training and the production of scientific knowledge".[10] Further evidence can be found in Warwick's *Masters of Theory*. Despite the indisputable value and originality of this work, it is regrettable that he did not take into account contemporary literature that in fact dealt with the organization of examinations and coaching practices in nineteenth-century France.[11] In fact, the comparative assessment of educational developments happening in Continental Europe was a well-established method in nineteenth-century Cambridge through the work of educationists such as Henry Latham, fellow and tutor of Trinity Hall, who published a well-informed monograph on the pedagogical role of examinations.[12] This literature could have taken his study further, and most importantly, it would certainly have contributed to erase the too often taken-for-granted Cambridge exceptionality, integrating it into a larger framework.

The history of education emerged as a discipline in the early nineteenth century as a subfield within studies on education or pedagogy, aiming at illuminating contemporary educational research and organization through a historical perspective. The reports on foreign education cited in Josep Simon's paper in this part are examples of the early literature in this field. This disciplinary emergence in a field with immediate practical applications offers interesting comparisons with the development of other historical specialisms, such as the history of medicine and the history of science. In this context, the use of international comparisons was considered a fundamental method. Despite different epistemological challenges to this approach, comparative education is today a well-established field. Its basic aim is to compare any activities associated with learning and teaching, whether the comparison acts in the same local or national context or across national boundaries. [13] Despite the major contemporary focus of this discipline, it has also produced valuable international studies of school systems in a historical perspective.[14] Hence, there are elements and approaches in history of education and education studies that historians of science could fruitfully exploit.

With *Pedagogy and the Practice of Science*, Kaiser's great achievement was gathering together an important and international team of historians of science, while offering an excellent account of current results and potential developments. Among the book contributors Kathryn Olesko, Graeme Gooday and the team constituted by José R. Bertomeu, Antonio García-Belmar and Bernardette Bensaude-Vincent had previously developed a long-standing research focus on science education.

Olesko has developed her research on educational organization and science pedagogy in nineteenth-century Germany since the 1980s, and communicated it through a series of excellent papers and a ground-breaking monograph on the Königsberg physics seminar, which, along with Warwick's book, constitutes an analytical and methodological model for future historians of science education and historians of science in general. [15] Despite the localized focus of her monograph study, she provided a general overview of German physics in its connection with the Königsberg seminar and its actors. Furthermore, she accurately integrated the international dimension of physics by studying the appropriation and communication of French physics performed by German physicists in reading, discussion and teaching.

Graeme Gooday is perhaps the author of the latest national account of British science education written by a historian of science. His PhD thesis mapped the rise of the physics teaching laboratory in nineteenth-century Britain. In subsequent years, he has decisively contributed to our knowledge of the practices of measurement in physics and electrical engineering by narrowing his focus through fine-grained case studies of laboratory practice, where pedagogy and training have an important place.[16] The national comparative focus of his earliest work has challenged the historiography of English experimental physics conventionally focused on developments at Cambridge. This has also been one of the aims and the major comparative stance in his editorship, in collaboration with Robert Fox, of a monograph on physics and its teaching in Oxford.[17] Despite the clear historiographical necessity of this comparative approach and the excellent overview of physics and pedagogy in the local context of Oxford offered by this work, it has inevitably contributed to reify a geography of

English physics that now assigns privileged predominance to Oxbridge in the history of the discipline in this country.[18] Their comparative effort could also have been taken beyond national borders. As argued in Josep Simon's paper in this part, educational reforms and the expansion of the sciences in the university and school curriculum took place in nineteenth-century Europe through international observation. Thus, for instance the British and French governments sent commissioners abroad to observe foreign educational systems before preparing or urging universities to undertake reforms. Furthermore, many of the physics textbooks used in nineteenth-century Oxford were French and German.[19] Understanding Oxbridge and British physics would certainly have gained from comparisons with French and German cases and the assessment of the role that international communication played in this context.

In the last decade, Bertomeu, García-Belmar and Bensaude-Vincent have developed a sustained program of recovery and study of nineteenth-century chemistry textbooks published in France and Spain. They have mapped the production of chemistry textbooks in this period and have highlighted the links between pedagogy, research and educational organization. They have coordinated international teams of historians of science and of education devoted to the use of this type of source. Drawing on the rich strand of the use of laboratory notebooks as sources, they have also begun to study student notebooks and connected them with their previous research on textbook writing.[20] Their work has dealt with the international communication of science through the study of the travels of Spanish chemistry students to France and of transnational figures.[21] Certainly they have still to challenge the Paris-centred perspective that characterizes most French historiography. Nonetheless, they are committed to undertaking comparative studies of teaching practices and textbooks through national case studies of countries belonging to the so-called European periphery.[22] However, this project is still under development.[23]

This variety of approaches was reflected in Kaiser's volume despite his ambitious editorial effort.[24] To strengthen the robustness and potential outreach of the book, he provided an introduction to the volume and – in collaboration with Andrew Warwick – a concluding essay. Their approach is based on two classical authors, Thomas S. Kuhn and Michel Foucault. Through the wide readership among historians of science of his *Structure of Scientific Revolutions*, Kuhn contributed to shape conventional ideas on scientific education in our discipline through the role he assigned to training and textbooks in the making of "normal science".[25] Foucault stressed the role of disciplinary regimes – such as those implemented in the school – in the interconnected formation of individuals and society.[26] Kaiser and Warwick propose to enlarge Kuhn's approach by considering the role of training regimes in which the use of textbooks was inscribed, complemented by Foucault's powerfully polysemous use of "discipline".[27] Despite the high value of their respective works, this approach is neither particularly original nor challenging, although in a pragmatic way it could still be useful in maintaining interest in science education among historians of science.

The approaches of Kuhn and Foucault have obvious shortcomings. As argued by García-Belmar, Bertomeu and Bensaude-Vincent, Kuhn's conception of textbooks as repositories of "normal science" is inaccurate in assuming that the making of textbooks is an uncreative task and that its role is restricted to maintaining the "paradigm" without contributing in any other way to the making of science.[28] On the other hand, Foucault's skilful connection of the individual, society and the state through discipline is indeed valuable, but his own experience as an intellectual in the Cold War and postcolonial period probably had an important role in his overemphasis on power as an overarching category.[29] In fact, his approach often contributes rather to obscure the study of pedagogy than to illuminate its everyday mechanisms. Indeed, what Kuhn and Foucault cannot provide is a genuinely historical approach to the study of science education.

By contrast, the work of Rudolf Stichweh and Kathryn Olesko has illuminated through historical inquiry the driving agency that pedagogy has in the constitution of scientific disciplines. In their respective studies, both showed the critical role that the training of secondary school teachers had in the formation of physics as a discipline in nineteenth-century Germany.[30] The stress of Kaiser and Warwick on the role that the circulation of students and teachers trained in particular pedagogical regimes, had in the configuration of this discipline,[31] contributes to strengthen this framework. Textbooks had an important role in this context, but also a wider range of pedagogical practices that Olesko has meticulously studied through examination of course programmes and notebooks.[32] In this sense, she has not only critically undermined Kaiser and Warwick's simplistic mapping of the field through Kuhn and Foucault, but their strong reliance on Polanyi's concept of "tacit knowledge". Olesko has repeatedly argued that although this concept could describe certain situations in scientific practice and pedagogy, many pedagogical processes were perfectly explicit and can be reconstructed through historical sources and techniques.[33] To the communicational obscurity involved in the concept of "tacit knowledge" – attributed by

Olesko to the Cold War culture in which it was conceived[34] – we can oppose instead a new focus on explicit communication.

Stichweh has defined "discipline" as a social system of communication articulated through particular pedagogical practices, and through the creation of new modes of communication such as the rise and consolidation of formal practices of scientific publication, and in particular scientific journals.[35] Analogously, in her paper in this part, Mónica Blanco has shown how different modes of communication (from conversation and correspondence to journal and textbook writing, publication, and international circulation) configured differential calculus as a discipline in eighteenth-century Europe. Moreover, Josep Simon has stressed the fundamental role that the multifarious processes of communication and appropriation of Adolphe Ganot's *Physique* played in the definition of nineteenth-century physics as a discipline in a Franco-British comparative perspective.

In her recent historiographical essay on science pedagogy, Olesko has emphasized the importance of the work of Ludwik Fleck in stressing the role of education as a social process in the making of science, and the role of publishing in shaping pedagogy and science.[36] In an equally thoughtful historiographical essay on science popularization, Jonathan Topham has also highlighted Fleck's work and the role of communication in science to obtain a better historical understanding of science popularization and its relation to the making of science.[37] As expressed in the introduction to this volume, Topham's rethinking of science popularization as a history of science communication is connected to a wider perspective formulated by James Secord, intending to make communication central to our analysis. Like Topham, Rudolph has noted the consonance between the study of education and of popularization, and the fundamental characterization of education as knowledge and skills' communication. This idea is also implicit in Jonathan Rose's recent call for the conception of history of education as a history of reading.[38]

Communication is a fundamental concept, especially if, for instance, we take into account the fact that orality has a primordial role in educational practices. As stressed by García-Belmar in his study of Louis-Jacques Thénard's chemistry lectures,[39] by Olesko on her analysis of Friedrich Kohlrausch's canonical textbook,[40] and by Mónica Blanco and Josep Simon in their respective studies of the development of differential calculus and physics, textbooks were often printed appropriations of oral lessons. Orality has received much attention from historians of popular culture, anthropologists, linguists and historians of reading, with a tendency to focus in contexts related to civilizations without writing, early stages of language learning, eighteenth and nineteenth-century social conversation, and informal education. [41] However, in an exceptional monograph – surprisingly, ignored by Anglo-American historians of science – Françoise Waquet has provided us with a long range account of the high status assigned to orality by science practitioners from the sixteenth to the twentieth century. Even in the context of nineteenth-century science – highly dominated by communication through printing and by organized systems of education – she shows that orality was considered by many scientists as the highest mode of communicating science, above its printed reproduction. [42]

In this sense, the study of communicational practices in education, extending James Secord's example of the role of literary replication, [43] can offer new avenues to study science education across different periods and national boundaries. The papers in this part are an example of this approach, in which the configuration of scientific disciplines is studied through the communication and appropriation of scientific knowledge into different formats and meanings (encompassing the oral, visual, manuscript and printed). Comparison of the eighteenth-century case study by Mónica Blanco and the nineteenth-century case study by Josep Simon gives some indication of the critical changes in Europe signalling the emergence of the journal as a mode of scientific communication, the expansion of scientific education and its state organization, the consolidation of textbooks as a form of pedagogical and scientific communication, and changes in authorship practices. Despite the differences informed by obvious contextual and temporal parameters, the two cases can be fruitfully handled through a focus on communication and appropriation, and they contribute to highlight common aspects characterizing the historical study of science education. For example, international communication is shown to have a fundamental role in the shaping of pedagogical and scientific practices, and textbooks are seen as fundamental agents of these across national and cultural boundaries. Booksellers have important agency in this circulation and – with teachers and textbook authors – they decisively contribute to shape scientific knowledge. Furthermore, textbooks are not repositories of "normal science". Their authors certainly try to reproduce scientific consensus and appropriate it for the purposes of pedagogical practices, but textbooks also display scientific disagreement and can be the subject of dispute or controversy. They contribute to illuminate the complex relations between teaching and research and they have a fundamental role in discipline building.

Hopefully, the combination of the fine-grained analysis of local cases, with tools offered by the disciplines which intersect across science education, the macroscopic approach of comparative education and the analysis of communication processes will contribute to strengthen the historical study of science education and further its research. The papers in this part are an attempt to move in this direction.

Notes

[1] Kaiser, D. (2005). *Drawing Theories Apart. The Dispersion of Feynman Diagrams in Postwar Physics*. Chicago: The University of Chicago Press, and, ed. (2005). *Pedagogy and the Practice of Science: Historical and Contemporary Perspectives*. Cambridge, Mass.: MIT Press; Warwick, A. (2003). *Masters of Theory: Cambridge and the Rise of Mathematical Physics*. Chicago: Chicago University Press.

[2] Kaiser, D. (2005). "Training and the Generalist's Vision in the History of Science". *Isis* 96: 244-51, pp. 250-1.

[3] Rudolph is the author of a magnificent account on the configuration of science education in Cold War America. Rudolph, J. L. (2002). *Scientists in the Classroom: the Cold War Reconstruction of American Science Education*. New York: Palgave.

[4] Rudolph, J. L. (2008). "Historical Writing on Science Education: A View of the Landscape". *Studies in Science Education* 44 (1): 63-82. I am grateful to John Rudolph for sending me a copy of his essay before publication.

[5] Olesko, K. M. (2006). "Science Pedagogy as a Category of Historical Analysis: Past, Present, and Future". In García-Belmar, A. et al. eds. "Textbooks In The Scientific Periphery". *Science and Education*. Special Issue 15 (7-8): 863-80.

[6] Brock, W. H. (1990). "Science Education". In Olby, R.; Cantor, G. N.; Christie, J. R. R., and Hodge, M. J. S., eds. *Companion to the History of Modern Science*. London: Routledge, pp. 946-59; Jenkins, E. W. (1985). "History of Science Education". In Husen, T., and Postlethwaite, T. N., eds. *International Encyclopaedia of Education: Researches and Studies*. Oxford: Pergamon, pp. 4452-56.

[7] In addition to their scholarly projects, this can be seen in some of their publications. See for example Brock, W. H. (1973). *H. E. Armstrong and the Teaching of Science, 1880-1930*. Cambridge: Cambridge University Press, and (1974). "From Liebig to Nuffield: A Bibliography of the History of Science Education, 1839-1974". *Studies in Science Education* 2: 67-99; Mac Leod, R., ed. (1982). *Days of Judgement : Science, Examinations and the Organization of Knowledge in Late Victorian England*. Driffield: Nafferton; Mac Leod, R., and Moseley, R. (1978). "Breadth, Depth and Excellence: Sources and Problems in the History of University Science Education in England". *Studies in Science Education* 5: 85-106; Jenkins, E. W. (1979). *From Armstrong to Nuffield: Studies in Twentieth-Century Science Education*. London: Murray, and, "History of Science Education".

[8] Compère, M.-M. (1995). *L'histoire de l'éducation en Europe: essai comparatif sur la façon dont elle s'écrit*. Paris: INRP. See also for example the history of education service of the French National Institute for Pedagogical Research (INRP). [http://www.inrp.fr/she/].

[9] Rudolph, J. L. (2006). " [Review of *Pedagogy and the Practice of Science* by David Kaiser ed.]". *History of Education Quarterly* 46 (4): 628-30.

[10] Kaiser and Warwick. "Kuhn, Foucault, and the Power of Pedagogy". In Kaiser. *Pedagogy and the Practice of Science*, pp. 393-409, on p. 393.

[11] See for example Belhoste, B.; Dalmedico, A. D., and Picon, A. (1994). *La formation polytechnicienne, 1794-1994*. Paris: Dunod; Belhoste, B. ; Picon, A., and Sakarovitch, J. (1990). "Les exercices dans les écoles d'ingénieurs sous l'Ancien Régime et la Révolution". *Histoire de l'éducation* 46 (mai): 53-63; Belhoste, B. (2001). "La préparation aux Grandes Écoles scientifiques au XIXème siècle". *Histoire de l'éducation* 90 (mai): 101-30, and Belhoste, B., ed. (2002). "L'examen: Évaluer, sélectionner, certifier. XVIe-XXe siècles". *Histoire de l'éducation* 94 (mai).

[12] Latham, H. (1877). *On the Action of Examinations Considered as a Means of Selection*. Cambridge: Deighton, Bell and Co.

[13] See Thomas, R. M. (1990). "The Nature of Comparative Education: How and Why are Education Systems Compared". In Thomas, ed. *International Comparative Education: Practices, Issues & Prospects*. Oxford: Pergamon Press, pp. 1-21, on pp. 1-2; Epstein, E. H. (1992). "The Problematic Meaning of 'Comparison' in Comparative Education". In Holmes, B., ed. *Theories and Methods in Comparative Education*. Peter Lang: Frankfurt am Main, pp. 3-23; Arnove, R. F. (1999). "Introduction: Reframing Comparative Education. The Dialectic of the Global and the Local". In Arnove, R. F., and Torres, C. A., eds. *Comparative Education: The Dialectic of the Global and the Local*. Lanham: Rowman & Littlefield Publishers, Inc., pp. 1-23.

[14] Although the only study of this type well-known to historians of science is represented by the work of Joseph Ben David (whose defects are pinpointed in the introduction to part V in this volume), there have been other relevant publications in this field such as Vaughan, M., and Archer, M. S. (1971). *Social Conflict and Educational Change in England and France, 1789-1848*. Cambridge: At the University Press; Archer, M. S. (1979). *Social Origins of Educational Systems*. London: Sage Publications Ltd, and Green, A. (1990). *Education and State Formation. The Rise of Education Systems in England, France and the USA*. Basingstoke: Macmillan.

[15] See Olesko, K. (1985). "The Mental World of Physiklehrer: Subject and Method in History of Mathematics". *Recherches en Didactique des Mathématiques* 6 (2-3): 347-62, (1989). "Physics Instruction in Prussian Secondary Schools before 1859". *Osiris* 2nd series 5: 94-120, (1991). *Physics as a Calling: Discipline and Practice in the Königsberg Seminar for Physics*. Ithaca: Cornell University Press, (1993). "Tacit Knowledge and School Formation". *Osiris* 2nd series 8: 16-29, and (2005).

"The Foundations of a Canon: Kohlrausch's Practical Physics". In Kaiser. *Pedagogy and the Practice of Science*, pp. 323-55.
[16] See Gooday, G. J. N. (1989). *Precision Measurement and the Genesis of Physics Teaching Laboratories in Victorian Britain*. Canterbury: University of Kent. unpublished PhD thesis, (1991). "Teaching Telegraphy and Electrotechnics in the Physics Laboratory: William Ayrton and the Creation of an Academic Space for Electrical Engineering in Britain, 1873-1884". *History of Technology* 13: 73-111, (1995). "The Morals of Energy Metering: Constructing and Deconstructing the Precision of the Victorian Electrical Engineer's Ammeter and Voltmeter". In Wise, N., ed. *The Values of Precision*. Princeton: Princeton University Press, pp. 239-82, and (2004). *The Morals of Measurement: Accuracy, Irony, and Trust in Late Victorian Electrical Practice*. Cambridge: Cambridge University Press.
[17] Fox, R. and Gooday, G., eds. (2005). *Physics in Oxford, 1839-1939. Laboratories, Learning and College Life*. Oxford: Oxford University Press.
[18] The contribution by Jeff Hughes is a single exception to this pattern. Hughes, J. (2005). "Redefining the Context: Oxford and the Wider World of British Physics, 1900-1940". In Fox and Gooday, pp. 267-300.
[19] Fox, R. (2005). "The Context and Practices of Oxford Physics, 1839-77". In Fox and Gooday, pp. 24-79, on pp. 71-2.
[20] See Lundgren, A. and Bensaude-Vincent, B. (2000). *Communicating Chemistry: Textbooks and Their Audiences, 1789-1939*. Canton, Mass.: Science History Publications; Bertomeu-Sánchez, J. R.; Garcia-Belmar, A., and Bensaude-Vincent, B. (2002). "Looking for an Order of Things: Textbooks and Chemical Classifications in Nineteenth Century France". *Ambix* 49: 227-250; Bensaude-Vincent, B. ; García Belmar, A., and Bertomeu Sánchez, J. R. (2003). *L'émergence d'une science des manuels: Les livres de chimie en France (1789-1852)*. Paris: Éditions des archives contemporaines; Garcia-Belmar, A. (2006). "The Didactic Uses of Experiment: Louis Jacques Thénard's Lectures at the Collège de France". In Bertomeu, J. R., and Nieto-Galan, A., eds. *Science, Medicine and Crime: Mateu Orfila (1787-1853) and His Times*. Sagamore Beach: Science History Publications, pp. 25-53, and García Belmar, A. and Bertomeu Sánchez, J. R. (2004). "Les cahiers d'élèves sources pour une histoire des contenus et des pratiques de l'enseignement de la chimie". In *Le cours magistral: modalités et usages* [http://www.inrp.fr/she/cours_magistral/expose_thenard/expose_thenard_complet.htm].
[21] García Belmar, A. and Bertomeu Sánchez, J. R. (2003). "Constructing the Center from the Periphery. Spanish Travellers to France at the Time of the Chemical Revolution". In Simoes, A.; Carneiro, A., and Diogo, M. P., eds. *Travels of Learning. A Geography of Science in Europe*. Dordrecht: Kluwer Academic Publishers, pp. 95-139; Bertomeu and Nieto-Galan. *Science, Medicine and Crime*.
[22] On the centre-periphery model see introduction to part V in this volume.

[23] Bertomeu Sánchez, J. R., et al. (2006). "Textbooks in the Scientific Periphery: Introduction". In García-Belmar, et al. "Textbooks In The Scientific Periphery", pp. 657-65.

[24] This was pointed out by knowledgeable reviewers like John Rudolph.

[25] Reference to Kuhn in this context is still conventional nowadays. Like Kaiser and Warwick, excellent scholars such as Robert Fox and Graeme Gooday refer to his work to stress the importance of studying training in science. Fox, R., Gooday, G., and Simcock, T. (2005). "Physics in Oxford: Problems and Perspectives". In Fox and Gooday, pp. 5-6; Kuhn, T. S. (1962). *The Structure of Scientific Revolutions*. Chicago: University of Chicago Press.

[26] Foucault, M. (1975). *Surveiller et punir: Naissance de la prison*. Paris: Gallimard.

[27] Warwick, A. and Kaiser, D. "Kuhn, Foucault, and the Power of Pedagogy". In Kaiser. *Pedagogy and the Practice of Science*, pp. 393-409.

[28] García-Belmar, A., Bertomeu-Sánchez, J. R. and Bensaude-Vincent, B. (2005). "The Power of Didactic Writings: French Chemistry Textbooks of the Nineteenth Century". In Ibid., pp. 219-51, on pp. 220-2.

[29] Seigel, M. (2005). "Beyond Compare: Comparative Method after the Transnational Turn". *Radical History Review* 91 (Winter): 62-90, p. 64; Olesko. "Science Pedagogy", p. 868.

[30] Stichweh, R. (1992). *Zur Entstehung des Modernen Systems Wissenschaftlicher Disziplinen: Physik in Deutschland*. Frankfurt: Suhrkamp, and (1994). "La structuration des disciplines dans les universités allemandes au XIXe siècle. *Histoire de l'éducation* 62 (mai): 55-73; Olesko. *Physics as a Calling*.

[31] Kaiser. *Drawing Theories Apart*, pp. 60-7; Warwick. *Masters of Theory*, pp. 254-64.

[32] See also Olesko. "The Mental World of Physiklehrer", and "The Foundations of a Canon".

[33] Olesko "Tacit Knowledge and School Formation", pp. 16-7, and "Science Pedagogy", n. 3 p. 878.

[34] Olesko. "Science Pedagogy", pp. 865, 871.

[35] Stichweh. *Zur Entstehung des Modernen Systems*, pp. 7-93, and (2003). "Differentiation of Scientific Disciplines: Causes and Consequences". In *Encyclopedia of Life Support Systems*. Paris: UNESCO.

[36] Olesko. "Science Pedagogy", pp. 872-4.

[37] Topham, J. R. (2008). "Rethinking the History of Science Popularization/Popular Science". In Papanelopoulou, F.; Nieto-Galan, A.; Perdiguero, E., eds. *Popularising Science and Technology in the European Periphery, 1800-2000*. Aldershot: Ashgate. I thank Jon Topham for sending me a copy of his paper before publication. See also Simon, J. (2008). "Circumventing the 'elusive quarries' of Popular Science: the Communication and Appropriation of Ganot's Physics in Nineteenth-century Britain". In Ibid.

[38] Rudolph. "Historical Writing on Science Education"; Rose, J. (2006). "The History of Education as the History of Reading". *History of Education* 36 (4): 595-605.

[39] Garcia-Belmar. "The Didactic Uses of Experiment".

[40] Olesko. "The Foundations of a Canon".

[41] See Goody, J. (1977). *The Domestication of the Savage Mind.* Cambridge: Cambridge University Press, and (1987). *The Interface Between the Written and the Oral.* Cambridge: Cambridge University Press; Olson, D. R. (1994). *The World on Paper: The Conceptual and Cognitive Implications of Writing and Reading.* Cambridge: Cambridge University Press; Burke, P. (1993). *The Art of Conversation.* Cambridge: Polity; Miller, S. (2006). *Conversation. A History of a Declining Art.* New Haven: Yale University Press; Secord, J. (2007). "How Scientific Conversation Became Shop Talk". In Fyfe, A. and Lightman, B., eds. *Science in the Marketplace.* Chicago: Chicago University Press, pp. 23-59. I am grateful to Jim Secord for sending me a copy of his paper before publication.

[42] Waquet, F. (2003). *Parler comme un livre. L'oralité et le savoir (XVIe-XXe siècle).* Paris: Albin Michel. This argument has also been recently put forward by James Secord in a vivid analysis of the role of conversation in science in early nineteenth-century England. Secord. "How Scientific Conversation Became Shop Talk".

[43] Secord, J. A. (2000). *Victorian Sensation: The Extraordinary Publication, Reception, and Secret Authorship of Vestiges of the Natural History of Creation.* Chicago: The University of Chicago Press, p. 126.

On How Johann Bernoulli's Lessons on Differential Calculus Were Communicated in Eighteenth-Century France and Italy

Mónica Blanco Abellán[*]

Over the past thirty years there has been a fair amount of historical work on eighteenth-century differential calculus, approaching the subject matter in a variety of ways. Some of these works insist on the discussion of foundations or methodology. Meanwhile, others discuss particular developments or the work of a single author. In any case, these studies regard Europe as a unit, wherein the communication of mathematical knowledge met no obstacles in its transit between countries. Even where the dissemination of differential calculus is investigated, there is no discussion of the context of the practice of mathematics. The emergence and application of differential calculus in well-defined contexts have been thus ignored, as have the interaction between such contexts.

In the history of mathematics it has been usual to assume that the mathematical communication between countries flowed without constraints, partly because of an "internalist" approach, which bases on intrinsic elements the explanation of the development of mathematical concepts.[1] Dissenting in this matter, Gert Schubring prefers "to analyze communication processes, to identify basic units of communication" for a common understanding of knowledge, and "to study interaction between such units".[2] A major example of such units are state educational systems, in which communication is potentially possible. This characterisation does not exclude that within a state boundaries several educational systems may coexist. Schubring proposes comparative analysis of textbooks as a means of examining the differences between national peculiarities with regard to style, meaning and epistemology, in considering that textbooks emerge from a specific national educational system.[3]

Insofar as communication contributed to the formulation of the ultimate discourse of differential calculus, it cannot be considered to be a neutral process. James Secord's proposal of conceptualizing science as a

form of communication offers in this context an interesting frame.[4]
"What" is being communicated can only be answered through the
understanding of "how", "where", "when" and "for whom".[5]
Unfortunately, while communication practices have especially attracted
the interest of early modern and Renaissance historians of science, only
few works have been devoted to its study in the eighteenth century.[6] In
spite of this, I believe that eighteenth century practices of communication
concerning differential calculus deserve an analysis in depth, both at
national and transnational level. In particular, the reading and writing of
educational books[7] represents only a part, however crucial, in the process
of communication. Taking Schubring's approach as starting point, in my
paper I intend to revise the comparative analysis of educational books
from a different point of view, as part of a more complex network of
diversified communicative practices.

The analysis of the reading practices of educational books, extended by
a survey of their actual uses, fits conveniently into Lissa Roberts'
approach in the sense that "the local particularities of education (...) help
to shape the context wherein that which circulates is taken up, interpreted
and put to use".[8] This naturally entails the study of local appropriation,[9]
which turns out to be essential since some of the readers later became
authors of new educational books. Furthermore, in this paper I take on
board Jonathan Topham's call for a more systematic survey of scientific
authorship by shedding some light on the authorship practices involved in
the writing of educational books on differential calculus in eighteenth-
century France and Italy.

In addition, my analysis of authorship practices supports Kathryn
Olesko's view on the close connection between research and teaching.[10]
Then it is easy to understand how important educational books appeared to
be in shaping differential calculus as a discipline, in the frame of Rudolf
Stichweh's theorization of this concept.[11] But, while Stichweh's approach
focused on the emergence of disciplines in the context of nineteenth-
century German universities, in my paper I intend to extend the
conceptualization of differential calculus as a discipline to contexts other
than the university, such as scientific academies, societies and institutions
offering more elementary education.

This paper opens with the examination of the different practices of
communication involved in the circulation of Johann Bernoulli's lessons
on differential calculus in eighteenth-century France. The study of the
book of the Marquis de L'Hôpital, an essential element in the
communication of differential calculus, originated from Bernoulli's
lessons, provides an exemplary case of circulation of knowledge in France.

Subsequently, I show how another educational book based in Bernoulli's lessons text, by the French Oratorian Charles Reyneau, was appropriated in Italy by Maria Agnesi, thus providing a transnational analysis of communication. Given the subject's novelty, educational works on differential calculus were needed to help both in learning and in teaching the new method. In the last section of this paper, the publication of L'Hôpital's work and its subsequent commentaries allows reconstruction of the discussion that started within this specific context regarding what was to be considered as "accepted" pedagogy, as David Kaiser puts it.[12]

Johann Bernoulli and the *Analyse des Infiniment Petits*

Several education systems coexisted in eighteenth-century pre-Revolutionary France. University education was mainly restricted to the *collèges*, run by religious orders, such as the Jesuits and the congregation of the Oratoire. In addition, by the 1750s a well-developed network of *écoles militaires* established in France, some of which, like the one in Sorèze, were formerly religious schools.

Involvement in educational matters was not among the founding principles of the Oratorians. However, the gathering of intellectuals around the Oratoire, on the one hand, and the contemporary needs for popularization of culture, on the other, led them to establish their own schools. While Jesuits established their *collèges* in big cities, the Oratoire developed its educational activity among smaller towns, at the request of town councils.[13] This preference was the consequence of the reluctance among the Oratorians to compete against the Jesuits, all the more so after conflicts over the control of the schools at Angers and Nantes. As a result, the Oratorian *collèges* accepted students from the *bourgeoisie* and artisan *élite*, that is to say, from a less favoured social *milieu* than the students at the Jesuit *collèges*.[14]

The social background of the students was not the only difference between Jesuit and Oratorian *collèges*. Their teaching methods differed profoundly, the Jesuits being definitely more conservative and austere in their teaching. Instead of dictating the courses – a standard practice for the latter – the Oratorians envisaged them as reading seminars, in which students were actively engaged, while the teacher guided their reading and work.[15] Latin had a dominant role in Jesuit education, contrary to the Oratorian tendency towards education in vernacular. It is, however, remarkable that mathematics was taught in French at the Jesuit *collèges* already in the seventeenth century, in view of the *"gens d'épée"*[16] who were enrolled in the courses and apparently did not master Latin.[17] As I will discuss later in this paper, the dichotomy between Latin and French appears closely connected with the process of communication of differential calculus.

In Paris, the congregation of the Oratoire was closely connected with the Académie des Sciences. The fact that the Oratorian Maison d'Institution was located opposite the Louvre – where the meetings of the Académie were held – is a significant circumstance, and it enhanced the communication between these two institutions. Among the Oratorians, Nicolas Malebranche (1638-1715) stands out for his extended "vocation académicienne", as Robinet puts it.[18] Although not officially an honorary member until 1699, Malebranche's involvement in the affairs of the Paris academy exerted a large influence on the development and spread of differential calculus in France through his circle of friends, *académiciens* and correspondents. This group included the Marquis de L'Hôpital (1661-1704), Pierre Varignon (1654-1722), Charles R. Reyneau (1656-1728), Louis Carré (1663-1711), and Pierre R. de Montmort (1678-1719), among others. In fact, it was not unusual for academicians to gather at Malebranche's room at the Oratoire for their scientific discussions.

Originally educated in Cartesianism, in the 1670s Malebranche shifted from Cartesian geometry to Leibnizian theory of the infinite.[19] Malebranche became acquainted with Gottfried Wilhem Leibniz (1646-1716) when the philosopher and diplomat visited Paris between 1672 and 1676. When Leibniz critized Malebranche's work for its Cartesian standpoint, the Oratorian, always inclined to revision and self-improvement, positioned himself at an ever-growing distance from his original Cartesianism and considered a revision of his work in the light of Leibniz's approach. The final break with Cartesianism came when one of Malebranche's theological works was banned by the Pope in 1690. It was then that he openly turned his interest exclusively to scientific matters, just when the calculus was in full swing.[20]

In 1684 Leibniz published a short seminal paper on the differential calculus in the journal *Acta Eruditorum*.[21] Leibnizian calculus had its origins in the method of finite differences he had elaborated in the 1670s. Having read Pascal's *Traité des sinus du quart de cercle*, he observed the behaviour of the sum and difference of the terms in the *arithmetic triangle*.[22] In this paper, Leibniz defined differentials, stated – without proof – the basic rules of differentiation, and applied them to problems on tangents and singular points. It is true that most of Leibniz's research was already known through his correspondence, as well as through his close connection with academies and societies. However, this short paper was certainly the first printed exposition of differential calculus.[23]

In his theory of discipline structuring, Stichweh stresses the publication of works in specialized journals as one of the factors shaping a discipline. Thus, Leibniz's paper was an early contribution to the configuration of differential calculus as a discipline. As essential for the communication of the differential calculus, the *Acta Eruditorum* deserves some lines, all the more so since the editor of the journal, Otto Mencke (1644-1707), maintained a frequent correspondence with Leibniz. In fact, both were members of the same local society in Leipzig. The production of such a journal had formerly been attempted by Leibniz himself.

The *Acta Eruditorum* was published in Leipzig monthly from 1692 to 1792. As a scientifically oriented journal, it was designed after the French *Journal des Sçavans*, the British *Philosophical Transactions* and the Italian *Giornale de'Letterati*. To attract a wider international audience, articles in the *Acta* were written in Latin. Here again we meet with the dichotomy Latin-French. Although French was gaining ground from about 1650, Latin was still the international language of science, and German was hardly spoken outside the German states. Not only were original articles to be published in the new journal, but also papers from other European journals, after being translated into Latin.[24] The distribution patterns of the journal's monthly issues varied. In the German states, they were distributed through a stable network of booksellers. Beyond German borders, Mencke tackled this task personally, through German correspondents living abroad, or through foreigners whom he had come to know in his travels around Europe. Most of the times the trade was established upon barter terms, that is to say, Mencke purchased books in exchange of issues of his journal. Likewise, Mencke obtained submissions for his journal through personal encounters and correspondents.[25] Given Leibniz's connection with the *Acta Eruditorum*, it is not surprising that he chose this journal to communicate his new method. Leibniz's himself, being a diplomat, contributed to the distribution of the journal abroad.

As an advocate of reform in the mathematical sciences, Malebranche had taken a special interest in fostering the study of Leibniz's articles. Yet, while acknowledging Leibniz as the master and inventor of the new calculus, he found no little difficulty in understanding Leibniz's paper, as he himself admitted in a letter to Leibniz:

> Could you not, Sir, give the public in more detail than you have done the rules for this calculus and the uses that one may take from it?[26]

Malebranche thus encouraged others of greater ability to deepen their understanding of the new calculus to spread it subsequently in a pedagogical way.[27]

One of Mencke's contacts in France was the Marquis de L'Hôpital, who supplied the Académie des Sciences with the monthly issues of his journal. Guillaume François Antoine de L'Hôpital, Marquis de Sainte-Mesme, Comte d'Entremont, Seigneur d'Ouques, with a taste for geometry, soon became interested in the Leibnizian calculus. A friend of Malebranche, and an honorary member of the Académie des Sciences since 1690, he was himself engaged in this same year in writing a manuscript on the arithmetics of the infinite. This project was encouraged by Malebranche, who wished to publish this manuscript at the end of L'Hôpital's work on conic sections. L'Hôpital's *Traité des sections coniques*, however, had to wait until 1707 for posthumous publication.[28]

In 1691 the Swiss Johann Bernoulli (1667-1748) visited Paris. Johann and his brother Jakob were considered to be the leading exponents of Leibnizian calculus at that time. It goes without saying that Johann's stay in Paris aroused intense interest in Malebranche's circle. Upon being introduced to L'Hôpital by Malebranche in his room at the Oratoire, Bernoulli challenged L'Hôpital on the resolution of a certain curve, concerning the osculating circle. Posing challenges to their colleagues on difficult problems, whose solution a *savant* had secretly arrived at, was a common practice to gain reputation within the intellectual sphere. Scientific journals proved to be an incomparable vehicle for making such challenges public. Bernoulli later described the episode in the following terms:

> After having employed nearly an hour to write drivel on several sheets of paper he finally found the exact value of the radius at the peak of this curve; but he was aware that I laughed at the horrible length of such a limited rule, and by telling him that the problem of determining the radius of the evolute for every kind of curve, either algebraic or transcendental, at the peak or at the other points, was nothing more than a children's game for us, that one could give a general formula from which one would find the radius in as many minutes as he had employed quarters of an hour...[29]

L'Hôpital counterattacked with a difficult curve, which Bernoulli immediately solved. According to Bernoulli's account of this episode, L'Hôpital was so fascinated by the new calculus that he wished to learn it from Bernoulli.

Johann and Jakob Bernoulli were Mencke's contacts in Switzerland. This was highly important because the trade between Germany and France was often interrupted due to war. In the meantime, the trade could continue through Switzerland, thanks to its neutral position. This explains why L'Hôpital was not in possession of the issues of the *Acta Eruditorum* containing Leibniz's articles on the new calculus and had to ask Johann

Bernoulli for a copy of them. On the margins of the articles L'Hôpital wrote down some notes expanding Leibniz's new theory in detail. Apparently, he intended to gather his notes as a book on the arithmetics of the infinite. In fact, the reading and discussion of Leibniz's papers were the starting point of Johann Bernoulli's lectures to the Marquis. L'Hôpital asked Johann Bernoulli to initiate him into Leibnizian calculus, and to that end invited Johann, both to his Paris home (end of 1691-July 1692) and to his estate in Oucques (August-October 1692). In Paris a friend of Johann Bernoulli, Johann Heinrich Stähelin (1668-1721),[30] was appointed to copy the lessons before submitting Bernoulli's original notes to L'Hôpital. During his stay in Oucques, however, Johann, lacking Stähelin's assistance, neglected duplicating his lessons, as he explained in a letter to Montmort.[31]

After Johann left France, the instruction continued in his subsequent correspondence with L'Hôpital. It is relevant here to note that while Bernoulli's lectures were written in Latin, their correspondence went on in French. I presume, therefore, that their discussions in Paris and Oucques were conducted in French too. The correspondence dealing with calculus ended in June 1694. In a letter dated August 22nd 1695, L'Hôpital told Bernoulli that he was about to get his work on conic sections printed, at the end of which he would include a small treatise on differential calculus. There, he would acknowledge Johann Bernoulli's contributions. The following year, L'Hôpital published what was considered, by contemporaries and subsequent historians, the first systematic work on differential calculus, *Analyse des infiniment petits pour l'intelligence des lignes courbes.*[32] Of course, this work largely relied upon the lectures that Johann Bernoulli gave L'Hôpital between 1691 and 1692.[33] In another letter L'Hôpital informed Johann Bernoulli of the publication of his book, and urged him to publish a similar book on integral calculus.

It is notable that the first edition of L'Hôpital's book was anonymous. In a letter of 1697 to L'Hôpital, Johann Bernoulli, after praising the *Analyse*, added that L'Hôpital should have his name written on the cover, in order to give more authority to the new calculus. The publisher of the first edition was the director of the Imprimerie Royale, Jean Anisson, who published the *mémoires* and *régistres* of the Académie des Sciences, as well as some works by Malebranche. It is therefore likely that Malebranche used his editorial contacts in the Académie to get L'Hôpital's book published. This anonymous edition seems to provide evidence that the role of the author was not yet well established in the late seventeenth century. If such was the case, then the Imprimerie Royale, through its connection with the Académie des Sciences, might have conferred authority on the book. That in 1715 Montalant published the second

edition under the name of L'Hôpital suggests a turn in authorship
practices.

Despite his few reflections on foundational aspects,[34] L'Hôpital's book
certainly contributed to the spread of Leibnizian calculus.[35] In the preface
to the *Analyse*, he claimed that with this new method the exact solution of
problems could be achieved in a shorter way than using Greek geometry.
Paradoxically, he turned to Greek geometry as a guarantee validating the
new calculus.[36] L'Hôpital's position on this point agreed with that of most
of the adherents to the new method.[37] In this struggle to seek public
recognition under the acknowledged standard of exactitude we can trace
the emergence of a discipline, turning again to Stichweh's conceptual
frame.

The publication of the *Analyse* raised a debate within the Académie des
Sciences (1700-1706) on the admissibility of the new calculus.[38] Some
académiciens, led by the conservative Cartesian Michel Rolle (1652-
1719), attacked the lack of rigour and the results achieved by means of the
new technique, supporting instead the classical techniques. Another group
of *académiciens* took the side of Leibnizian calculus; their defence was led
by Varignon between 1700 and 1701. The subsequent attempts to provide
calculus with an accurate methodology only reinforces the notion of a search
for recognition, a characteristic trend of an emerging discipline, as has
been argued before.

In 1742 Johann published the *Lectiones de calculo integralium*, a
compendium of his lessons to L'Hôpital concerning integral calculus. In a
note, Johann lamented that his lessons on differential calculus were in fact
contained in L'Hôpital's *Analyse*. But it was not until L'Hôpital's death,
that he openly claimed the authorship of the *Analyse*. The timing of
Bernoulli's reaction is significant, given that he had established a
particular economic and intellectual exchange with L'Hôpital, which was
only broken by his death, as I explain below. In 1922, Paul Schafheitlin
edited Johann's *Lectiones de calculo differentialium*, which was written
between 1691 and 1692 for L'Hôpital's benefit.[39] However, the source of
L'Hôpital's work became evident when, in 1955, Otto Spiess published
Johann's correspondence.[40] In a letter dating from March 1694, L'Hôpital
offered Johann a yearly income for life in exchange for communicating his
discoveries exclusively to him.[41] Although Johann's answer could not be
found, a subsequent letter made plain that he had accepted the deal. In
another letter L'Hôpital admitted that, given the scarcity of elementary
treatises on the new method, he was planning to publish a work on
differential calculus, where he would acknowledge Johann's contributions.
It was intended as the introduction to the treatise on integral calculus that

Leibniz apparently meant to prepare. In fact as we have seen above, L'Hôpital's intention dates back to 1690.[42]

The accomplishment of the *Analyse* was clearly the result of the communication and pedagogical interaction between L'Hôpital and Johann Bernoulli in a context where teaching and research did not have clear boundaries. Indeed, according to Stichweh the interaction between research and teaching is another factor defining a discipline. The first four chapters of the *Analyse* present the same structure as Bernoulli's text, namely: introduction of basic definitions and rules, tangents, extreme values and inflection points. In his preface to his edition of Bernoulli's *lectiones* Schafheitlin pointed out several examples where L'Hôpital would have corrected some innacurate results caused by Bernoulli's inattentiveness.[43] Likewise, more essential errors in the *Analyse* would have already been corrected by Johann himself in his subsequent papers. On the other hand, the correspondence with L'Hôpital made possible the communication of Johann Bernoulli's discoveries at another stage of the story. Insofar as L'Hôpital provided Johann with new problems, his research advanced thanks to the interaction with L'Hôpital, who challenged him with new questions and pointed out aspects that needed revision. In turn the correspondence between L'Hôpital and Bernoulli was the main source of the *Analyse*'s remaining chapters.

However, Bernoulli was not the only reader of L'Hôpital's work (and vice versa). It is therefore worth discussing how the *Analyse* and its author were regarded after its publication. L'Hôpital became a representative of the new calculus in France not only because of his scientific contributions to the *Journal des Sçavans*, the *Mémoires de l'Académie des Sciences* and even the international *Acta Eruditorum*, but also because of his correspondence with Leibniz, Jakob and Johann Bernoulli and Christian Huygens. This view was shared by Bernard de Fontenelle (1657-1757), permanent secretary of the Académie des Sciences:

> It would take too long to explain here all the masterpieces on Geometry with which M. de l'Hôpital, & the small group of his fellows have embellished the Journals of either Germany or France.[44]

Alongside Fontenelle, subsequent historians of mathematics such as Montucla and Bossut, praised L'Hôpital's work, all the more so since the author put an end to the mystery shrouding the new calculus and revealed all the secrets concerning the geometry of the infinite.[45] For instance Fontenelle stated in his *éloge* that:

The Geometry of the infinitely small quantities was still but a kind of mystery, &, so to speak, a cabalistic science held by five or six people.[46]

The *Analyse des infiniment petits* was widely read during the eighteenth century, and not only in France. Non-French authors of similar works on calculus cited L'Hôpital's book.[47] It was reissued several times in Paris throughout the eighteenth century, and also published in Avignon (1768).[48] Stone translated it into English (1730), though he rewrote, on Newton's behalf, L'Hôpital's text in terms of fluxions, with an extra section on the integral calculus. Stone's version was in turn translated into French in 1735. In Vienna there were also two Latin versions of the *Analyse* (1764, 1790).

In this first section we have met a range of communication practices in France, used by a whole gamut of actors: the inventor (Johann Bernoulli), the writer of educational books (L'Hôpital), the copyist (Stähelin), the chronicler (Fontenelle), the journal editor (Mencke) and his journal distributors abroad (Leibniz, L'Hôpital, Bernoulli). At the centre of this communication network, controlling all these practices, the figure of Malebranche, with his kaleidoscopic personality, stands out: the advocate of the new calculus, the promoter of a pedagogical vehicle for communicating it, the editor of manuscripts, and the man who exploited personal encounters, hence favouring the circulation of knowledge. Furthermore, some of these practices have led us into the Latin-French dichotomy. The choice of Latin appears to be essential for reaching a wider international audience, as in the case of the *Acta Eruditorum*, while in France the use of French was awarded a growing importance at the level of both the popular and the cultural *élite*. Finally, we have identified some decisive factors pointing to the shaping of the new calculus as an emergent discipline.

Johann Bernoulli's lessons were also the source which inspired another French author to publish a work on calculus in the early eighteenth century. The next section is devoted to this other work, and also emphasizes how it was appropriated by an Italian author some forty years later.

Charles R. Reyneau and Maria G. Agnesi:
How Johann Bernoulli's lessons got to Italy

It was Malebranche who in 1698 encouraged his friend the Oratorian Charles René Reyneau (1656-1728) to compose a work on the new calculus intended for beginners.[49] Reyneau met Malebranche at the Maison d'Institution in Paris. In 1682 he was appointed professor of mathematics at the Oratorian *collège* of Angers. However, he had to give up teaching in 1705 due to his increasing deafness, and he moved to Paris, where he was elected *associé libre* of the Académie des Sciences in 1716. To accomplish Malebranche's request, Reyneau managed to get a copy of Johann's lessons, worked them out and finally published his *Analyse démontrée* in 1708,[50] with the help of the Oratorians Louis Byzance and Claude Jaquemet. How Reyneau got the copy of Johann's lessons offers us the chance to reflect on practices of communication and authorship. Again the role of Stähelin turns out to be relevant. In October 1718 Montmort, in a letter to Johann, made it plain:

> It is from a friend [Stähelin] who was in Paris with you, & who copied your lessons for M. de l'Hospital that Father Reyneau has got his manuscript, in which I have noticed some small scraps in his book Analyse démontrée: Father Bizance had also one. When I urged M. de l'Hospital to lend it to me; he gave me a letter to Father Bizance, in which he asked him to lend me his; but apparently the request was in vain, because I did not get it; Father Reyneau lent me his around a year later.[51]

During Johann's stay in Oucques, Stähelin apparently lent his copy to the Oratorian father Louis Byzance, who ordered Carré to make a new copy, and finally Johann's lessons reached father Reyneau.[52] Once again the transcription and subsequent circulation of manuscripts appears to have been a common practice within the Oratorian context. They copied manuscripts and papers from journals, then studied them, writing their comments and corrections in the margins. These copies were passed around the members of Malebranche's group. It is worth stressing here that the work developed by Malebranche and his group can be regarded as "teamwork", in that it was not unusual that an original text would be revised by several anonymous authors. Malebranche himself filled the margins of the *Analyse* with comments and revisions.[53]

The full title of Reyneau's work, *Analyse démontrée ou la Méthode de résoudre les problèmes des mathématiques, et d'apprendre facilement ces sciences*, reveals that Reyneau's intended readers were beginners. However, it was not written for any kind of reader, but "pour les Lecteurs qui sçavent au moins médiocrement la Géometrie ordinaire".[54] To offer a completely up-to-date work on the new mathematics, he even got involved in the Rolle-Varignon debate.[55] He was actually entrusted with the reports of the debate, gathering Varignon's answers to the objections raised by Rolle to the differential calculus (1700-1701).[56] Once more it is remarkable that there is a link between teaching and research concerning the communication of differential calculus.

In 1716 Reyneau became a member of the Académie des Sciences.
Therefore, upon his death Fontenelle wrote his *éloge*, wherein he praised
the encyclopaedic nature of his work:

> ... he attempted to gather, into one work for his students' use, the principal
> theories found in Descartes, in Leibniz, in Newton, in Bernoulli, in the
> Acta of Leipzig, in the Mémoires de l'Académie...[57]

Concerning the reception of Reyneau's work, Fontenelle added that, at
least in France, Reyneau's book could be taken as an introductory guide to
the modern geometry, which prepared the reader for more complex
works.[58] Even Johann Bernoulli praised Reyneau's book in a letter to the
Oratorian in 1714.[59] Although Reyneau's book relied upon Bernoulli's
lessons on calculus, the fact is that, in this case, he did not claim
authorship.[60]

Reyneau's work turned out to be crucial in the systematization of the
differential calculus in eighteenth-century Italy. Despite being split into
several states and city-states, religious homogeneity prevailed in
eighteength-century Italy, where the Catholic faith was accepted in every
state. Catholic institutions, specifically the Jesuit ones, were entrusted with
the educational system, within which mathematics was taught at an
elementary and traditional level, displaying no specialization.[61] Academies
and societies were more active in mathematical research, in general, and in
the study of Leibnizian calculus, in particular, than the universities. Even
at military schools mathematics was taught at a higher level than at
universities. Leibnizian calculus was not introduced at the University of
Padua until some thirty years after Leibniz's paper of 1684.[62] Yet, it had
already been studied, especially in northern Italy, by individuals and
groups, such as Jacopo Riccati in Venice, Giulio Fagnano in Senigallia
and the brothers Manfredi in Bologna.[63]

In the early and mid-eighteenth century, Italy witnessed a reformist
Catholic movement, referred to by Mazzotti as the "Catholic Enlightenment",
which attempted at a reform of knowledge but within the orthodox Catholic
framework.[64] Enlightened Catholics showed a concern for the pedagogy of
the modern sciences, but without altering their religious framework. This
group consisted of authoritative ecclesiastics with an interest in modern
scientific practices, who were mostly university professors. Thus, modern
scientific culture was promoted within the framework of the traditional
Catholic system of knowledge. Unlike the Jesuits, the reformers supported
education in vernacular, thus rejecting the centrality of Latin. Here again,
Jesuit education was connected with social dominance. One of the strongest
focus points of this movement was Milan, where Maria Gaetana Agnesi

(1718-1799) was born. She is believed to be the first woman to publish a mathematical work, namely, the *Instituzioni Analitiche* (1748). As she stated in her preface, this work on modern analysis was written in Italian on behalf of young Italians. In the eighteenth century, women rarely committed themselves to scientific matters, yet, within this reformist context, talented women were socially accepted.[65] According to Truesdell[66] only this trend can account for Agnesi's popularity in the mathematical scenery of the century. However, more recent studies (such as Mazzotti's essay) offer a more positive portrait of Agnesi and her work.[67]

As a means to improve his social status, Agnesi's father held cultural and scientific meetings at his *palazzo*. He provided Agnesi with tutors, most of them ecclesiastics with an interest in modern science. Among these the Olivetan Benedictine monk Ramiro Rampinelli (1697-1759) played an important role in the production of Agnesi's book. Rampinelli taught at religious colleges in Bologna and Milan, and from 1747 he held a position at the University of Pavia, where he had been lecturing on calculus. According to Mazzotti, a few outdated works were the only introductions to infinitesimal calculus printed in Italy.[68] Moreover, Agnesi noted that the relevant materials were spread and scattered in journals and books by various authors.[69] The Italian skepticism towards printing scientific results in late seventeenth century could account for this reluctance to publish. Until the 1660s the concern in preserving priority in discovery was largely safeguarded by the correspondence between *virtuosi*,[70] which, among other things, was immune to the ecclesiastic authorities. As a result Italian science lacked basic and specialized treatises, to the extent that *virtuosi* looked down on writers of educational books. Around the 1660s the first foreign scientific journals began to arrive in Italy, disseminating the idea of scientific journalism as the best possible means to secure priority in discovery. In 1668 the first Italian scientific journal was published, the *Giornale de'Letterati*, following the *Journal des Sçavans* and the *Philosophical Transactions* in style. Scientific journalism fostered not only international scientific relations, so far neglected, but also a new technique for social advancement. Through publishing, the *virtuosi* aimed at winning the esteem of the great. Insofar as they had to reach a broader audience, they tried to appeal to the public sphere by encouraging the notion of the utility of science and, hence, boosting the production of educational books.[71]

In this context, Agnesi undertook a systematic, educationally oriented, introduction to algebra, Cartesian analysis (analytic geometry) and calculus. Although the social and cultural conditions favoured the association of women with the learned and academic community, in most cases they were discouraged to enter the university because lectures and discussions were conducted in Latin.[72] Though it is not mentioned, this situation might have carried weight on Agnesi's decision to write her book in Italian. Through Rampinelli, Agnesi corresponded with Jacopo Riccati (1676-1754), with whom she held discussions on topics related to calculus in view of her prospective work, mostly of a pedagogical nature. On Rampinelli's recommendation, she also studied Reyneau's *Analyse démontrée*, which was her most important source.[73] That Agnesi accepted this recommendation is hardly surprising since her philosophical background, shared by Reyneau, was Cartesianism filtered through Malebranche's interpretation. It is generally established that her work relied largely on Reyneau's. However, she did not simply translate his work; for instance, Agnesi's procedure did not follow Reyneau's for the differentiation of the product. In fact, being herself a follower of Newtonian natural philosophy, she might have been aware of Berkeley's criticism on Newton's compensation of errors and avoided this proof in her text.

In his paper of 1989 Truesdell critized Agnesi for not having included rational mechanics and experimental physics in her work.[74] However, Mazzotti[75] has shown that Agnesi essentially viewed mathematics and geometry as a way to discover absolutely certain truths, influenced by Malebranchian philosophy. Accordingly, she deliberately excluded physical topics from her work and focused on geometry to preserve the traditional framework of religious and metaphysical knowledge. [76] Mazzotti stresses that in centres of Leibnizian tradition, such as Basel and northern Italy,[77] the development of calculus was closely related to applied problems in dynamics, mechanics, and hydraulics. However, British practitioners, influenced by Newton, centered their interest on the geometrical nature of calculus and its application to celestial mechanics. As a result, Agnesi's geometrical point of view rendered the English translation of her work (1801) a suitable introduction to algebra and calculus in Newtonian context.

To round off this communication network, it is worth tracing how Johann's lessons returned to France through Agnesi's work. At the Académie des Sciences, Jean-Jacques Dortus de Mairan (1678-1771), who succeeded Fontenelle as permanent secretary, and Étienne Mignot de Montigny (1714-1782), an associate geometer at the time, reported favourably on the *Instituzioni* in 1749 before the *académiciens*.[78] Montigny praised Agnesi's work not only for its clearness and thoroughness, but also for having gathered the methods and conclusions of several geometers. It is not unlikely that Montigny introduced her work in

France, as it can be inferred from a passage in a letter to Agnesi: "I had the pleasure of having introduced into my country an extremely useful work".[79]

In 1775, a comission of the Académie des Sciences deemed it worth translating Agnesi's second volume – on differential calculus – into French. This version was recommended at institutions such as the royal military schools of Brienne (1782) and Sorèze (1784). The latter, a former Benedictine *collège* transformed into *école militaire* in 1776, offered a course on differential and integral calculus as early as in 1772. The review to the French translation of Agnesi's work in the *Journal des Sçavans* sheds some light on the actual audience of this work:[80]

> This work can be used to understand our learned articles spread by the hydrodynamics & mechanics of M. l'Abbé Bossut, & consequently to be a part of his great course for engineers.[81]

Before closing this section it is worth adding that Agnesi's undertook the printing of her book privately. Actually, she had the press brought into her father's house, in order to supervise the process personally. This practice stands in stark contrast to the printing and publishing procedures concerning the works of L'Hôpital and Reyneau.

To sum up, the network embracing the Académie des Sciences and the Oratoire boosted the interaction of a wide variety of actors, which in turn moulded the picture of Paris as a centre of scientific communication. By contrast, in the Enlightened Catholic context such an interaction might have been hampered by the existence of several small centres scattered over northern Italy.

On the other hand, in this section I have stressed how differently Agnesi and Reyneau appropriated Johann Bernoulli's *lectiones*. Even so, Bernoulli's text seems to have had a common structure that rendered it recognizable for both authors, despite belonging to different social worlds. For this reason, Johann Bernoulli's text can be viewed as a "boundary object" between France and Italy, in the sense used by Roberts in relation to knowledge circulation.[82]

The books discussed in the previous sections were not regarded as immutable things, but rather as objects that could be revised and improved. In the next section I provide a survey of the commentaries which appeared in eighteenth-century France on the subject of L'Hôpital's book, with the aim of producing more pedagogical texts.

What was accepted as a "textbook" on differential calculus in eighteenth-century France?

So far I have deliberately avoided using the word "textbook" on differential calculus, since the status of the educational books on calculus was not yet well established. However, the publication of L'Hôpital's *Analyse* revealed that a debate on "accepted" pedagogy was brewing, to such an extent that the role of authors of educational works was shaping up beside the so-called *inventors* and their *original* discoveries. [83] Likewise, the structure of such books started to take shape. In the first four chapters of L'Hôpital's *Analyse* we observe the same structure as in Bernoulli's *Lectiones*, namely, introduction to basic concepts and rules, tangents, extreme values, and points of inflection. This structure is retained in several subsequent educational books on calculus in the eighteenth century.

From the correspondence between L'Hôpital and Johann Bernoulli one could infer that Johann approved of the *Analyse des infiniment petits*. However, in a letter to Leibniz, he lamented that L'Hôpital's most remarkable merit was having translated his lessons into French in an accurate and well ordered manner. This kind of author, Johann proceeded further, differed greatly from those whose books introduced the reader to new discoveries. [84] In our time, Truesdell has held the same argument when he referred to Agnesi's work as lacking "any specific invention", the most outstanding contribution being its clear presentation and order. [85]

Yet, L'Hôpital's *Analyse des infiniment petits* received a favourable reception by the *Journal des Sçavans*, a periodical whose main intention was to "faire savoir ce qui se passe de nouveau dans la *République des Lettres*", [86] and, among other things, to provide an exact catalogue of the books printed in Europe. The *Journal*'s review of the *Analyse* provides some hints to define what could be labelled as "accepted" pedagogy. Among other aspects, it praised the *Analyse*'s generality and conciseness:

> It is worth noticing that in the whole work there are but very few propositions: but they are all general, and like so many methods which are easy to apply to as many particular propositions as one may wish. [87]

Beside these features Fontenelle, in L'Hôpital's obituary *éloge*, pointed out the significance of the work's form:

> There are, especially in Mathematics, more good books, than there are well made ones, that is to say, one sees several which can be used for instructing, & few which instruct with a certain method, & so to speak,

with a certain charm. It is well to have a good subject, but one neglects the form. M. de l'Hôpital has given a book as well made as good; he had the skill to gather an endless number of things into a quite small volume; he wrote with this brevity & this neatness so delightful for the mind, the order & precision of ideas...[88]

Three commentaries appeared in France as a supplement to L'Hôpital's book, namely, the *Commentaire sur l'Analyse des infiniment petits* (Paris, 1721) by Jean-Pierre de Crousaz (1663-1750), the *Éclaircissemens sur l'Analyse des infiniment petits* (Paris, 1725) by Pierre Varignon, and the *Analyse des infiniment petits, suivie d'un nouveau commentaire pour l'intelligence des endroits les plus difficiles de cet ouvrage* (Avignon, 1768) by Aimé-Henri Paulian (1722-1802). They commented on L'Hôpital's work point after point, that is, they kept an identical structure. Though they thought highly of L'Hôpital's work in general, Crousaz and Paulian objected that it was intended for *savants*, and their commentaries aimed at filling in the pedagogical gap. In this sense, the analysis of their prefaces proves to be revealing.

The Swiss Jean-Pierre de Crousaz studied theology and philosophy at Lausanne, Leiden and Paris. During his stay in Paris he met Malebranche. A former Calvinist deacon in Lausanne, from 1700 to 1724 he held a professorship of philosophy and mathematics at the Academy of Lausanne and for a short period was appointed professor of philosophy at the University of Groningen (1724-1726). The preface of Crousaz's *commentaire* opens with some criticism of L'Hôpital's *Analyse*. Crousaz disapproved of an excessive use of *donc*,[89] which might convey the idea that a line was the immediate consequence of the previous one. To reach such a result, however, the reader needed to be acquainted with geometry, algebra and differential calculus. Had L'Hôpital included some references, he would have guided the reader through his *Analyse*. Crousaz objected to L'Hôpital's laconism, and therefore attempted to write a more extended and thorough work. So much so, in fact, that he asserted that readers with a knowledge only of elementary algebra and geometry would easily follow his text. Nevertheless, he acknowledged L'Hôpital's merit in having systematically collected a number of hitherto scattered studies[90] – all the more since the publication of the *Analyse* had put an end to the secrecy shrouding the new calculus. Only by not revealing their discoveries, Crousaz added, could the inventors' reputation be safeguarded. Crousaz sent an exemplar to Johann Bernoulli, who did not think much of the work, on the grounds of the amount of errors it contained. When Crousaz won the prize contest of the Académie des Sciences in 1720, Johann went as far as dismissing him as a mathematician in a letter to Varignon.[91]

Crousaz's preference for Newton's philosophy, instead of Leibniz's, and Johann's own personal dispute with L'Hôpital might have brought on such a reaction. But, as it is claimed in his *éloge*,[92] Crousaz was also known for other mathematical works, pedagogically oriented, such as his *Réflexions sur l'utilité des Mathématiques, & sur la manière de les étudier, avec un nouvel essai d'Arithmétique démontrée* (1715) and his *Traité de l'Éducation des Enfans* (1722).

The edition and publication of Varignon's commentary followed a different path. In fact, it made plain "the role of the publisher in determining what texts appeared in print".[93] The Jesuit Pierre Varignon was educated at the Jesuit *collège* in Caen. From 1688 to his death he was professor of mathematics at the *collège* Mazarin, and in 1694 he held in addition a chair of Greek and Latin philosophy at the *Collège Royal* in Paris.[94] A member of the Academies of Sciences of Paris, London and Berlin, he became part of Malebranche's circle, along with L'Hôpital and Reyneau. Varignon's commentary contained not only the explanations of – what he considered to be – some obscure points and a revision of L'Hôpital's text, but also contributed new propositions, additional problems, rules, constructions, and different methods, which he had worked out. While he dictated and edited his courses at the Jesuit *collège* Mazarin in Latin,[95] his commentary of L'Hôpital's *Analyse* was written in French. As has been hinted above, the mathematical courses at the Jesuit *collèges* were conducted in French, and this fact might account for his decision to write his commentary in French. Varignon intended to publish his notes annexed to L'Hôpital's text, but never did so. Some of Varignon's papers were bequeathed by Fontenelle, but some of the most important ones were sold in auction and bought by Guillaume Debure, a *marchand libraire* who inaugurated a dynasty of book traders in Paris. The publisher Jacques Rollin found out that this *marchand* was in possession of Varignon's manuscript and bought it.[96] In the foreword addressed to the reader, Rollin explained that, when he became aware of the existence of such notes, he decided to publish them for the benefit of the public. Otherwise, Varignon's text would have ended up in private hands and would have been left unknown. In fact it was the publisher himself who gave the title to the commentary, *Éclaircissemens*, and decided to publish it independently from the *Analyse*:

> Since the Analyse des Infiniment Petits has already come out several times, & in France & in foreign countries is very widespread, & all learned people are supplied with it, I thought that I would please them if I could save them the expense of buying it a second time, & that it would be no less convenient to read the Commentaire in a separate volume, which one

could easily have open at the same time as the Analyse which it clarifies & explains, than to have them both in a single book.[97]

It was not only the errors in Crousaz's *commentaire*, but also the difficulty of Varignon's *éclaircissements*, excessive for a beginner, which compelled Aimé-Henri Paulian to publish a new commentary on the *Analyse des infiniment petits* in the third edition of 1768, as he explained in its preface. A Newtonian Jesuit,[98] Paulian taught physics at several French Jesuit *collèges*. In his opinion, Varignon had just attempted to make clear those aspects that he himself had not grasped completely, but his *éclaircissements* failed to introduce beginners to the subject.[99] He acknowledged the positive contributions of both Varignon and Crousaz notwithstanding. His criticism of L'Hôpital's work turned essentially to its too concise rules and rather complicated developments. His *commentaire* included what Paulian considered as necessary prerequisite knowledge to tackle the study of the differential calculus. A new edition of Paulian's work appeared in 1781, enlarged and revised by Louis Lefèvre-Gineau, who held a chair of experimental physics at the *Collège Royal* from 1786 to 1829.

The prefaces of the commentaries to L'Hôpital's *Analyse* allow us to discuss what a book with educational purposes should be like according to historical actors, and, on the way, to describe another factor characterizing the emergence of the differential calculus as a new discipline. Broadly speaking, an educational book on differential calculus should avoid complicated developments and laconic explanations; furthermore, it was most desirable that it included references to other works, as well as some hints on the prerequisite knowledge required to fully understand the book. In this section I have shown Malebranche's inclination towards working in a team, revising and enlarging original works, and shown that the line between public and private knowledge became increasingly blurred. Of course, this practice might not be ascribed exclusively to the Oratoire, since several members of Malebranche's circle, like Varignon himself, belonged to the Jesuit order. Whether this was an individual choice or a more generalized practice would merit more detailed study.

Alongside the figure of the commentator, we find another actor with a crucial role: the publisher Jacques Rollin. His zeal in acquiring Varignon's manuscripts illustrates the importance of the publisher's role in deciding which works to publish, how to publish and present them and, in this case, in communicating Leibnizian calculus.

Conclusions

In this paper I have analyzed the practices of communication involving Johann Bernoulli's lessons on differential calculus in Paris and, to a certain extent, in northern Italy. Taking a transnational perspective, I have presented the different ways in which differential calculus was communicated in both settings and between them. Besides Malebranche's commitment to Leibnizian calculus, the combination of the Oratorian educational practices with Malebranche's connection with the Académie des Sciences seems to have contributed to the emergence of Paris as a focus of scientific socialization, essential to the transnational communication of the new calculus. This context by no means displays a homogeneous picture, but one in which the disputes between Leibnizians and Cartesians took place, and where the Oratorian pedagogical activities coexisted with the more conservative Jesuit ones.

Meanwhile, in northern Italy it was Catholic reformists who undertook the introduction of differential calculus in the educational context. In Catholic Enlightened Italy learned women were encouraged to get involved in scientific matters. Their active commitment to scientific societies and academies, and in few cases even to the university, had no counterpart elsewhere in Europe. Of course Paris and northern Italy shared some communicative practices, namely correspondence, patronage and personal encounters. Yet, in northern Italy I have found no proof of teamwork activities similar to the ones developed in Paris around Malebranche's circle. In contrast to Paris, the fact that Agnesi personally undertook the printing of her book may point to a lack of publishers willing to publish the new method, unless it was simply a personal decision on Agnesi's side. In any case, the absence of a center of socialization in northern Italy as significant as Paris might have slowed down the interaction between research and teaching. However different it was, this paper makes plain that communication between Paris and northern Italy was indeed possible. Not only did Reyneau's book travel to Italy, where it was appropriated by Agnesi, but Agnesi's book was also later translated into French and introduced before the Académie des Sciences in Paris.

Scientific communication was conducted either in Latin or in the vernacular, depending on the context and the intended audience. The *Acta Eruditorum* was written in Latin because this language was still considered to be the most suitable vehicle to reach a broad international readership. Alongside it, French was gaining ground among cultural *élites* and in the public sphere, even beyond the French borders. In the communicative practices considered in this paper the progressive advance of French was no doubt due to Malebranche's reputation and his interest in communicating the research on calculus in a pedagogical way. As far as Italy is concerned, there are two points worth commenting. On the one hand scientific journals in Italy were written in Italian, hence targetting a local audience. On the other, Latin was the language used at the university. So as not to deter learned women from the study of differential calculus, Agnesi might have decided to publish her work in Italian.

Throughout the paper the evolution of communicative practices can be traced. The episode on the publication in the *Analyse* of part of the correspondence between L'Hôpital and Johann Bernoulli illustrates the progress of private knowledge towards public knowledge. Again, the circulation and appropriation of Johann Bernoulli's text allows us to reflect on authorship practices in the eighteenth century. The teamwork activities involved in the writing, revision and circulation of manuscripts – at least within the Oratorian context – seem to reveal that in the late seventeenth century and early eighteenth century, the figure of the author was not as yet clearly defined. The anonymous edition of L'Hôpital's book only confirms this statement. But the decision of the publisher of the second edition under the name of L'Hôpital, suggests some initial steps towards a turn in the author's status.

This leads me to defend the role of some traditionally neglected actors, such as the publisher and the journal editor. Often overlooked by the role of the inventors and their discoveries, these actors had an important role in the communication of the new calculus. Writers of educational books belong to this category too. Educational books have been neglected by earlier historians of mathematics. The most obvious example is Truesdell's consideration of Agnesi's work as lacking originality. However, the authors of educational books did provide original contributions, and the interaction between research and teaching turned out to be an especially relevant factor in the configuration of differential calculus as a discipline.

In short, the study of the wide range of practices of communication considered in this paper has shown that it is fundamentally important to single out the variety of factors determining the ways and forms of the shaping of differential calculus into a discipline in the eighteenth century.

Notes

[*] To Gert Schubring and Josep Pla.

[1] Schubring, G. (2005). *Conflicts between Generalization, Rigor, and Intuition.* New York: Springer, p. 2.

[2] Schubring, G. (1996). "Changing Cultural and Epistemological Views on Mathematics and Different Institutional Contexts in Nineteenth-Century Europe". In Goldstein, et al., eds. (1996). *Mathematical Europe. Myth, History, Identity.* Paris: Éditions de la Maison des sciences de l'homme, pp. 361-88, on p. 363.

[3] In this point Schubring's position concurs with that of David Edgerton discussed in the introduction to part IV in this volume.

[4] Secord, J. A. (2004). "Knowledge in Transit". *Isis* 95 (4): 654-72.

[5] Ibid., pp. 663-4

[6] Topham, J. R. (2000). "Scientific Publishing and the Reading of Science in Nineteenth-Century Britain: A Historiographical Survey and Guide to Sources". *Studies in History and Philosophy of Science* 31 (4): 559-612, p. 562. See for instance Avramov, I. (1999). "An Apprenticeship in Scientific Communication: the Early Correspondence of Henry Oldenburg (1656-63)". *Notes and Records of the Royal Society* 53 (2): 187-201; Dooley, B. (1995). "The Communications Revolution in Italian Science". *History of Science* 33 : 469-96; Daston, L. (1991). "The Ideal and Reality of the Republic of Letters in the Enlightenment". *Science in Context* 4 (2): 367-86.

[7] Since the word "textbook" was not defined in the eighteenth century yet, my preference here is to consider them as educational books or "books employed for educational purposes", as they are referred to in Bertomeu Sánchez, J. R.; García Belmar, A.; Lundgren, A., and Patiniotis, M., eds. (2006). "Textbooks in the Scientific Periphery". *Science and Education* 15 (7-8), p. 658.

[8] Roberts, L. (2005). "Circulation Promises and Challenges". *The Circulation of Knowledge and Practices: The Low Countries as an historical laboratory.* Workshop, Woudschoten, 27-28 May, p. 9. [http://www.gewina.nl/werkgroep/]

[9] On appropriation and reading practices see also Topham. "Scientific Publishing"; Secord, J. A. (2000). *Victorian Sensation. The Extraordinary Publication, Reception, and Secret Authorship of Vestiges of the Natural History of Creation,* Chicago: The University of Chicago Press, and Simon, J. (2008). "Circumventing the 'elusive quarries' of Popular Science: the Communication and Appropriation of Ganot's Physics in Nineteenth-century Britain". In Papanelopoulou, F.; Nieto-Galan, A., and Perdiguero, E., eds. *Popularising Science and Technology in the European Periphery, 1800-2000.* Aldershot: Ashgate.

[10] Olesko, K. M. (2006). "Science Pedagogy as a Category of Historical Analysis: Past, Present and Future". *Science & Education* 15: 863-80.

[11] Stichweh, R. (1994). "La structuration des disciplines dans les universités allemandes au XIXe siècle". *Histoire de l'éducation* 62 (mai): 55-73.

[12] Kaiser, D. (2005). "Training and the Generalist's Vision in the History of Science". *Isis* 96: 244-51, p. 250.

[13] Taton, R. ed. (1986). *Enseignement et diffusion des sciences en France au dix-huitième siècle*. Paris: Hermann, pp. 67-9.

[14] Ibid., p. 69.

[15] Ibid., p. 81.

[16] The aristocracy.

[17] Taton. *Enseignement et diffusion*, pp. 51-2.

[18] Robinet, A. (1970). *Malebranche de l'Académie des sciences: l'œuvre scientifique, 1674-1715*. Paris : J. Vrin.

[19] Ibid., p.9.

[20] For a thorough account of this episode see ibid.

[21] Leibniz, G. W. (1684). "Nova methodus pro maximis & minimis, itemque tangentibus...". *Acta Eruditorum*. In the same journal he published "De Geometria recondita et analysi indivisibilium atque infinitorum" in 1686, a paper of similar format dealing with the integral calculus.

[22] Bos, H. (1974). "Differentials, Higher-Order Differentials and the Derivative in the Leibnizian Calculus". *Archive for the History of Exact Sciences* 14: 1-90, p. 16.

[23] One can find a list of locations with complete sets of *Acta Eruditorum* where consequently Leibniz's seminal paper may have reached in the appendix 3 in Laeven, H. (1990). *The* Acta Eruditorum *under the Editorship of Otto Mencke (1644-1707): The History of an International Learned Journal between 1682 and 1707*. Amsterdam & Maarssen: APA-Holland University Press.

[24] Mencke's sources were essentially the *Journal des Sçavans* and the *Philosophical Transactions*. Not always an easy task, the translation of the articles was endeavoured by Mencke's collaborators, a sort of journal editorial board.

[25] In Laeven. *Acta Eruditorum*, p. 139, there is an exhaustive report on the distribution of Mencke's correspondents and submitters of articles, between 1682 and 1706, in Germany, Low Countries, Switzerland, Italy, Britain, France and Sweden.

[26] Robinet. *Malebranche*, p. 58. All translations from French are by the author of this paper.

[27] In fact, in the inventory of Malebranche's group all the papers from the *Acta Eruditorum* regarding the new calculus are arranged in such a homogeneous way that Robinet believed that was a proof of Malebranche's educational intention. See Robinet, A. (1960). "Le groupe malebranchiste introducteur du Calcul infinitésimal en France". *Revue d'Histoire des Sciences et de leurs applications* XIII, n. 2, p. 292.

[28] Cantor, M. (1880-1908). *Vorlesungen über Geschichte der Mathematik*. Leipzig, p. 222.

[29] Robinet. *Malebranche*, p. 56.

[30] I became aware of the role played by Stähelin regarding Johann Bernoulli's lessons on calculus when I first read Spiess' preface to Johann Bernoulli's correspondence, in Bernoulli, J. (1955). *Der Briefwechsel von Johann Bernoulli, Herausgegeben von der Naturforschenden Gesellschaft in Basel*. Ed. O. Spiess.

Basel: Birkhäuser. Stähelin was a physician and an outstanding botanist, professor
of rhetoric, anatomy and botany at the University of Basel, of which he became
rector in 1719. See the *Dictionnaire Historique et Biographique de la Suisse*, vol.
VI, p. 335.

[31] Johann Bernoulli to Montmort, in a letter of May 21st 1718, quoted in Radelet-
de Grave, P. (2004). "L'édition des figures manuscrites des Bernoulli".
[http://www.brickscommunity.org/material/RadeletAbstract.doc].

[32] Bossut, C. (1802). *Essai sur l'histoire générale des mathématiques*. Paris: Chez
Louis. Vol II, p. 138.

[33] Bernoulli, J. (1922). *Lectiones de calculo differentialium (1691-92)*. Basel: ed. P.
Schafheitlin. There is a comparative analysis of L'Hôpital's *Analyse* and Johann
Bernoulli's *Lectiones* in Blanco, M. (2001). "Análisis de la discusión L'Hôpital-
Bernoulli". *Cronos* 4 (1-2): 81-113. See also Schafheitlin's preface to Bernoulli's
Lectiones (1922). From an epistemological point of view how L'Hôpital appropriated
Bernoulli's lessons deserves to be analyzed in more detail than is possible here.

[34] Contrary to what it has been traditionally accepted, L'Hôpital introduced some
innovations, like some reflections on the concept of variable concept. See
Schubring. *Conflicts*, p. 190.

[35] At the time the nature of infinitesimals was not yet clear. Leibniz himself did not
mention "infinitely small quantities" in his papers in the *Acta Eruditorum*.
However, since this conception appeared in L'Hôpital's book, unmistakably
borrowed from Johann Bernoulli's lessons, the group around Malebranche
appropriated and communicated Leibnizian calculus this way. On this point, see
Schubring. ibid., pp. 186-7.

[36] L'Hôpital, G. F. A. de (1696). *Analyse des infiniment petits pour l'intelligence
des lignes courbes*. Paris: Imprimerie Royale, Préface.

[37] For instance, the fist volume of Maclaurin's *Treatise of Fluxions* (1742) was an
attempt to validate the new calculus by means of Greek geometry entirely.

[38] On this episode see Blay, M. (1986). "Deux moments de la critiques du calcul
infinitésimal: Michel Rolle et George Berkeley". *Revue d'histoire des sciences* 39:
223-53, and Mancosu, P. (1989). "The Metaphysics of the Calculus: A
Foundational Debate in the Paris Academy of Sciences, 1700-1706". *Historia
Mathematica* 16: 224-48. On Leibniz's discussion on foundations see also Bos.
"Differentials".

[39] Bernoulli. *Lectiones*.

[40] Bernoulli. *Briefwechsel*.

[41] Being a wealthy man, it is rather unlikely that L'Hôpital's interest in keeping
Bernoulli silent stemmed from financial needs. I feel more inclined to believe that
he wished to secure a reputation within the intellectual sphere. This behaviour is in
contrast to that – according to Lorraine Daston – characterizing the Republic of
Letters, as represented by the financial relation between the Abbé de Saint Pierre
and Varignon. Daston. "The Ideal and Reality of the Republic of Letters", p. 371.

[42] See Robinet. *Malebranche*, p. 295.

[43] Bernoulli. *Lectiones.*

[44] "Éloge de M. le Marquis de l'Hôpital". In Fontenelle, B. le B. de (1740). *Éloges des académiciens avec l'histoire de l'Académie des Sciences de Paris*, I, p. 54. This comment could serve as a contemporary proof of the dominant role played by France and Germany in spreading Leibnizian calculus in the eighteenth century.

[45] Montucla, J. E. (1758). *Histoire des mathématiques.* Paris: Chez H. Agasse, II, p. 397; Bossut. *Essai*, pp. 162-3.

[46] *Éloges*, I, p. 57.

[47] See Blanco, M. (2004). *Hermenèutica del càlcul diferencial a l'Europa del segle XVIII: de l'*Analyse des infiniment petits *de L'Hôpital (1696) al* Traité élémentaire de calcul différentiel et de calcul intégral *de Lacroix (1802).* Barcelona: Universitat Autònoma de Barcelona. unpublished PhD thesis.

[48] In Paris there was a second edition in 1715 published by Montalant, who also reprinted it in 1716. This same edition was reprinted in 1725 by Rollin. The third edition, published in Avignon, was reprinted in 1768 in Paris and was later enlarged and revised by Lefèvre in 1781.

[49] See Gillispie, C. C., ed. (1970-). *Dictionary of Scientific Biography.* New York: Charles Scribner's Sons.

[50] The publisher, Jacques Quillau, published works by other members of Malebranche's circle, namely Guisnée and Montmort. A second edition of Reyneau's *Analyse*, enlarged by Varignon, came out in 1714.

[51] Quoted in Radelet. *Figures manuscrites.*

[52] It is not clear whether Reyneau used the original copy or the Carré-Byzance copy. Finally, apparently Reyneau passed the papers on to another member of Malebranche's circle, Pierre R. de Montmort, and in this process they strayed partly. See Radelet, ibid., n. 20, and Spiess' preface in Bernoulli. *Briefwechsel.* The manuscript that Schafheitlin published in 1922 was the only manuscript of the lessons on differential calculus preserved. It contains both the differential calculus and the first eleven lessons on the integral calculus. This manuscript, handwritten by Nicolaus (I) Bernoulli, dates from 1705. Since Nicolaus was born in 1687 it seems most likely that when he visited his uncle in Groningen to be lectured on mathematics, Johann lent him Stähelin's duplicate and Nicolaus rewrote it.

[53] See also Robinet's account of the manuscripts found in the Oratorian library. See Robinet. "Groupe malebranchiste", p. 290.

[54] "To readers who have at least an average knowledge of the ordinary Geometry". Reyneau, C. R. (1708). *Analyse démontrée, ou la Méthode de résoudre les problèmes des mathématiques et d'apprendre facilement ces sciences*, Préface.

[55] In the *Analyse démontrée* Reyneau proved the elimination of *dxdy* in the product rule by compensation of errors, exactly in the same way as Newton did in his *Principia Mathematica.* In doing so, Newton apparently avoided the use of second order fluxions. This approach was strongly critized by Berkeley in 1734. Reyneau followed very closely the Rolle-Varignon debate, centered on the nature of infinitesimals. Since his text was published after this episode, but before

Berkeley's attack, Reyneau might have attempted to avoid second order
differences resorting to Newton's apparently epistemologically unproblematic
proof. See Newton, I. (1687). *Philosophiae Naturalis Principia Mathematica.*
London, book II, section II, lemma II.

[56] See Blay. *Deux moments*, n. 14, p. 227; Mancosu. *Metaphysics*, p. 230.

[57] "Éloge du Père Reyneau in Fontenelle". *Éloges*, II, p. 346. Fontenelle's remark
is not fully exact. Since in 1705 Reyneau gave up teaching at the *collège* in Angers
and moved to Paris, it is rather unlikely that his intended audience were his own
disciples: I think thus his text targetted a more general audience.

[58] Ibid., p. 348.

[59] See Greenberg, J. L. (1986). "Mathematical Physics in Eighteenth-Century
France". *Isis* 77 (1): 59-78, p. 66.

[60] I believe a detailed analysis of Johann's reaction on the reception of Reyneau's
book would be enlightening for a better understanding of questions of authorship
and private and public knowledge.

[61] See Bottazzini, U. (1994). "The Italian States". In Grattan-Guinness, I. ed.
(1994). *Companion Encyclopedia of the History and Philosophy of the
Mathematical Sciences.* London: Rontledge.

[62] Ibid., vol. 2, pp. 1497-8.

[63] See Schubring, G. (2002). "Mathematics between Propaedeutics and
Professional Use: A Comparison of Institutional Developments". *Enciclopedia
Italiana.* Rome: Istituto dell'Enciclopedia Italiana, VI.

[64] Mazzotti, M. (2001). "Maria Gaetana Agnesi. Mathematics and the Making of
the Catholic Enlightenment". *Isis* 92: 657-83, pp. 660-1.

[65] On the subject of women involved in scientific fields in Enlightenment Italy,
especially in Bologna, see Logan, G. B. (2003). "Women and the Practice and
Teaching of Medicine in Bologna in the Eighteenth and Early Nineteenth
Centuries". *Bulletin of the History of Medicine* 77 (3): 506-35; Cavazza, M.
(1997). "Dottrici e lettrici dell'Università di Bologna nel settecento". *Annali di
storia delle università italiane* 1: 109-25, and (1995). "Laura Bassi e il suo
gabinetto di fisica sperimentale: Realità e mito". *Nuncius: Annali di Storia della
Scienza* 10: 715-53; Findlen, P. (1993). "Science as a Career in Enlightenment
Italy: the Strategies of Laura Bassi". *Isis* 84: 440-69.

[66] See Truesdell, C. (1989). "Maria Gaetana Agnesi". *Archive for History of the
Exact Sciences* 40: 133-42.

[67] See also Kleinert, A. (1990). "Maria Gaetana Agnesi und Laura Bassi: Zwei
italienische gelehrte Frauen im 18. Jahrhundert". In Schmidt, W. and Scriba, C. J.,
eds. (1990). *Frauen in den exakten Naturwissenschaften.* Stuttgart: Steiner, pp.71-
85; Vettori-Sandor, C. (1988). "L'opera scientifica e umanitaria di Maria Gaetana
Agnesi". In *Alma Mater Studiorum: La presenza femminile dal XVIII al XX secolo.
Richerche sul rapporto donna/cultura universitaria nell'Ateneo bolognese.*
Bologna: Cooperativa Libraria Universitaria Editric Bologna, on pp. 105-18.

[68] Such as Guido Grandi's *Quadratura circuli et hyperbolae per infinita hyperbolae & parabolas geoemtrice exhibita* (Pisa, 1703) and Gabriele Manfredi's *De constructione aequationum differentialium primi gradus* (Bologna, 1707). Mazzotti. "Catholic Enlightenment", p. 674. It only recently came to my knowledge that, apparently, the first Italian work entirely devoted to calculus was *De'Calcoli differenziale e integrale memoria analitiche* (1743-44), written by Domenico Corradi d'Austria (1677-1756), a mathematician, engineer and chemist born in Modena. Unfortunately, I have not had the chance to discover why Mazzotti considers these works to be outdated.

[69] Mazzoti. ibid., p. 674. Regarding other publications on differential calculus in Italy, *The Giornale de'Lletterati in Italia* published the research papers of individuals such as Riccati and Fagnano, thus spreading the methods of Newton and Leibniz among educated Italians, for some thirty years. See Bottazzini. "Italian states", p. 1497, and the section on Italy in Schubring. "Institutional developments".

[70] This term could encompass the learned community in Italy.

[71] On the subject of scientific communication in Italy see Dooley. "Communications Revolution".

[72] See Logan. "Women and Medicine in Bologna", pp. 512, 533.

[73] I deem relevant that a bookseller in Venice, Francesco Pitteri, published in 1739 in French two works by Reyneau previously published in Paris: the *Analyse démontrée* (1705) and *La science du calcul des grandeurs en général, ou Les élémens des mathématiques* (1714-1736).

[74] Truesdell. "Agnesi", p. 135.

[75] Mazzotti. "Catholic Enlightenment".

[76] On the dichotomy between "mixed mathematics" and "pure mathematics" see part I in this volume.

[77] Mazzotti. "Catholic Enlightenment", p. 675. When Leibniz visited Italy in 1689-90, especially the northern cities, he was interested in contact the local mathematicians and scholars to communicate his calculus.

[78] On the subject of similar practices in the Royal Society of London see the works of Iordan Avramov and Maria Boas Hall.

[79] Quoted from Sartori, E. (2006). *Histoire des femmes scientifiques de l'Antiquité au XXᵉ siècle*. Paris : Plon, pp. 226-7.

[80] Taton. *Enseignement et diffusion*, pp. 117, 534. The works of L'Hôpital and Reyneau were also to be found in the library of several *écoles militaires*. Inventory of the École de Valence (1785) and École Royale d'Artillerie de Strasbourg (1789). See Taton. ibid., pp. 539-540. For an analysis of the increasing institutionalization of calculus it would be worth tracing the Italian libraries which contained Agnesi's book in their catalogues at that time.

[81] *Journal des Sçavans* (1776), p. 125.

[82] Roberts. "Circulation".

[83] On scientific authorship see Topham. "Scientific Publishing", pp. 591ff.

[84] Schafheitlin's preface in Bernoulli. *Lectiones*, pp. 4-5.

[85] Truesdell. "Agnesi", p. 27. As it has been mentioned above, the Italian *virtuosi* held the same position around the 1660s.

[86] "To communicate the novelties which take place in the *Republic of Letters*". L'Imprimeur au Lecteur in the *Journal des Sçavans* (1665). This review dates from 1696. It is worth observing that the second part of the Rolle-Varignon debate was fought publicly in the *Journal des Sçavans*, initiated by the Jesuit fathers Gouye and Bignon, who took Rolle's side in 1702.

[87] *Journal des Sçavans* (1696), p. 427.

[88] Fontenelle. *Éloges*, I, p. 59.

[89] In English "thus".

[90] Crousaz, J.-P. (1721). *Commentaire sur l'Analyse des infiniment petits*. Paris: Chez Montalant, Preface, p. 6.

[91] Greenberg. *Mathematical Physics*, p.75.

[92] In 1772 Crousaz was appointed associate member of the Paris Academy of Sciences. See *Histoire de l'Académie Royale des Sciences. Année 1750*. Paris.

[93] Topham. "Scientific Publishing", p. 581.

[94] Taton. *Enseignement et diffusion*, p. 277.

[95] Ibid., p. 143.

[96] On the role of Rollin in the publishing of Varignon' *éclaircissements*, see the introduction to Bernoulli. *Der Briefwechsel von Johann Bernoulli*, Band 3, pp. 18-23.

[97] "Le Libraire au Lecteur". In Varignon, P. (1725). *Éclaircissemens sur l'Analyse des infiniment petits*. Paris: Chez Rollin.

[98] Taton. *Enseignement*, p. 49.

[99] The definition of "beginner" seems to be not yet established, since Montucla believed Paulian's commentary was only useful for "slow beginners"! See Montucla. *Histoire*, II, p. 398.

THE FRANCO-BRITISH COMMUNICATION AND APPROPRIATION OF GANOT'S *PHYSIQUE* (1851-1881)

JOSEP SIMON

In 1851, Adolphe Ganot (1804-1887) published in Paris his *Traité élémentaire de physique expérimentale et appliquée.*[1] The book was the result of twenty years' experience in a science-teaching career, first as a student at the École Normale,[2] then as a teacher in a French provincial *collège*, and finally in a private school in Paris.[3] The *Traité* met with rapid success, running through eight editions in eight years. In 1859, he published another textbook, the *Cours de physique purement expérimentale*, intended for a different readership.[4] Ganot produced successive editions of his two books until 1881, when he handed them over through contract to Hachette, the leading French publisher of secondary school textbooks. According to Ganot, the last editions of the *Traité* (18th, 1880) and of the *Cours* (8th, 1881), that he prepared himself, had print runs of 20,000 and 13,000 copies, respectively.[5] By then he claimed to have produced 204,000 copies of the former (since 1851), and 64,500 copies of the latter (since 1859).[6]

Furthermore, in this period, the *Traité* was translated into Italian (1852), Spanish (1856), Dutch (1856), German (1858), Swedish (1857-60), Spanish (Paris, 1860), English (1861-3), Polish (1865), Bulgarian (1869), and Turkish (1876). The *Cours* was translated into English (New York, 1860), Dutch (1862), Italian (1868), English (1872) and Spanish (1873).[7] Although the translation of French physics textbooks into other languages was common in this period,[8] Ganot's textbooks were certainly amongst the most widely translated and read, and as such made a major contribution at an international level to the configuration of physics as a discipline. By the 1880s, they were considered standard works of physics by a wide range of readers across the social and cultural spectrum in France, as well as in countries such as Britain.[9] As I have argued elsewhere, this made them central to French and British culture.[10]

However, Ganot's textbooks are seldom (if at all) cited in the history
of nineteenth-century physics literature, and they have only been the
object of a very limited number of publications.[11] Furthermore, when they
are cited, it is often in dismissive terms, informed by an historiographical
prejudice that still too often categorizes pedagogy as a clearly distinct
practice inferior to research.[12] The reasons for Ganot's absence in history
can be clarified if we analyse the current major trends in the
historiography of nineteenth-century physics.[13] Some of these can in fact
be found to some degree in all major syntheses published in the last three
decades.[14]

One of these major trends is the peripherality assigned to educational
structures. Educational qualifications are usually mentioned and
sometimes inserted in the discussion of the work of physicists, but it is the
selection of the latter, along with their research work, that drives historical
narratives. Furthermore, only higher education institutions are taken into
account, and considered to be the only places where research and creative
work is performed. By contrast, the work of Rudolf Stichweh and Kathryn
Olesko has shown that, in Germany, the training of secondary school
teachers and students had a fundamental role in the constitution of physics
as a discipline, through its defining role in the physics community, its
professional practices and the university curriculum.[15] Olesko and Andrew
Warwick have also manifested the importance of considering pedagogical
practices in relation to discipline building.[16]

In addition, physics is conventionally portrayed as a well-defined
institutional field, clearly distinct from other disciplines. Thus, for
example, in spite of the sensitivity of Iwan Morus' research to the role of
medicine in the practice of electricity,[17] he has failed to make any
connection between the rise of physics as a teaching discipline and
medical education. By studying the place of Ganot's physics in the French
and British educational systems, I argue in this paper that the intersection
of secondary schooling, university medical and science education, and
textbook writing had a fundamental role in shaping physics as a discipline
in France and Britain.

Another major trend concerns three interconnected aspects, namely
periodization, nationality and conceptual unification. The canonical
periodization used by historians of nineteenth-century physics is
characterized by a strong focus on late nineteenth- and early twentieth-
century developments, a particularly important period for the constitution
of physics as the discipline we know today. This periodization is
structured through alternation of national hegemony. Developments in
physics are considered to be predominantly French until the 1830s, and

British and German during the second half of the century. Accordingly, early French physics is often seen as a mere contribution to subsequent developments produced in Germany and especially Britain, and scientific communication between these countries is only examined in the particular moments when leadership transfer is considered to happen.

National developments are considered to be structured by conceptual unification programmes leading towards the *fin de siècle* consolidation of the discipline. Thus, physics is considered to have been defined first by the French Laplacian programme, and subsequently, in succession, by the correlation, conversion and conservation of forces in Britain, the conservation of energy in Britain and Germany, and Maxwell's field theories. As expressed by Rudolf Stichweh, this element of discontinuity in the historicization of physics is an attractive solution, as it allows presenting "physics" as an "invention", thus making the contingency of its origin a central object of discussion.[18]

However, this three-dimensional framework of time, space and epistemological foundation has in fact favoured narrative linearity and contributed to narrow the historical and ontological diversity of the field.[19] In terms of time, this narrative is teleological in assuming, more or less explicitly, that nineteenth-century developments add up to form twentieth-century physics. In terms of space, it is geographically narrow, implicitly assuming a centre-periphery model in which physics is defined by one or two countries being central in a certain period, and diffusing knowledge to the rest.[20] What happens in peripheral countries, or in those slipping from their central position in periods of loss of hegemony, is not considered relevant. It is inaccurately assumed that the local practices and conceptual frameworks in force at the centre are predominant everywhere. In terms of epistemological foundation, it is assumed that the different conceptual unification processes in physics can be subject to early disagreement and debate, but that they end up in generalized consensus and acceptance everywhere.

In this perspective, books like those of Ganot would therefore not belong to an hegemonic period of French physics and can thus be ignored. By contrast, as argued by Faidra Papanelopoulou in this volume, the practices defining nineteenth-century physics were diverse within the same country and in the international context. Furthermore, she has shown the relevance of studying French physics in periods outside of the aforementioned canonical national periodization.[21] Ganot's books offer a big picture of nineteenth-century physics in which mid-century developments are core. The study of their French and English editions pinpoints the diversity of epistemological frameworks and practices coexisting in physics in different periods and places. Thus, for example, both gave distinct (but in both cases small) relevance to the principle of the conservation of energy in a period in which it is conventionally assumed that this conceptual framework pervaded physics. In fact, a careful analysis of nineteenth-century physics textbooks in different countries is still necessary in order to determine how common the unified picture of physics was, in the form of conceptual frameworks such as that of the conservation of energy. Due to their international relevance, Ganot's textbooks are a good starting point to undertake this task.

Surprisingly, historians of nineteenth-century physics have in general failed to assess the relevance of the fact that, during most of the second half of the century, French textbooks had a fundamental role in the education of British school and college physics students.[22] Communication regarding educational matters between these two countries had an important role in this process. A transnational analysis of French and English science education allows the dismissal of many stereotypes and exceptionalities that have been built on the basis of national histories. In this context, physics textbooks are privileged sources as the meeting point of the spaces defined by physics research and teaching, educational organization and curriculum design, instrument making and publishing.[23] Furthermore, this case study allows the implementation of James Secord's proposal of making communication central to history of science narratives by emphasizing the connections between pedagogical, research and popularizing practices, between textbook and journal science, and lecturing, and between oral, visual and printed communication, in international perspective.[24]

My aim in this paper is to show different ways in which making textbooks central can contribute to improve the historiography of nineteenth-century physics. Accordingly, I will first introduce a case study of the genesis and communication of Ganot's physics textbooks in France. Hence, I will consider the role played by Ganot and his pedagogical practice in the rise of physics in French education, in the context of conceptual and political tensions related to the configuration of this subject as a discipline. By studying the structure of the French educational system, I will highlight the role of medical faculties, secondary education and examinations in the shaping of physics and in the concomitant success of Ganot as an author. Furthermore, by examining the communication and appropriation of Ganot's physics from France to Britain, I argue that booksellers and students had an important role in this process, and, consequently, that they are actors to be considered in a study of the constitution of physics as a discipline. By analysing the role of Ganot's textbooks' English translator, I also provide a comparative perspective on physics and the organization of science education in England and France, and will study its contribution to the definition of distinctive features differentiating the English translations from the French originals.

Ganot's authorship and the rise of physics in French education

The inauguration of Ganot's role as an author came together with four events of extreme relevance for his book.[25] After the revolution of 1848, the Association Philotechnique had been founded as an institution offering free courses to the working class in Paris. Ganot was an original member and taught physics in the school. The Philotechnique was a split from the Association Polytechnique – founded in 1830 by students of the École Polytechnique – constituted by members of the latter that considered that its courses were too elitist for the workers of the Parisian workshops.[26] In 1850, after fifteen years experience as a teacher of mathematics and the sciences in the private school of the chemist Alexandre Baudrimont, Ganot opened his own school in Paris. His entrepreneurial action was favoured by the Falloux law that year, promoting private initiative through educative freedom. In spite of its major aim to favour the Catholic Church regaining control over education, this law also boosted the opening of secular private schools like Ganot's in Paris, preparing students for scientific careers.[27]

As in Baudrimont's institution, the aim of Ganot's school was to prepare candidates for university and engineering studies and for the *baccalauréat ès-sciences*. This examination was the scientific equivalent of the *baccalauréat ès-lettres*, both created in the early nineteenth century with the Napoleonic national reforms of education. During most of the century, the *baccalauréat ès-lettres*, sanctioning classical studies, had ruled education, as the requirement for the culmination of secondary education, and also for accessing any literary or scientific studies in higher education. The *baccalauréat ès-sciences* was merely optional.

However, this examination was about to acquire an unprecedented relevance. In the 1830s, it had acquired further importance by being requirement for enrolment in the medical faculties. In the late 1840s, this examination – administered by the French state network of faculties of science[28] – had also for the first time been made a requirement for candidates for the *Écoles du Gouvernement*.[29] Its definitive expansion came in 1852, when the reform called the *bifurcation* created two parallel specializations in secondary education, giving equal status to the two

baccalauréats. As a consequence, student numbers for the *baccalauréat
ès-sciences* were decisively boosted, equalling those of the classical
baccalauréat.[30] The number of hours assigned to physics in the secondary
school curriculum doubled, and consequently there was an important
increase in the number of teachers, strengthening the identity of this
professional collective.[31]

The rise of the *baccalauréat ès-sciences* and the faculties of science
was the result of the confrontation between two different approaches to
education and to physics, played in the battlefield of the Conseil de
l'instruction publique – the highest state organ in educational matters. The
position of mathematics and the École Polytechnique was represented by
Siméon-Denis Poisson who believed in the mathematical character of
physics, and thus coupled the teaching of these two subjects, against the
position of physics as experimental science and the faculty of science,
represented by Louis-Jacques Thénard who stressed its links with
chemistry. Due to Thénard's actions, in 1840, for the first time the
agrégation – an examination designed on a German model to provide
teaching positions in secondary and higher education – was split in two
specializations: mathematics and sciences.[32] Thénard's efforts were
decisively continued by his assistant Jean-Baptiste Dumas, a student
trained in Geneva and Paris, who succeeded him in all his positions,
representing through his chemistry professorships the link between the
French faculties of medicine and of sciences, and the experimental
approach to physics. Dumas was the major actor in the design and
promotion of the *bifurcation*.[33]

Several factors converged in its successful promotion. On the one
hand, there was the rise of the faculty of sciences against the École
Polytechnique, and the consequent disciplinary rise of physics and
chemistry against mathematics. On the other, there was the related strategy
of promotion of the sciences and scientific teaching careers by the
members of this faculty and of the École Normale – a special institution in
charge of training science teachers – as a way to strengthen their position
in French society. French scientists argued for the need for more science
teaching in order to improve the country's industrial development.
England was cited as an example of the major role given to the
"application" of the sciences to industry in education, and Germany, as a
country which had already started to follow this direction.[34] Although
incubated during the previous decade, the *bifurcation* was implemented
only a year after the Great Exhibition, and similar arguments were critical
to debates on middle-class education promoted then in England by British
scientists. These, however, used France as the model to follow in the

organization of science teaching and research, and also cited Germany in similar terms.[35] Finally, after the coup d'état of Louis-Napoléon, the same year, the *bifurcation* also aimed at regaining state control over education by counteracting the promotion of religious schools by the Falloux law, taking into account the latter's usual specialization in the classical curriculum.

Teachers and authors like Adolphe Ganot were eager to exploit the possibilities offered by this new framework. Since the 1830s, he had taught both mathematics and the sciences (especially chemistry and physics). In the 1840s, while maintaining his profile as a science teacher, he increasingly specialized in physics, a process culminating in his authorship of the *Traité*. In the same period Louis Pasteur – like Ganot a student of the École Normale – went against the advice of his father, who tried to orient him towards mathematics because of its perceived prestige. On the contrary, he decided to specialize in sciences and, in 1846, he got through the *agrégation* his first appointment as secondary school physics teacher, which was the start of a successful career in science.[36]

The aims and pedagogical structure of Ganot's *Physique*

In 1855, Ganot's school had 180 students and he expected an increase in numbers of approximately one-third every year.[37] Most of the students studied medicine and pharmacy, with less than five candidates per year going on to the military École de Saint-Cyr and to the École Polytechnique.[38] However, the publication of his textbook allowed him to expand and to diversify his audience beyond his school classroom. After its first edition (1851), the subtitle of his book targeted any student following courses in secondary schools, science faculties and engineering schools, as well as those not engaged in formal education, but interested in the "applications" of physics.[39] Ganot also showed that he expected his book might be read by foreign authors, translators and publishers, since he inserted a note concerning international prosecution for piracy.

With his book, Ganot aimed at reproducing in printed form, the oral, visual, and manuscript form defining the physics courses he gave in his school.[40] Thus, he stressed that the numerous illustrations in the *Traité* were reproductions in perspective of the instruments of his cabinet, by first-rate artists, whose work he had carefully supervised. The quality and up-to-date character of the *Traité*'s illustrations were stressed by reviewers such as the abbé Moigno, who complained of the common use of old eighteenth-century and early nineteenth-century illustrations in many contemporary textbooks.[41] Indication of instrument size accompanied

Ganot's illustrations, and the use of letters and numbers assisting their description in the text completed this printed replication of the classroom experience. Illustrations of recently designed instruments available in the shops of leading Parisian makers completed the *Traité*'s instrumental display. Ganot often mentioned the makers and his visits accompanied by a draughtsman to their workshops – most of which were close to his school – in order to prepare printed replicas of their products.

The final high quality product was obtained through collaboration with the printer Jules Claye, renown in France and abroad for his aesthetic taste, technical skill and mechanical inventions, which allowed him to print superb wood engravings cheaply. After Claye was awarded a Prize Medal at the London Great Exhibition, the major scientific book he displayed in 1855, at the subsequent Paris Exhibition, was Ganot's *Traité*.[42] Indeed, scientific collections were considered by Ganot a fundamental aspect in advertising his school. His cabinet of physics was presented as holding around three hundred pieces of apparatus, all "amongst the most modern",[43] roughly the same number as represented through illustration in his *Traité*. His collection seems to have been displayed in glass-fronted cupboards around the lecture hall of his school.[44] The large investment required for such collection might have been facilitated by his inheriting the collections of Alexandre Baudrimont's school after he departed Paris to take a professorship at the faculty of sciences of Bordeaux. The spread of his investment in the production of the first edition of his book was secured by publishing it in two parts – a common financial strategy in the French book trade.[45]

Although I have not found direct evidence of Ganot's school lessons, a certain amount of his pedagogical practices can be reconstructed indirectly. Ganot's school certainly had a direct connection with the school of Baudrimont, in which he previously worked. As already mentioned, he probably inherited the school's equipment and certainly preserved part of its staff. The calendar design of Ganot's courses was similar to that of Baudrimont's,[46] and, arguably, he used the pedagogical experience he had acquired in his school, and adapted it to a new educational framework, characterized by an increase in students, the possibility of entering in direct competition with the state secondary schools, and the increase of the status of physics.

A published description of Baudrimont's school courses helps us to imagine what would have been included in Ganot's teaching. According to Baudrimont, the short length of his courses in relation to the normal curriculum taught at the state schools allowed students to go into greater depth in the most relevant questions without loosing a general picture of the subject. Every lesson in his school consisted of an exposition for one hour, followed by an interrogation of fifteen minutes. The student was thus placed in similar conditions as in the examinations they were preparing for. In chemistry and physics, experimental demonstrations were performed by the teacher in front of the students. Collections of physical instruments, chemical products, medicine and zoology were kept in glass-fronted cupboards, in the lecture hall to allow the students to see them continually and to remember more easily the subjects explained using them.[47]

In fact, this picture coincides with that provided by Jean-Baptiste Dumas in his report leading to the *bifurcation* reform. Dumas praised the pedagogical methods used in many private preparatory schools and recommended their implementation in the state schools. He remarked that, in contrast with the latter, the competence of the preparatory schools' headmasters and teams of specialized teachers, the close supervision of students, and the reproduction of examination conditions through regular interrogation was superior. Moreover, he stressed the indispensability of implementing pedagogical practices and material resources, articulating the teaching of science through demonstration, experimental manipulation and observation, and illustrating it through its major current applications.[48] As I have suggested, these ideas informed the organization and pedagogical practice of Ganot's physics courses and their oral, visual and manuscript form, which were then replicated in printed form through the publication of the *Traité*.

Thus, for instance, from its second edition, the book included an appendix of questions and problems based on the *baccalauréat ès-sciences* examination questions. Ganot's was not the first book of physics problems published in nineteenth-century France, but he certainly was pioneer in condensing in a book all the pedagogical practices defining the teaching and learning of physics in this context, including interrogation and examination. This characteristic is related to its origin in the context of French private preparatory schools, and was subsequently generalized in most French physics textbooks, and concomitantly in French state schools.

With the publication of successive editions of his book Ganot aimed at expanding his readership beyond the walls of his classroom. Accordingly, through collaboration with his printer, he used a special system of asterisks and small size type to introduce selective discrimination of the contents of his book. This allowed him to target readers enrolled in different educational structures, covering the whole spectrum of science education in the critical area of access to higher education. Furthermore, the small size type system distinguished the basic contents of the book from those

directed to more advanced readers. These often contained more advanced
algebraic calculus, contrasting with the main text, which was characterized
by its mathematical simplicity. Indeed, the *Traité*'s major emphasis was on
instrument design, and experimental procedures. In addition, small size
type was also used to mark accounts of physics developments that,
because of their novelty and Ganot's perception of the relative consensus
surrounding them, were not considered ready to be part of the main course.
In subsequent editions, some of these contents could be integrated in the
main body of the text, or conversely be eliminated, when the author judged
they had been superseded or ruled out.

These were the only instances in which Ganot explicitly cited his
sources. Accordingly, we know that he read periodicals such as the abbé
Moigno's *Cosmos*, Germer Baillière's *Revue des cours scientifiques*, and
the more elitist *Annales de chimie et de physique* and *Journal de physique*,
and books such as John Tyndall's *La Chaleur*, Rudolf Clausius' *Théorie
mécanique de la chaleur*, Angelo Secchi's *L'unité des forces physiques*,
William Grove's *Corrélation des forces physiques* and Auguste de la
Rive's *Traité d'électricité*. He thus kept abreast of developments in
physics through the accounts and lectures by French and foreign authors
published in French scientific periodicals, and through more advanced
treatises published in French, and through translations of foreign books
into French. A relevant number of these translations were performed by
authors inhabiting the same pedagogical, scientific and social space as
Ganot in Paris, and whom he befriended, such as the abbé Moigno and the
Dr. Deslechamps,[49] so he may also have known about foreign research
through conversation with them. Analogously, he informed himself about
new developments in instrument design by visiting international leading
instrument makers' workshops in Paris, most of which were located in the
vicinity of his school.[50] In addition, he had access to the numerous World
Exhibitions celebrated in Paris during the second half of the century, and
thus obtained illustrations and descriptions for instrument patents recently
presented.

If Ganot used periodicals and books produced in Paris to update
successive editions of his *Traité*, the process of communication and
appropriation also worked in the other direction. Thus, for instance, in
1867 the abbé Moigno's account in his journal *Les Mondes* of the World
Exhibition, held in Paris that year, reproduced a large number of
descriptions and illustrations of the instruments presented there from the
thirteen edition of the *Traité*. With Ganot's permission, they appeared a
few months before the publication of his book.[51]

Through his editorship of the journals *Cosmos* and *Les Mondes* between 1852 and 1881, François-Napoléon Moigno had an important role in the communication of foreign science in France and vice versa. His journals offered regular accounts on foreign research, and had an international readership and correspondence. They also are a monumental testimony of his frantic activity as journal editor and man of science.[52] For instance, he regularly attended the meetings of the British Association for the Advancement of Science, and accompanied Parisian instruments makers like Jules Duboscq to participate in public demonstrations in London at the Royal Polytechnic Institution.[53] Furthermore, he contributed to international scientific communication through his translation into French of works from the English of John Tyndall, William R. Grove, August Wilhelm von Hofmann and Ebenezer C. Brewer and the Italian of Angelo Secchi, some of which were read by Ganot.

In turn, Moigno published in his journals praiseworthy reviews of Ganot's books.[54] Moreover, through his collaboration with Molteni, a firm making projection lanterns, he designed a large collection of projection slides for teaching purposes. This collection included a physics course with 138 slides all based on the illustrations of Ganot's *Cours*.[55] From the inception of *Cosmos* onwards, Moigno considered his journal as only one side of his educational mission. He expressed his intention of completing the journal's written teaching with "an even more efficient teaching, that which comes through the senses of hearing and sight". Accordingly he intended to establish a hall in Paris for the teaching of science through lectures with the aid of projection lanterns using electric light.[56] Moigno had in fact been impressed by this technique, that he had witnessed in London at the Royal Polytechnic Institution, and through his collaboration with Molteni and Duboscq he pioneered its use in popular education in France.[57]

In fact, one of Ganot's major aims in writing the *Cours* had been to provide an elementary physics book for those readers who had no access to real instrument collections, which were very expensive. This purpose was achieved by including a large number of illustrations which, by contrast with those of the *Traité*, included human figures manipulating instruments. Illustrations were thus supposed to act as substitutes for real instruments and experimental demonstrations whenever those were not available.[58]

By rewriting the *Traité* in a different form through the publication of his second textbook, the *Cours*, Ganot further expanded his readership, targeting readers enrolled in the higher ranks of female primary education, in the secondary school literary classes, and in the context of informal

education and social conversation. [59] Furthermore, as shown in the
introduction of this paper, Ganot's textbooks were read beyond France.
The next sections are devoted to discuss how Ganot's *Traité* was
communicated and appropriated through its English translation.

International booksellers and the communication and appropriation of Ganot's *Physique* in England

By the mid nineteenth-century, Paris, Leipzig and London formed a major
triangle in the international book trade. In addition to the large
concentration of French booksellers in Paris, supplying the faculties,
engineering schools and other teaching and research institutions, the city
hosted a large number of foreign booksellers. The latter had typically first
been trained in their native countries and subsequently came to Paris to
complete their education before establishing businesses. Some of them
stayed, and their experiences allowed them to serve as focal points of the
two cultural spaces formed by their native and adoptive countries.[60]

The medical and science faculties in Paris also attracted a considerable
number of foreign students. Around one fifth of the foreign students
registered at the medical faculty in this period were British. [61] On
completion of their degrees, they returned to their home countries, and
acted as mediators between French science and medicine and their own
national research and teaching contexts, often engaging in making foreign
works available through translation and through reviews and regular
accounts in journals. This was often made possible through collaboration
between students and foreign and native booksellers.

French booksellers had a leading role in the international book trade
during the nineteenth century. Early in the century, different Parisian firms
opened branches in other countries in order to expand their market, as well
as to protect their national businesses from bankruptcy due to
overproduction, and from piracy – the most common reasons for crisis at
the time. During the century, England was the main market – after
Belgium – for French books, and French international booksellers
typically structured their businesses by establishing branches in at least the
three major leading metropolises of the book trade, including London.[62] In
this context, the Baillières were arguably the most important international
medical and scientific publishers and booksellers operating in mid-
nineteenth-century Britain. The bookshop and publishing business
established by Jean-Baptiste Baillière in 1818 in Paris acquired leading
status in France over the following decade through acting as bookseller to
the Academy of Medicine. In the following decades the house expanded

abroad through the work of his brothers and nephews. By the 1860s, the family had successfully established bookshops and publishing businesses in London, Madrid, New York and Melbourne.[63]

In this period, J.-B. Baillière published a bibliographical catalogue of French and foreign medical and scientific works, fulfilling the role of promoting bibliographical practices and advising men of science on recent publications, and also advertising the books available at his shop. As a well-known book in the Parisian medical and scientific context, successive editions of Ganot's *Traité* were included in the Baillière catalogue as soon as they appeared or even some months before.[64] At the same time, the book was advertised from the early 1850s in the catalogues of Jean-Baptiste's brother, Hippolyte, director of the London and New York branches.[65] Thus, from an early stage, the Baillières made Ganot's *Traité* available in Britain and America.

Although Jean-Baptiste and his brother Germer (who had opened another bookshop in Paris in 1830) specialized in medicine, they also published books on chemistry, physics and natural history, in alignment with the role that the sciences played for medical students in relation to the *baccalauréat ès-sciences*, and the teaching of these subjects as "preliminary" sciences in the curriculum of the French medical faculties. However, they could not compete in this field with other strictly scientific booksellers in Paris.

By contrast, when Baillière started to publish, in London, English translations from French and German authors, as well as works by British authors, Hippolyte soon detected the lack of elementary treatises in physics, thanks to his Parisian experience,[66] and identified the potentially emergent market of scientific secondary education.[67] In 1847, he published Johann Müller's *Principles of physics and meteorology*, a translation of a German work that was itself a short version of a translation of a textbook by Claude-Servais-Mathias Pouillet from French into German.[68] The second physics textbook published by Baillière, between 1861 and 1863, was *Elementary Treatise on Physics Experimental and Applied*, a translation of Ganot's *Traité*.[69] As already explained, this work, had its origins at the crossroad of French medical and secondary school education, a context well known to Hippolyte and his brothers.

Both Ganot's and Müller's books were part of a *Library of Illustrated Standard Scientific Works* launched by H. Baillière, which also included works by British authors such as the chemist Thomas Graham and the histologist and microscopist John Quekett. J.-B. Baillière was renowned in Paris for the quality of the illustrations in his publications, and his brother Hippolyte aimed to reproduce this standard in England by designing the aforementioned collection in addition to other projects such as anatomy, surgery, botany and geology atlases.[70] Ganot's care for the illustrations of the *Traité*'s together with Claye's first-rate professional contribution made it an excellent work to be integrated in Baillière's *Library*.

In keeping with Baillière's standard commercial practices, the *Treatise* was issued in England in monthly parts. As already mentioned, this practice allowed publishers to spread their investments over a longer period, as well as to attract subscribers. It was also used to fight piracy, as subscription sales allowed market share to be fully taken up before the whole work was completed. In addition, it facilitated distribution, as book parts could be sent by post like newspapers. This practice was characteristic of the publication of cheap literature and of novels appearing in the periodical press, but also of large encyclopaedic works or dictionaries and of expensive volumes with large numbers of illustrations, in publishing which J.-B. Baillière had a long experience.[71] Hence, Baillière's commercial strategies made the publishing format of the periodical press and of large encyclopaedic works cohere with that of textbooks.

Ganot's identification in the title page of the *Traité* as "professeur" – a term used in France for any teacher in secondary and higher education (in line with the idea of "université") – was transformed by Baillière in "professor" – used in Britain only to strictly designate university positions – thus, enhancing the author's authority. Furthermore, the *Traité*'s system of asterisks and small size type was completely eliminated in the *Treatise*. This printed feature of the highly stratified French educational system was difficult to adapt to the emerging English system, and in addition, apparently, English readers did not like it.[72] As we will see, this modification had important consequences in the configuration of the book as a new product through the work of its translator.

Edmund Atkinson and the teaching of physics in mid nineteenth-century England

Through the recommendation of one of his peers, the London publisher William Francis, Baillière assigned the translation of the *Traité* to Edmund Atkinson,[73] a young chemist and active Fellow of the Chemical Society, whose journal and proceedings Hippolyte published. Francis, who had previously been in charge of this task, knew Atkinson through his regular contribution to his *Philosophical Magazine*, with translations and accounts of recent Continental researches in chemistry, especially German and French.[74] Previously, after being educated at Owens College, Manchester, under Edward Frankland, Atkinson had, like Francis, studied in Germany, where he took a PhD in chemistry. After a subsequent research stay in Adolphe Wurtz's chemistry laboratory at the faculty of medicine in Paris, he had returned to England and engaged in teaching and journal writing.[75] In the scientific and medical context of Paris, Atkinson was likely to have noticed the success of Ganot's *Traité*, and in the preface to its first English edition, he stated that his high regard of the book was informed by the previous use he had made of it in teaching.[76]

Atkinson's teaching career started in the 1850s – after returning from his training in Germany and Paris – at Queenwood College, a pioneering science school in which Frankland and John Tyndall had met and taught a decade earlier, before themselves going on to do postgraduate research in Germany. When he started to work on the translation of Ganot's *Traité*, he was Lecturer in Chemistry and Physics at Cheltenham College. In 1863, when the translation was completed, he transferred with the same title to the Royal Military College, Sandhurst.

In the 1860s British education was assessed by Jean Demogeot and Henri Montucci, two French commissioners chosen for this task by the French minister of public instruction. Their survey was part of a larger preparatory enterprise organised on the eve of reforms in education in France, in which commissioners were also sent to Prussia, Austria, Sweden, Switzerland and the USA.[77] After being submitted to the minister, Demogeot's and Montucci's report was printed in a high-quality edition by the French national press. Although it received some criticisms in Britain, it was in general praised for its meticulosity and accuracy together with its gentlemanly respect. And British writers considered it to be a Continental standard reference work on British education.[78]

For Demogeot and Montucci, science education was underdeveloped in England, in comparison with France. In certain occasions, in journals and newspapers, English reviewers explicitly pinpointed this comparative state of affairs as the reason for the success of French and German textbooks in England.[79] Only schools such as Cheltenham, Marlborough and a few other institutions in industrial towns such as Liverpool were considered by Demogeot and Montucci to match the level attained by French education through the *bifurcation*. In this sense, they considered that the development of national systems of examinations such as those of the civil and military service, the London university matriculation, the Oxford and Cambridge Locals and the Society of Arts, and those for teachers held by the College of Preceptors, had had a fundamental role in starting to raise the quality of secondary education and in the introduction of the sciences in this level of education. However, English education was still considered to be particularly oriented towards the classical curriculum, and the national standard of education to be irregular and heterogeneous.[80] In addition to the examinations they mentioned, it is worth to pinpoint the powerful action exerted by the Science and Art Department, a state department funded through the benefits of the Great Exhibition. Through its system of teacher training and school examinations, and the pedagogical and political action of teachers such as Edward Frankland, Frederick Guthrie and Thomas H. Huxley, the Department exerted a comparable – although not equally omniscient – function to the École Normale in France.[81]

Cheltenham College, together with the colleges at Marlborough and Wellington, was one of a new type of large proprietary school founded in the 1840s and 1850s, and that soon acquired the first grade status of public schools. They were, however, distinct from traditional public schools in that they developed differentiated curricula which included the teaching of the sciences for the preparation of candidates for the army and civil service. Thus, they had a close relationship with the Royal Military Colleges at Sandhurst and Woolwich. By contrast, the old public schools – with the exception of Rugby – kept their curricula in the classical ideal for decades.

Through his pedagogical practice at Chelthenham and Sandhurst, and his translation of the *Traité*, and subsequently of the *Cours*, Edmund Atkinson had a prominent role in the development of physics as a school subject, and concomitantly as a discipline in England. When he died in 1901, his colleagues George Carey Foster and Hugo Müller remembered him as one of the first teachers who had taught physics systematically in a large public school.[82] Trained as a chemist, he successfully moved to physics, decisively contributing to it through his teaching and textbook writing. He was not the only mid-nineteenth-century physicist with this profile. In France, Ganot had taught chemistry before publishing his *Traité*, although he did not produce research in this field, and he also taught mathematics in the particular context of the French school curriculum. But another English chemist trained in Germany, Frederick Guthrie, had also a fundamental role in the configuration of physics, through his position at the Department. Analogously, George Carey Foster, professor of physics at University College, London, was originally a research chemist who had finished his training in German, Belgian and French laboratories. Atkinson's translations of Ganot's books were used by generations of teachers and students, and were recommended in a wide range of English examinations (including those of the Science and Art

Department and London University), contributing to the shaping of physics as a discipline in Britain and abroad.[83]

Edmund Atkinson's appropriation of Ganot's *Physique*

Atkinson's translation of the *Traité* transformed the book in several ways. First, together with the publisher, he helped shape their intended audience. The book was addressed to "Colleges and Schools", and it was certainly used in many British colleges, schools and medical faculties during this period and was standard for the preparation of a wide range of science examinations.[84] Furthermore, he reshaped the contents and form of the text in significant ways, according to his pedagogical practice as a teacher, and in alignment with the evolving context of scientific education in England. This process was constrained by the internal mechanics of the book that, as I explain later, can be considered as vademecums, in the sense coined by Ludwik Fleck.[85] Finally, he appropriated Ganot's section of problems as an author.

In the preface to the *Treatise*, Atkinson valued the book for its large number of editions and translations, its "clearness and conciseness" and "methodical arrangement", and the quality of its illustrations. However, because of its close link to the "French systems of instruction", he thought necessary to make "alterations and additions" to meet the needs of the English student. In its first edition, his translation was often literal and, in general, did not significantly supplement the *Traité*'s contents. However, it was characterized by more synthetic sentences, shorter historical introductions, different examples, more algebraic formulae (still, simple ones), a more quantitative and mathematical approach, the recalculation for London of observational data given for Paris, and a stronger anti-realist approach in relation to the physical agents and the theories of electricity. Both Ganot and Atkinson explicitly discriminated new conceptual or theoretical frameworks by sometimes choosing new theories for pedagogical reasons, such as prioritizing simplicity and tradition for the sake of readers. However, as we will see, Atkinson was more sensitive, for instance, to the doctrine of the conservation of energy, showing the emerging importance that this framework had in Britain.

In subsequent editions, Atkinson introduced new contents and significantly changed some sections. The first parts to be completely reshaped were the introductory chapters on mechanics, which Ganot had limited to a minimum, due to the greater independence in relation to physics of this subject in France, by contrast with Britain.[86] In addition, Atkinson introduced new contents related to recent research conducted in

Britain, as well as results originally published in English and German. Sometimes additional illustrations were added, often referring to local instrument makers.

New articles were typically introduced at the end of chapters, keeping intact the general structure of the book. The *Traité* included results produced in France, but also in England, Germany, Italy and other countries. However, as already mentioned, Ganot's knowledge of foreign research was based on its appropriation by French journal writers, and translators. Following Fleck's distinction of "journal" and "vademecum" science, Ganot's book was not a mere aggregation of journal articles but, a closed organized system, a vademecum. The tight internal mechanics of the structure of the *Traité*, and perhaps the possibility of saving labour on the basis of Ganot's regular new editions,[87] are factors which configured Atkinson's appropriation. Hence, he respected the general structure and most of the contents of the *Traité*'s successive editions, and in general, introduced additions only at the end of chapters. Thus, for instance, Atkinson never completely reconfigured the *Treatise* in terms of the driving concept of energy conservation, promoted in Britain from the 1860s; however, he introduced an article on this subject at the end of the first book of the *Treatise* as early as 1868.

In spite of the general praise received by Ganot's English translations, they were sometimes criticized and the role of the conservation of energy in the structure of the book was a topical question. In 1872, George Rodwell – physics professor at Marlborough College – regretted the limited space devoted to this doctrine, considering it had a fundamental role in defining current and future physics.[88] A year later, Atkinson expanded his treatment of this topic by including a lengthy discussion of work, energy and the principle of the conservation of energy. By 1880, Ganot's account was instead based on the earliest formulation of the correlation and conversion of forces. However, Atkinson still respected the main structure of Ganot's book and usually added new matter at the end of chapters.

The close links between translation and authorship are especially noticeable for the case of Atkinson's appropriation of the *Traité*'s problems. Atkinson included this section from the *Treatise*'s seventh edition (1875), arguing that teachers and other users of the book had conveyed to him the need. He found many of Ganot's problems devoid of interest in having only algebraic or geometrical solutions, so he added new problems focusing on the use of physical principles, based on his teaching experience and that of his colleagues.[89] In addition, the appendix was published separately in 1876, with Atkinson identified as the author.

After Hippolyte Baillière's death in 1867, Atkinson proposed translating Ganot's *Cours* to Longmans, the most important London publisher in this period.[90] Its first edition, published in 1872, as *Natural Philosophy for General Readers and Young Persons*, was addressed to students at a more elementary level of instruction.[91] It was therefore a priority to eliminate the *Treatise*'s mathematical formulae. Considering it would be difficult to produce a coherent work by expurgating the *Treatise*,[92] Atkinson preferred translating Ganot's *Cours*, which he knew had already had an extensive circulation in France. His translation introduced modifications aimed at targeting students in the English upper classes of boys' and girls' schools, and candidates for the University of London matriculation examination. It was also considered suitable for the general reader wishing to acquire knowledge of the main physical phenomena and laws in "familiar language".[93]

The London examinations included a fair amount of science and, from the 1860s medical and science students sat for the same papers in the first stage of their university education. As a traditional faculty, medicine attracted a large number of students, strengthening the presence of the sciences in the university curriculum. Furthermore, these examinations also had an important role in articulating the teaching of science in England as they became a certificate sought by students aiming either to follow a career in science or simply to crown their school education with a certificate, and thus they directed the curriculum and performance of a certain number schools.[94]

The *Traité* and the *Treatise* shared the aim to be reference textbooks for students enrolled in the last courses of secondary school and in the first courses of higher education, and for students preparing for major science examinations giving access to higher education in France and England, respectively. The *Cours* was characterized by its target in the higher ranks of primary education teacher training and the candidates to the classical *baccalauréat*. By contrast, *Natural Philosophy* was built as a treatise aimed at the preparation of university access examinations, so it developed as a book targeting students in the higher levels of formal education. In spite of this, both the *Cours* and *Natural Philosophy* shared their target of self-taught readers and those bringing science into social conversation, an intention certainly favoured by their use of the same illustrations.

Conclusion

The international presence of Ganot's textbooks during the second half of the century challenges the canonical periodization and national mapping of physics in this period. A preliminary comparative analysis of these textbooks and their English translations suggests that by the 1880s, although the doctrine of the conservation of energy had an increasing role in the perception that physicists had of their emerging discipline, it did not have the general consensus that is usually attributed to it. Furthermore, the increasingly important status it had in Britain was not equally shared in other countries such as France.

Concomitantly, there were other fundamental aspects that contributed to shape physics as a discipline. The important role that the sciences had in the preliminary education of medical students and the support that the well-established medical faculties provided for the teaching of these subjects was decisive in France and in England for the expansion of the public and professional community of physics. The experimental connection of physics and chemistry and its convergence in the medical curriculum contributed to the emergence of a community of physicists whose focus was more experimental than mathematical. On the other hand, the development of secondary education, through the establishment of schools, national systems of examinations and teacher training programmes was also essential for the rise of physics and contributed to shape this discipline, both in its contents and its form. As a meeting point of these various factors, Ganot's textbooks illuminate the different actors and phenomena that contributed to configure physics.

International communication played a fundamental role in this context, and students, booksellers and journal editors were important actors in this phenomenon. France and England observed each other through politics, educational organisation, scientific practice and technological innovation. The observation of the other went beyond rhetoric and was used in the efforts to promote pedagogical reforms, scientific practice and emerging disciplines such as physics. Internationalism was present, to different extents, in the work of textbook authors and teachers such as Ganot, Atkinson, Guthrie and Carey Foster, journal editors such as Moigno and Francis, and publishers such as Jean-Baptiste and Hippolyte Baillière. Their appropriation of practices and knowledge through their experience of foreign cultures contributed to the shaping of pedagogy, educational organization, and physics in particular, and science and medicine in general, in their respective countries.

As emphasized in this paper, literary genres had fluid boundaries in this period and physics was communicated and appropriated in a wide range of forms including oral and visual communication, teaching, public display and textbook and journal reading. Textbooks are an important source to attempt to recover this diversity defining science in this period.

Comparing the French and English Ganot also allows us to highlight the common importance of national examinations in shaping pedagogical practices and physics as a discipline. The French model of state educational control has conventionally been opposed to the English *laissez faire* tradition and national heterogeneity. However, the role of Ganot's textbooks and his school highlight the importance that the interaction of private and state initiative had in the configuration of the sciences and its teaching in France. The role of Atkinson and his textbooks suggests the importance of the new public schools and the military academies in the promotion of the science curriculum in England. Moreover the different examination systems instituted in England in this period had a national coverage – although it was heterogeneous and dependent on local initiative – and institutions such as the Science and Art Department were funded by the state, suggesting potential comparisons with the French École Normale.

Notes

[1] Ganot, A. (1851). *Traité élémentaire de physique expérimentale et appliquée.* Paris: Chez L'Auteur, Éditeur.

[2] Moigno, F.-N. (1868). *Cosmos* 16 (janvier-avril): 306.

[3] Archives Nationales (Paris), F17 20793, *Ganot* file.

[4] Ganot, A. (1859). *Cours de physique purement expérimentale*. Paris: Chez L'Auteur-Éditeur.

[5] Archive of the House of Longmans, *Atkinson File* (reel n°64, N107).

[6] Data consigned on the back page of Ganot (1880). *Traité.*

[7] Dates between brackets indicate the year of first editions. In most cases there was more than one. The Spanish and English editions were almost as numerous as the French.

[8] Paul, H. W. (1980). "The Role and Reception of the Monograph in Nineteenth-Century French Science". In Meadows, A. J., ed. *Development of science publishing in Europe*. Amsterdam: Elsevier, pp. 123-48.

[9] In fact, Ganot's books received more criticism from the 1880s onwards than in previous decades. However, this was due to changes in pedagogical thought that considered examinations and teaching through textbooks as traditional and deficient methods which did not lead to real learning. At the core of the pedagogical kingdom of physics, Ganot's textbooks were likely to be attacked as the symbols of the prevalent educational regime. See Anon. (1880). "Recent Electrical Researches". *The Times* Aug. 6: 3; I thank Graeme Gooday for pointing out this review to me.

[10] Simon, J. (2008). "Circumventing the 'elusive quarries' of Popular Science: the Communication and Appropriation of Ganot's Physics in Nineteenth-century

Britain". In Papanelopoulou, F.; Nieto-Galan, A., and Perdiguero, E., eds. *Popularising Science and Technology in the European Periphery, 1800-2000.* Aldershot: Ahsgate.

[11] Takata, S. (1987). "Ganot's Textbooks of Physics introduced into Japan". *Historia Scientiarum* 33: 25-41; Khantine-Langlois, F. (2006). "Un siècle de physique à travers un manuel à succès: le traité de physique de Ganot". *SFC*; Newton, D. P. (1983). "A French Influence on nineteenth and twentieth-century physics teaching in English secondary schools". *History of Education* 3 (12): 191-201. I would like to thank Seiji Takata and Françoise Khantine-Langlois for sending me their papers.

[12] Tannery, P. (1880). " [Review of *La Philosophie Scientifique* by H. Girard]". *Revue Philosophique* 9 (janvier-juin): 338-50; Fox, R. (2005). "The Context and Practices of Oxford Physics, 1839-77". In Fox, R. and Gooday, G., eds. *Physics in Oxford, 1839-1939. Laboratories, Learning and College Life.* Oxford: Oxford University Press, pp. 24-79, on p. 72.

[13] Some of the major synthesis produced in the last decades in this field are Harman, P. M. (1982). *Energy, Force and Matter: The Conceptual Development of Nineteenth-Century Physics.* Cambridge: Cambridge University Press; Purrington, R. D. (1997). *Physics in the Nineteenth Century.* New Brunswick: Rutgers University Press; Buchwald, J. Z. and Hong, S. (2003). "Physics". In Cahan, D., ed. *From Natural Philosophy to the Sciences: Writing the History of Nineteenth-Century Science.* Chicago: University of Chicago Press, pp. 163-95; Nye, M. J., ed. (2003). *The Cambridge History of Science. The Modern Physical and Mathematical Sciences.* Cambridge: Cambridge University Press; Morus, I. R. (2005). *When Physics Became King.* Chicago: The University of Chicago Press.

[14] Many driving concepts can be found in a seminal paper by Thomas S. Kuhn, although they have subsequently been developed in various directions. In the last account which appeared (*When Physics Became King*) Iwan Rhys Morus has contributed to enlarge the scope of most histories of nineteenth-century physics by recasting traditional accounts in the richer framework provided by cultural history. He acknowledges the international dimension of physics, but particularly focuses on Britain, and builds his account around national blocks corresponding to the latter, namely France and Germany. The book is original in its stress of the role of public display and instrument design. However, it also shares core historiographical tenets with its predecessors. T. S. Kuhn. (1975). "Tradition mathématique et tradition expérimentale dans le développement de la physique". *Annales. Economies, sociétés, civilisations* 30: 975-98, republished as (1976). "Mathematical vs. Experimental Traditions in the Development of Physical Science". *Journal of Interdisciplinary Science* 7: 1-31.

[15] Stichweh, R. (1992). *Zur Entstehung des modernen Systems wissenschaftlicher Disziplinen: Physik in Deutschland.* Frankfurt: Suhrkamp; Olesko, K. (1991). *Physics as a Calling. Discipline and Practice in the Königsberg Seminar for Physics.* Ithaca: Cornell University Press.

[16] Warwick, A. (2003). *Masters of Theory: Cambridge and the Rise of Mathematical Physics*. Chicago: Chicago University Press.

[17] See for example Morus, I. R. (1998). *Frankenstein's Children: Electricity, Exhibition, and Experiment in Early-Nineteenth-Century London*. Princeton: Princeton University Press, and (2006). "Bodily Disciplines and Disciplined Bodies: Instruments, Skills and Victorian Electrotherapeutics". *Social History of Medicine* 19 (2): 241-59.

[18] Stichweh. *Zur Entstehung des modernen Systems wissenschaftlicher Disziplinen*, p. 98.

[19] Despite Morus' insistence on the historical contingency of the facts presented, the periodization and national base of his approach makes difficult to avoid a linear reading of his account. See Morus. *When Physics Became King*.

[20] On centre/periphery see part V in this volume.

[21] Papanelopoulou, F. (2004). *The Emergence of Thermodynamics in Mid-Nineteenth-Century France*. Oxford: University of Oxford. unpublished D.Phil. thesis, and (2006). "Gustave-Adolphe Hirn (1815-90): Engineering Thermodynamics in Mid-Nineteenth-Century France". *British Journal for the History of Science* 39: 231-54.

[22] The major physics textbooks used by British students were those by Adolphe Ganot, Augustin Privat-Deschanel and Jules-Célestin Jamin. The first two were translated into English. These books remained canonical for the British student until the end of the century, but from the 1870s they started to compete with some books by British authors such as those by Balfour Stewart.

[23] Choppin, A. (1992). *Les manuels scolaires: histoire et actualité*. Paris: Hachette; Bertomeu Sánchez, J. R. ; García Belmar, A. ; Lundgren, A., and Patiniotis, M. (2006). "Textbooks in the Scientific Periphery: Introduction". *Science and Education* 15 (7-8): 657-880; Olesko, K. M. (2006). "Science Pedagogy as a Category of Historical Analysis: Past, Present, and Future". Ibid.: 863-80.

[24] See general introduction and introduction to this part in this volume.

[25] Before publishing his *Traité*, Ganot had contributed to a collective handbook aimed at students preparing for the *baccalauréat ès-sciences*, in which he wrote the sections on mathematics and part of the sections on chemistry. D'Orbigny, Ganot; Leblond; Rivière (1838*). Manuel a l'usage des aspirans au grade de bachelier es sciences physiques*. Paris: Bechet jeune.

[26] Pressard, A. (1899). *Histoire de l'Association Philotechnique*. Paris: Association Philotechnique.

[27] Belhoste, B. (2001). "La préparation aux Grandes Écoles scientifiques au XIXème siècle". *Histoire de l'éducation* 90 (mai): 101-30.

[28] By the 1850s there were sixteen faculties of science in France, one for each *académie* into which France was academically divided (Paris, Aix, Besançon, Bordeaux, Caen, Clermont, Dijon, Douai, Grenoble, Lyon, Montpellier, Nancy, Poitiers, Rennes, Toulouse and Strasbourg). A secondary school was created in the capital city of each of the 83 French *départements*, as well as in other important

towns able to provide the necessary funds and infrastructure for its organization.
Fox, R. and Weisz, G. (1980). "Introduction: The Institutional Basis of French
Science in the Nineteenth-Century". In Fox and Weisz, eds. *The Organization of
Science and Technology in France, 1808-1914*. Cambridge: Cambridge University
Press, pp. 1-28, on p. 5.

[29] Covering the whole spectrum of French engineering schools, including the École
Polytechnique.

[30] In 1842, the number of candidates attending the *baccalauréat ès-sciences* was
one sixth of that of the *baccalauréat ès-lettres*. In the 1850s and early 1860s, both
examinations had approximately the same number of candidates, of around 4000
candidates each. Gerbod, P. (1965). *La Condition universitaire en France au XIXe
siècle, Etude d'un groupe socio-professionnel. Professeurs et administrateurs de
l'enseignement secondaire public de 1842 à 1880*. Paris: PUF, pp. 357, 384.

[31] Fournier-Balpe, C. (1994). *Histoire de l'enseignement de la physique dans
l'enseignement secondaire en France au XIXe siècle*. Paris: Université Paris XI.
unpublished PhD. thesis, p. 88.

[32] Anon. (1847). *Projet de Loi sur l'enseignement et l'exercice de la médecine et de
la pharmacie...*. Paris: Union Médicale-Victor Masson, p. 62.

[33] See Anderson, R. D. (1975). *Education in France, 1848-1870*. Oxford:
Clarendon Press.

[34] Dumas, J. B. (1847). "Rapport sur l'état actuel de l'enseignement scientifique
dans les collèges, les écoles intermédiaires et les écoles primaires, ... (extraits) ".
In Belhoste, B.; Balpe, C.; Laporte, T., eds. (1995). *Les sciences dans
l'enseignement secondaire français. Textes officiels*. Paris: INRP-Éditions
Economica, pp. 207-23.

[35] Gooday, G. (2000). "Lies, Damned Lies and Declinism: Lyon Playfair, the Paris
1867 Exhibition and the Contested Rethorics of Scientific Education and Industrial
Performance". In Inkster, I.; Griffin, C.; Hill, J., and Rowbotham, J., eds. *The
Golden Age. Essays in British Social and Economical History, 1850-1870*.
Aldershot: Ashgate, pp. 105-20.

[36] Balpe, C. (1997). "L'enseignement des sciences physiques: naissance d'un corps
professoral (fin XVIIie-fin XIXe siècle) ". *Histoire de l'éducation* 73 (janvier): 49-
85, on pp. 62, 70; Chervel, A. (2004). "Lauréats des concours d'agrégation de 1821
à 1900". Paris: INRP. [http://www.inrp.fr/she/chervel_laureats1.htm] (accessed 10
September 2007).

[37] An increase that he probably managed to approximately maintain only during
the 1850s, due to the partial dismissal of the *bifurcation* framework in the early
1860s, and the fierce competition between private schools.

[38] Ganot, A. (1856). *A messieurs les membres du jury de l'expropiation pour cause
d'utilité publique ... Audience du onze février 1856*. Paris: Henri Plon, pp. 4-9.

[39] Such as barometry, steam machines, medical optics and electricity, and electrical
"applications" such as telegraphy and lighting.

[40] As suggested by its wide intended readership and by Ganot's statements in an advertisement of his work inserted in its first edition. Ganot (1851). *Traité.*

[41] Moigno, F.-N. (1853). "Traité élémentaire de physique expérimentale et appliquée et de météorologie, par M. A. Ganot". *Cosmos* 3 (II): 513-4.

[42] Robin, C. (1855). *Histoire illustrée de l'exposition universelle.* Paris: Furne.

[43] Bottin (1851). *Annuaire et almanach du commerce, de l'industrie, de la magistrature et de l'administration.* Paris: Firmin Didot Frères.

[44] Ganot. *A messieurs*, pp. 3, 6.

[45] Ganot was probably prudent and produced an initial small print run of the *Traité*, since this, coupled with the rapid success of the book had him preparing a second print run a year later (I have found copies of the *Traité*'s first edition printed in 1852 instead of 1851, although they seem to be rare).

[46] Baudrimont offered a course of ten months, which could be taken in two parts of five months each, and quarterly courses starting every month. Ganot also offered quarterly courses.

[47] Baudrimont, A. (1836). *École spéciale de chimie théorique et pratique*, Paris : Paul Renouard, and (1837). *Enseignement préparatoire aux études médicales,...,* sous la direction de M. A. Baudrimont, ..., n.p.

[48] Dumas. "Rapport sur l'état actuel de l'enseignement scientifique".

[49] Ganot (1866). *Traité* , p. 3; Ganot (1870). *Traité*, pp. 3-4.

[50] See series of papers on French instrument makers by Paolo Brenni in *Bulletin of the Scientific Instrument Society* (1993-6).

[51] Moigno (1867). *Les Mondes* 15 (août-décembre): 364-75.

[52] Redondi, P. (1988). "Physique et apologetique. Le *Cosmos* de l'abbé Moigno et de Marc Seguin". *History and Technology* 6: 203-25; Crosland, M. (2001). "Popular science and the arts: challenges to cultural authority in France under the Second Empire". *British Journal for the History of Science* 34 (3): 301-322.

[53] Moigno, F.-N. (1854). "Faits Divers. Nouvelles d'Angleterre". *Cosmos* 5 (28 juillet): 85-7.

[54] Moigno. "Traité élémentaire" (see n. 41).

[55] Moigno, F.-N. (1873). *Les Mondes* 30 (janvier-avril): 3-6, 667-8.

[56] Moigno, F.-N. (1852). *Cosmos* 1, pp. iii-iv, 1-3, and (1854). "Avis". *Cosmos* 5 (18 août): 201-2.

[57] Mannoni, L. (1995). *Le grand art de la lumière et de l'ombre. Archéologie du cinema.* Paris: Nathan, pp. 249-57.

[58] Advertisement inserted in Ganot. *Cours.* (1859).

[59] Simon "Circumventing the 'elusive quarries' of Popular Science".

[60] Most French scholarship has focused on Franco-German cases, but the international dimension of the Parisian book trade was larger, including booksellers of other nationalities such as Spanish, Italian or Swedish. Kratz, I. (1992). "Libraires et Éditeurs Allemands installés à Paris, 1840-1914". *Revue de synthèse* 1-2: 99-108; Barbier, F. (1988). "Les échanges de librairie entre la France et l'Allemagne, 1840-1914". In Espagne, M. and Werner, M., eds. *Transferts: les*

relations interculturelles dans l'espace franco-allemand (XVIIIe et XIXe siècle).
Paris: Editions Recherche sur les civilisations, pp. 231-60.
[61] Caron, J.-C. (1991). *Générations romantiques, les étudiants de Paris et le Quartier latin, 1814-1851*. Paris: A. Colin; Desmond, A. (1989). *The Politics of Evolution*. Chicago: The University of Chicago Press; Warner, J. H. (1998). *Against the Spirit of System. The French Impulse in Nineteenth-Century American Medicine*. Baltimore: The Johns Hopkins University Press.
[62] Barber, G., ed. (1994). "Treuttel and Würtz: Some Aspects of the Importation of Books from France c. 1825". *Studies in the Booktrade of the European Enlightenment*. London: The Pindar Press, pp. 345-352, on p. 381; Barbier, F. (1981). "Le commerce international de la librarie française au XIXe siècle". *Revue d'Histoire Moderne et Contemporaine* 27: 94-117; Martin, O. and Martin, H. J. (1985). "Le monde des éditeurs". In Martin, H. J., ed. *Histoire de l'édition française*. Paris: Promodis, pp. 159-215, on pp. 172, 176; Feather, J. (1994). *Publishing, Piracy and Politics. An Historical Study of Copyright in Britain*. London: Mansell.
[63] Simon, J. (2008). "The Baillières: The Franco-British Book Trade and the Transit of Knowledge". In Fox, R., and Joly, B., eds. *Franco-British interactions in science since the seventeenth century*. Paris: Vuibert.
[64] Baillière, J.-B. et Fils. (1860). *Bulletin bibliographique des sciences physiques, naturelles et médicales* 1, p. 61, and (1861). *Bulletin bibliographique des sciences physiques, naturelles et médicales* 4, p. 117.
[65] Baillière, H. (November 1853). *Catalogue of Scientific Books. Medicine, Natural History, Chemistry and Mathematics. American, French and German*. London and New York: H. Baillière, p. 15; (February, 1856). *Mr. H. Baillière's Catalogue of Medical, Chemical and Scientific Works*. London and New York: H. Baillière, p. 15, and (1858). *H. Baillière's Catalogue of Recent Foreign Books on Chemistry, Electricity, Physics, Meteorology, &c., &c*. New York: H. Baillière.
[66] Anon. (1851). "Elementary Works on Physical Science". *The North American Review* 72: 358-98. I would like to thank Jim Secord for pointing me at this review.
[67] Newton. "A French Influence".
[68] Lind, G. (1992). *Physik im Lehrbuch, 1700-1850. Zur Geschichte der Physik und ihrer Didaktik in Deutschland*. Springer-Verlag: Berlin, pp. 235, 381.
[69] The fourth edition (1870) of the book had a print run of 5,500 copies, increasing by 1879 to around 7,000 in its ninth edition, which sold 4,000 copies in the first ten months. Archive of the House of Longmans, *Atkinson file*.
[70] Simon. "The Baillières".
[71] Barbier. "Les marchés étrangers"; Feather, J. (1988). *A History of British Publishing*. London: Croom Helm, pp. 114-5; Zachs, W. (1998). *The First John Murray and the Late Eighteenth-Century London Book Trade*. Oxford: Oxford University Press, pp. 68-9.
[72] Anon. (1870). "Fernet's Elementary Physics". *Nature*, 3 (November): 23-4.

[73] Brock, W. H. (1996). *Science for All: Studies in the History of Victorian Science and Education*. Aldershot: Variorum, p. 197.

[74] Brock, W. H. and Meadows, A. J. (1998). *The Lamp of Learning. Two Centuries of Publishing at Taylor & Francis*. Bristol: Taylor & Francis, pp. 135, 138-9.

[75] [George Carey Foster and Hugo Müller]. (1901). "Obituary Notices". *Journal of the Chemical Society, Transactions* 79: 888-9.

[76] Ganot (1861). *Treatise*.

[77] Hippeau, C. (1872). *L'instruction publique en Angleterre*. Paris: Didier et Cie.

[78] Anon. (1868). "[review of Demogeot's and Montucci's report]". *The Quarterly Review* 125: 473-90; Todhunter, I. (1873). *The Conflict of Studies, and other Essays on Subjects Connected with Education*. London: Macmillan and Co.

[79] See for example Anon. (1871). "Christmas Books and Annuals. Ganot's Elementary Treatise on Physics". *The Leeds Mercury* December 7 (10502).

[80] Demogeot, J. and Montucci, H. (1868). *De l'enseignement secondaire en Angleterre et en Écosse. Rapport adressé a son Exc. le Ministre de l'Instruction Publique*. Paris: Imprimerie Impériale.

[81] Butterworth, H. (1970). "The Department of Science and Art (1853-1900) and the Development of Secondary Education". *History of Education Society Bulletin* 6: 34-43; Gooday, G. J. N. (1989). *Precision mesurement and the genesis of physics teaching laboratories in Victorian Britain*. Canterbury: University of Kent. unpublished PhD thesis, pp. 1/50-4, 8/1-54.

[82] [Foster and Müller]. "Obituary Notices".

[83] The English editions of Ganot's books were reprinted in the USA and were also used in countries such as India and Japan.

[84] Newton. "A French Influence"; Fox. "The Context and Practices of Oxford Physics, 1839-77".

[85] Simon. "Circumventing the 'elusive quarries' of Popular Science"; Fleck, L. (1979). *Genesis and Development of a Scientific Fact*. Chicago and London: The University of Chicago Press.

[86] Crosland, M. and Smith, C. (1978). "The Transmission of Physics from France to Britain: 1800-1840". *Historical Studies in the Physical Sciences* 9: 1-61.

[87] Ganot sent–at least in certain occasions–a copy of his book to its English publisher, in order to help the work of the translator, and apparently, he was regularly sending engraving stereotypes for each new edition. Archive of the House of Longmans, *Atkinson File*.

[88] Rodwell, G. F. (1872). "Ganot's Physics". *Nature* 5 (8 Feb.): 285-7.

[89] *Preface* in Ganot (1875). *Treatise*.

[90] Topham, J. R. (2000). "Scientific Publishing and the Reading of Science in Nineteenth-Century Britain: A Historiographical Survey and Guide to Sources". *Studies in History and Philosophy of Science* 31 (4): 559-612, on p. 584.

[91] Its second edition (1875) had a print-run of 5,000 copies and sold around 2,600 copies in the first seven months. The print-run of the third edition (1878) increased

to 7,000 copies and the fourth (1881) and fifth (1884) attained 9,000 and 10,000 copies respectively. Archive of the House of Longmans, *Atkinson File*.

[92] As previously pinpointed Fleck's concept of 'vademecum' helps us to understand the fact that the *Traité* was a closed organized system.

[93] "Preface" in Ganot (1872). *Natural Philosophy*.

[94] Mansell, A. L. (1982). "Examinations and Medical Education: The Preliminary Sciences in the Examinations of London University and the English Conjoint Board, 1861-1911". In MacLeod, R., ed. *Days of Judgement: Science Examinations and the Organization of Knowledge in Late Victorian England*. Driffield: Hafferton Books, pp. 87-107.

PART III

POPULARIZATION OF SCIENCE

POPULARIZATION OF SCIENCE

PEDRO RUIZ-CASTELL

Interest in the popularization of science has been promoted in the last decades from very different historiographical agendas. For example, the development in the 1970s of the Sociology of Scientific Knowledge contributed to promote the focus on science popularization through the lens of the social construction of scientific knowledge.[1] These studies generally focused on the role of popularization of science as a tool to claim authority and legitimate the professional, social and political interests of science and its practitioners.[2] In the 1980s, the study of the interaction between science and its publics received further impulse from the Public Understanding of Science movement, which involved both political and professional interest-groups and civic movements. According to the famous report of the Royal Society of 1985, the crucial role played by scientific knowledge in the progress of modern societies and its influence over policy-making demanded public participation in debates about the production of scientific research. This context, in which new funding opportunities were created, favoured the development of projects on popularization amongst historians of science, who aimed to understand, in new critical ways, both the social structure of scientific practice and the complex relations between science and the public. In fact, communicating science not only involves educational centres and international meetings of experts, but also the interaction between scientists and society in a broad sense. Therefore, the study of this relationship between scientific elites and lay people comprises not only the production and formulation of scientific knowledge, but also its appropiation by audiences with various cultural, social and expertise profiles.[3]

Building on Habermas' assertion of the rise of a bourgeois public sphere in eighteenth-century Europe,[4] several historians of science have identified this period as that of the creation of a public space for natural philosophy.[5] However, this idea of science in the public sphere in modern societies is far from what is currently conventionally understood as the popularization of science, since the construction of the former space was intended for elites rather than for lay people. As the first of the papers in

this section shows, it was not until the nineteenth century that the word
"popularization" was used in Spain to deal with the spread of knowledge
amongst a broad general public, a semantic development in accordance to
the use of terms such as *science populaire, popular science,* and *scienza
popolare* in French, English and Italian respectively.[6]

All three papers in this section are, therefore, devoted to the study of
the "popularization of science" as a mass phenomenon, being its focus
those scientific products accessible to or intended for the lay public during
the nineteenth and twentieth centuries. The great development of scientific
instruction in this period was very much linked to factors such as the
growth of an educated middle class and the invention of cheaper printing
technology.[7] New scientific educational projects were carried out, and
changes in the popularization of science had a strong effect through social
and individual initiatives, including the direct intervention of governments.
Examples such as the university extension movement and the foundation
of science museums following international exhibitions are to be found in
many European countries, as in the case of the foundation of the Science
Museum in London after the Great Exhibition of 1851.[8] In addition,
publishing houses became more and more involved in marketing scientific
works for the lay people. The high demand for these works, together with
important economic interests, meant that science became part of the
common work of publishing houses, editors and authors, who made great
efforts to engage with a broad range of readers.

The historical study of popularization of science was originally
understood by scholars on the basis of a diffusionist model, which made a
clear distinction between the making of scientific knowledge and its
subsequent translation to the general public. However, social historians of
science revisited the problem, considering that popularization of science
acquired significance as a rhetorical tool for public authority, used to
legitimate science and to obtain economic and social support.[9] In fact,
some nineteenth-century initiatives in science education, such as the
foundation of Mechanics' Institutes in Britain, are seen as mainly
motivated by interests in social control, i.e., as attempts to modify the
consciousness of the working classes.[10] Moreover, studies on the social
dimension of the practice of science have shown how scientific activities
were enmeshed within the complexities of collaboration between scientists
from different social backgrounds, the division of labour being distributed
according to social status.[11] In addition, studies on nineteenth-century
astronomy have also challenged the traditional view of scientific practice
and its communication by showing the important role played by amateurs
and women in the development of this discipline.[12]

During the last decades, historians of science have gradually stretched the use of popularization of science to embrace an extensive set of actors, sites and practices.[13] Popularization of science has been understood not only as a medium of acquiring authoritative knowledge but also as a way to mobilize public participation, underlining the popular awareness of scientific knowledge, its interest, and the public and open character of science.[14] In the history of science and technology, voice has been given not only to scientists and popularizers, but also to publishers, audiences and users.[15] At the same time, new agendas have been brought into scene: popularization of science also served political, metaphysical and economic plans. Moreover, the influence of specific local conditions has been considered to determine the different creative processes of appropriation of scientific knowledge. In other words, local factors have been proved very important in what images of science were spread and in the singular evolution of programs of popularization of science at different locations.[16]

Nineteenth-century commentators deliberately displayed both popular science works and popularization of science to society as the neutral explanation of a coherent body of knowledge. The three articles presented in this part show, however, that popularization activities also supported specific moral, social and political agendas, which may have varied in each country and cultural context. For instance, the two papers of this part dealing with Spain show that science was seen during the twentieth century by Spanish elites as an engine for industrial development and technological modernization. It was believed that such modernization of the nation would be easier with the spread of applied and utilitarian scientific knowledge amongst the population. Popularization of science was thus considered to be a tool to build up scientific culture from a national and international perspective.

The complex relationships between the creation of scientific knowledge, its communication and its appropriation have recently generated some discussion on the convenience of using the concept "popularization". In fact, several other historiographical tags such as "public science" and "expository science", which coexist in a rather ambiguous way with those of "popular science" and "popularization of science", have recently set up a historical and epistemological debate around what these categories are. These discussions focus not only on the meaning of the words "popular" and "popularization", but also on the terms "public", "understanding" and "science": the problem lies in understanding what the "public" (whoever that may be) should, might or would have liked to understand about science – whatever "to understand" means and however "science" is defined.[17] For example, some historians of science have demonstrated the absence of clear boundaries between scientists and the lay public prior to professionalization of scientific disciplines in the nineteenth century.[18] At the same time, both experts' and laymen's accounts have been analysed in terms of a *continuum* of communication strategies, which included everything from research journals to popular texts, in order to understand the different kinds of liaison between these actors.[19] Furthermore, drawing on the legacy of historical studies on popular culture, it has been acknowledged that "popular" and "popularization" have changing meanings in specific contexts and cultures, being this essential to unveiling both the social and political agendas behind popularization programmes and the place of science in culture.[20]

As a consequence of these complex semantic ambiguities, some historians have lately called for abandoning the use of the terms "popular science" and "popularization of science", arguing that they are no longer valid in articulating what has become a rather confusing model. Instead, they suggest new historiographic tools dealing with aspects such as the communication of knowledge, which might help to remove the distinction between the production and the consumption of scientific knowledge, as well as between education and popularization. In this sense, as argued in the introduction to this book, new proposals such as the development of a wider history of communication and knowledge in science may contribute to renew the interest on science popularization and make it central to the discipline.[21] Nonetheless, as shown by the three papers presented here, "popularization of science" remains a very appealing and useful category for the historian of science who aims to understand and to better explain, not only why historical actors were so eager to employ such a concept, but also how it was a crucial tool in the dynamic dialogue between scientists, political bodies, and society during the nineteenth and twentieth centuries.[22]

All three papers in this part provide an extensive analysis of popular scientific literature for a period of approximately a century, from the late nineteenth to the late twentieth century, engaging with critical questions on popularization of science from the inspection of a wide range of sources, including books, leaflets, newspapers and periodicals. In particular, these articles explore the difficulties of the analysis of books and periodicals, highly influenced by ideological and economic agendas, and the social forces involved in the production of popularization of science projects. For example, the first two articles are devoted to the last years of the nineteenth and early decades of the twentieth century, a period in which popularization of science programmes may be seen, in a broad sense, as part of different movements towards the constitution of national identities in several countries.[23] As it will be pointed out, and similarly to what happened in countries such as Italy – where science popularization initiatives coincided during the nineteenth century with the process of national construction – science popularization in Spain was expected to play an active role in the programme of modernization of the country, which aimed to bring its international prestige back, while popularization of medicine during the late nineteenth century was in Finland part of a middle-class project to constitute a national identity on the basis of specific social, moral, cultural and political principles. Furthemore, from the late 1970s, a critical period of transition to democracy in Spain, science popularization in the newspaper *El País* had a prominent role in the changing values associated to models of national economical development. This involved privileging private initiatives in front of state ones, and a change of national reference models, from France to the USA.

The articles in this part not only show how popular works intended to legitimate social groups and create professional boundaries in different European countries, but also analyse in detail the strategies by which audiences were captivated in terms of leisure, economic profit, tourism, and political and ideological agendas. For example, Pedro Ruiz-Castell analyzes the growth of popular works and science popularization strategies used to promote and legitimate the development of science in Spain – in terms of national pride and prestige – on the occasion of the total solar eclipses of 28 May 1900 and 30 August 1905, two events which attracted international interest and engaged a large audience. Promoted and celebrated by the Spanish intellectual elites, the study of these events allows exploring the different nature of and collaboration between actors involved in the popularization and commercialization of eclipses: professional and amateur astronomers, journalists, publishers, instrument makers, travel and leisure entrepreneurs, lecturers, etc. The central role of popularizers in the constitution of scientific diciplines and the legitimization and consolidation of professional scientists and disciplines is clear from the claims of Spanish elites with occasion of these astronomical events.

Similarly, the paper by Matiana González brings to light the hidden agendas of the most influential Spanish newspaper of the last quarter of the twentieth-century. In fact, her analysis of this publication's coverage of genetics issues uncovers the interests, mainly economic (and related to private funding of scientific research), of popularization of science strategies, problematizing the role of media as a space for public debate in the twentieth century. Moreover, she unveils the strategic role of popularization in the constitution and consolidation of a community of geneticists in Spain. The analysis of spaces for public debate is also present in the article by Tiina Männistö, which focuses on how popularization of medicine in Finnish young women's guidebooks may be seen as part of a broader project to build an independent national identity for Finland, on the basis of specific moral principles promoted by middle-class women. Her work shows, indeed, not only how these popular texts lay on the ambiguous boundary between the private and the public, as the private health of young women became the concern of a whole society, but also how the domestic space in which medicine had to be learnt was understood as a fundamental part of that political space in which the Finnish state was being constituted. Moreover, these guidebooks promoted a hygienic view of the body in which the organs and their functions were the basis of mental functions, subordinating females' body to the authority of doctors, whose crucial role in society was then legitimated.

In summary, as the papers in this part reveal, studies of popularization of science – with their breadth in considering historical actors and sources – allow scholars to reconstruct the interactions between a wide range of different perceptions built around science. This may include central problems such as what activities may be legitimately considered as *science*, and who has the authority to decide and rule over this. As the three papers of this part show, popularization of science can be understood as part of a broad public space in which citizens, both experts and lay people, play an active role in society through expressing their attitudes and opinions, crucial for policy-making and other discourses such as the construction of national and professional identities.

Notes

[1] Bloor, D. (1976). *Knowledge and Social Imagery*. London: Routledge & Kegan Paul; Shapin, S. and Barnes, B. (1977). "Science, Nature and Control: Interpreting Mechanic's Institutes". *Social Studies of Science* 7: 31-74.
[2] Cooter, R. and Pumfrey, S. (1994). "Separate Spheres and Public Spaces: Reflections on the History of Science Popularization and Science in Popular Culture". *History of Science* 32: 237-67, on pp. 241-2.
[3] Borgna, P. (2001). *Immagini pubbliche della scienza. Gli italiani e la ricerca scientifica e tecnologica*. Torino: Edizioni di Comunità.
[4] Habermas, J. (1989). *The Structural Transformation of the Public Sphere: An Inquiry into a Category of Bourgeois Society*. Cambridge: Polity Press.
[5] See for instance: Golinski, J. (1992). *Science as Public Culture: Chemistry and Enlightenment in Britain, 1760-1820*. Cambridge: Cambridge University Press;

Stewart, L. (1992). *The Rise of Public Science. Rhetoric, Technology and Natural Philosophy in Newtonian Britain, 1660-1750.* Cambridge: Cambridge University Press.

[6] Govoni, P. (2002). *Un pubblico per la scienza. La divulgazione scientifica nell'Italia in formazione.* Roma: Carocci, p. 20; Bensaude-Vincent, B. and Rasmussen, A. (1997). "Introduction". In Bensaude-Vincent and Rasmussen, eds. *La science populaire dans la presse et l'édition, XIXe et XXe siècle.* Paris: CNRS Éditions, pp. 13-30; Topham, J. R. (2008). "Rethinking the History of Science Popularization/Popular Science". In Papanelopoulou, F.; Nieto-Galan, A., and Perdiguero E., eds. *Popularising Science and Technology in the European Periphery, 1800-2000.* Aldershot: Ashgate.

[7] Topham, J. R. (2000). "Scientific publishing and the Reading of Science in Nineteenth-Century Britain: A Historiographical Survey and Guide to Sources". *Studies in History and Philosophy of Science* 31: 559-612; Turner, F. M. (1980). "Public Science in Britain, 1880-1919". *Isis* 71: 589-608; Bensaude-Vincent, B. and Blondel, C., eds. (2002). *Des savants face à l'occulte 1870-1940.* Paris: Éditions la découverte.

[8] Bennett, T. (1995). *The Birth of the Museum: History, Theory, Politics.* London: Routledge; Brain, R. (1993). *Going to the Fair. Readings in the Culture of Nineteenth-Century Exhibitions.* Cambridge: Whipple Museum of the History of Science; Butler, S. V. F. (1992). *Science and Technology Museums.* Leicester: Leicester University Press.

[9] Shapin, S. (1984). "Pump and circumstance". *Social Studies of Science* 14: 481-520; Hilgartner, S. (1990). "The dominant view of popularization: conceptual problems, political uses". *Social Studies of Science* 20: 519-39.

[10] Shapin and Barnes, "Mechanic's Institutes".

[11] See for instance Secord, A. (1994). "Science in the Pub: Artisan Botanists in Early Nineteenth-Century Lancashire". *History of Science* 32: 269-315; and (1994). "Corresponding interests: artisans and gentlemen in nineteenth-century natural history". *British Journal for the History of Science* 27: 383-408.

[12] Lankford, J. (1981). "Amateurs and Astrophysics: A Neglected Aspect in the Development of a Scientific Specialty". *Social Studies of Science* 11: 275-303, and (1981). "Amateurs versus Professionals: The Controversy over Telescope Size in Late Victorian Science". *Isis* 72: 11-28; Schaffer, S. (1988). "Astronomers Mark Time: Discipline and the Personal Equation". *Science in Context* 2: 115-45; Ogilvie, M. B. (2000). "Obligatory amateurs: Annie Maunder (1868-1947) and British Women Astronomers at the Dawn of Professional Astronomy". *British Journal for the History of Science* 33 (1): 67-84.

[13] Pyenson, L. and Sheets-Pyenson, S. (1999). *Servants of Nature: A History of Scientific Institutions, Enterprises, and Sensibilities.* New York: W. W. Norton & Company; Cooter and Pumfrey, "Separate Spheres".

[14] Walters, A. N. (1999). "English Broadsides of Solar Eclipses". *History of Science* 37: 1-43.

[15] Topham. "Scientific publishing".

[16] Roqué i Rodríguez, X. (1995). "Premsa i cultura de la ciència a Catalunya". In Puig-Pla, C. et al., coords. *Actes de les III Trobades d'Història de la Ciència i de la Tècnica.* Barcelona: Societat Catalana d'Història de la Ciència i de la Tècnica, pp. 47-60; Lightman, B., ed. (1997). *Victorian Science in Context.* Chicago: University of Chicago Press; Bensaude-Vincent and Rasmussen. *La science populaire.*

[17] Gregory, J. and Miller, S. (1998). *Science in Public: Communication, Culture, and Credibility.* New York: Plenum.

[18] Roberts, L. (1999). "Science becomes Electric: Dutch Interaction with the Electrical Machine during the Eighteenth Century". *Isis* 90: 680-714; Secord, "Science in the Pub".

[19] Shinn, T. and Whitley, R., eds. (1995). *Expository Science. Forms and Functions of Popularization.* Kluwer: Dordrecht; Bensaude-Vincent and Rasmussen, *La science populaire.*

[20] Topham. "Rethinking".

[21] Secord, J. A. (2004). "Knowledge in Transit". *Isis* 95: 654-672; Topham. "Rethinking".

[22] Bensaude-Vincent, B. (2000). *L'opinion publique et la science: A chacun son ignorance.* Paris: Sanofu-Synthélabo, and (2003). *La science contre l'opinion: histoire d'un divorce.* Paris: Les Empêcheurs de penser en rond.

[23] On the the relations between science and the nation, see part IV in this volume.

A "NATIONAL FIESTA":
TOTAL SOLAR ECLIPSES AND POPULARIZATION OF ASTRONOMY IN EARLY TWENTIETH-CENTURY SPAIN

PEDRO RUIZ-CASTELL

The public image and understanding of science plays a crucial role both in the development of science in a country and in the progress of a country itself. Indeed, the development of science and social change within a community is depicted by how science is interpreted in it.[1] Nonetheless, the language in which scientists deal with scientific issues is not easy to understand for the general public, who usually need a translation (which does not necessarily include all the troubles and uncertainties) of the subject into comprehensible words. *Popularization of science*, a term first coined and frequently used in the nineteenth century, has traditionally been identified with this process of explaining scientific knowledge to general audiences.

As Shapin has suggested, the various languages of popularization may be understood as special *occasional* modes of speech used when external support (mainly financial and moral) or subvention is required.[2] As the first part of this paper shows, this was partly the case for Spanish intellectual elites, who used the total solar eclipses of 28 May 1900 and 30 August 1905 to claim for more official and social support. Moreover, this text provides some evidence on how popularization of science should not be understood exclusively in terms of a unidirectional process, from elite to lay people, but as a concept and a category which helps to better understand the concept of *public sphere*, as defined by Habermas,[3] referring to a public space in which citizens interact and enrich societies' policies through rational and critical attitudes and opinions.[4] This political public debate in which popularization of science was engaged, as this text will show, was encouraged by the Spanish press – largely dominated during these years by a liberal culture[5] – which heavily promoted science on the occasion of these eclipses, impressing on the population the importance of scientific knowledge.

Moreover, commercial interests meant that booksellers supplied the country with leaflets, descriptions, and plates on the nature of and on how to observe the eclipses. These popular works were affordable to people for few centimes. But in a country that was yet to be modernized and with high levels of illiteracy, the fact is that even when the price of these publications indicates that they could have been intended for popular audiences, they mainly reached the upper classes. Nevertheless, the public image and understanding of astronomy was shaped through many other initiatives that were altogether broad in their targeting of various audiences in Spanish society, such as the organization of special eclipse trips to the path of totality, the public projection of films of eclipses, and the public interest that arose the visit to Spain of so many eminent astronomers to observe these solar events. Therefore, popularization of astronomy in Spain during the late nineteenth and early twentieth centuries has to be understood in terms of the different ways in which the several sectors of Spanish society (not only the upper classes, but also the working classes, women, etc.) interacted with astronomical knowledge in a dynamic and enriching way for both the configuration of the practice of science and its appropriation.

The nature and characteristics of astronomical phenomena – and particularly eclipses – have been shown to be especially relevant as potential sites for the popularization of science to a wide range of audiences across society. Their significance in culture, linked with their ready observability and the traditional coexistence of professional and amateur practitioners in this field, secured a broad and diverse phenomenon of integration of solar eclipses in early twentieth-century Spanish culture.

The role of foreign travellers and foreign scientific societies – in particular those promoting the practice of astronomy – in this cultural process which took place in Spain, was especially relevant. Being the only European country from which the total solar eclipses of 28 May 1900 and 30 August 1905 were observable, Spain became a centre of international attention. Special trips to Spain to observe the eclipses were promoted in countries such as Britain, France and Italy. The major trigger of these organized trips was astronomical observation, but they also included tourism and were an opportunity for Spaniards to display their cultural heritage in an international context.

Foreign astronomers used these astronomical expeditions for specific interests in their own countries – the study of this phenomenon would need another whole paper though, and it will not be fully explored in here. Nevertheless, it is certainly clear that internationally famous astronomers and popularizers such as Camille Flammarion used their visits to Spain both to recruit local adepts to their own national scientific societies and journals and to implement their popularizing program for a universal practice of astronomy and their ideal of cooperation between amateur and professional scientists. In this sense, channels of communication between France and Spain proved especially important, as the French astronomical milieu was already a visible model for Spanish practitioners of astronomy, who subscribed in great number to its journals and societies, and intervened in both amateur and professional fields. Consequently, the visit of Flammarion and other internationally celebrated astronomers served the purposes of Spanish professional and amateur astronomers, instrument makers, and booksellers in their aim to promote the practice and consumption of astronomy and to achieve a higher status for the discipline in Spanish society.

Solar eclipses reports: popular works or scientific news?

As Alice Walters has explained for the astronomical broadsides produced in Britain during the eighteenth century, popular astronomical prints constituted not only an important medium for the consumption of authoritative knowledge, but also mobilized public participation in the observation of an event like an eclipse.[6] This public participation was always declared of great importance in several matters such as refining eclipse theory. For example, by the turn of the twentieth century, the different values used for the Moon's semi-diameter had led to discrepancies in calculated results based on observations made at totality.[7] To measure accurately the contact times and the duration of totality was one of "the most useful observations that could be made by the average person" in order to help to solve this problem.[8]

Consequently, lay people were encouraged to take observational data and pursue astronomical tasks during totality in order to contribute their results to the progress of scientific knowledge. At the same time, the cultural legitimacy of astronomy and astrophysics was established, at least to a certain extent, by making eclipses commercially and intellectually accessible. In the new and vast literary market of the second half of the nineteenth and the early twentieth centuries, total solar eclipses were publicized and popularized in a wide range of publications, including newspapers, magazines, books and scientific journals. These popular reports not only emphasized the public and open character of astronomy, but also were excellent occasions to promote the popular awareness of and interest in the field.

In Spain, the passion for astronomy exploded on the occasion of the
total solar eclipses of 1900 and 1905:

> The press, the opportunism of publishers, and commercial temptation have
> spread leaflets, descriptions, prints, and optical devices all over the country
> to enable the progress of the eclipse to be observed in all its detail – all its
> rays, protuberances, and coroniums ... That is how one has seen interesting
> popular texts such as those by Comas Solá, good versions of the very
> valuable works by Flammarion, and other similar ones purchasable for a
> few centimes ... Some of this will remain in the people's souls, and may
> ferment into higher instruction. And even if it does not, it will always exalt
> and strengthen us, and brighten our national pessimism, like an aurora, to
> see people, even if just circumstantially, becoming involved with Science.[9]

In other words, the convergence of commercial, scholarly, and
educational aspects was seen by Spanish elites as a positive indicator and
the first step on the path to the instruction of the people and, consequently,
to the progress of the country itself. Furthermore, it is interesting to note
that the term employed by contemporary authors to deal with explanatory
texts addressed to lay people and written by prestigious astronomers such
as the French Cammille Flammarion (1842-1925) and the Catalan Josep
Comas i Solà (1868-1937) was that of *trabajos divulgadores*.[10]
    The verb *divulgar*, was in use in Spain by the mid-eighteenth century,
meaning the action of "publishing, extending, spreading something, telling
it to many people and in many places".[11] Its associated terms *divulgador*
and *divulgación* were also in use in this period. Only towards the end of
this century was the process given a more specific target, the *vulgo*,
defined as "the majority of lay people or the plebs".[12] By the 1830s, the
vulgo had been substituted by the more general concept of *public*, a
definition preserved in Spanish dictionaries until the twentieth century.[13]
Although a serious study of the changing uses of these terms in Spanish
society still needs to be pursued, the use of dictionaries published in
different periods allows us to see a hint of a certain correspondence with
the language used in Italy for this purpose, while the English term
popularization fits into the new meaning given to *divulgación* by the late
eighteenth century. [14] However, from the 1830s, the language of
popularization in Spain was directed towards a general public –
independent of social class – a meaning in harmony with the contemporary
use of the terms *science populaire*, *popular science*, and *scienza popolare*
in French, English and Italian respectively.[15]
    Like in the case of the boom of science popularization in Italy, which
took place in a period of special political, social, and cultural relevance for

that country,[16] popularization of the 1900 and 1905 eclipses happened at a time of important transformations in Spain. Indeed, in a country immersed in a process of modernization which aimed at emulating more developed European countries, the language of *divulgación* (and therefore that of popular science) was considered both as a way to obtain economic benefits, and as a tool to achieve two different objectives: on the one hand, to engage people in scientific practice and to awaken people's interest for science; on the other, as starting point to draw a line between experts and consumers of science in a period in which professionalization of astronomy was being consolidated.[17]

Popular works on the total solar eclipses of 1900 and 1905 were also found in Spanish newspapers, whose editorial staff invited prominent scientists and astronomers, both local and foreign, to explain in their publications the importance of astronomy (particularly of total solar eclipses) and of their astronomical observations.[18] Nonetheless, these explanatory texts were slightly different from the news that journalists published in these same newspapers dealing with the solar eclipses of 1900 and 1905. For example, the many different reports which appeared in the Spanish press written by non-scientists mainly included anecdotes and minor topics from a scientific point of view, such as the social activities of prestigious foreign astronomers and the events organized to honour them, which were recorded in all Spanish newspapers during the previous and subsequent months to the eclipses. Although it is tempting not to consider relevant, from the scientific point of view, the information reported in this kind of articles, there is no doubt that these works have to be considered as part of a science popularization programme set up to coincide with the total solar eclipses. In fact, even today, newspaper science writers need, in order to attract and hold readers' attention, "to sugar-coat the pill – that is, to "dress up" the story with catchy phrases and melodramatic anecdotes".[19]

Many publications sent their correspondents to different towns to write about the solar events. The following table is a good example of the large impact that the eclipses of 1900 and 1905 had on the Spanish press. It shows the unusual large number of contributors that the Madrid-based newspaper *El Liberal* had all around the path of totality of the eclipse of 1900 (see Table 1).[20]

Contributor	Location
Bernardino Torres González	Manzanares
Jose María López Campello	Elche
Joaquín Dicenta Benedicto	
"Orozco"	Santa Pola
Miguel Aguilar	Plasencia
Carlos Puente	
Carlos del Río	Argamasilla de Alba
Mariano Martín Fernández	
Carlos Miranda	Alcázar de San Juan
"Niño"	Daimiel
Eduardo Rosón	Navalmoral de la Mata
Antonio R. Lázaro	

Table 1. Contributors to *El Liberal* for the eclipse of 28 May 1900

Many newspapers contacted local relevant figures who acted as amateur journalists in order to write articles and notes for them. A good example is that of José María López Campello, physician at the city of Elche who hosted in his house several foreign and local astronomers.[21] Other publications sent their regular contributors to the path of totality. This was the case of Joaquin Dicenta Benedicto (1862-1917), one of the most relevant Spanish poets and playwrights, and Mariano Martín Fernández (1866-1940), a well-known journalist and lawyer who later became a member of the Spanish Parliament and Senator.[22]

Many periodicals from all over Spain, with different ideologies ranging from the pro-republican daily *El Liberal* (mainly read by a liberal non-conservative bourgeoisie, cultivated and open to ideas from outside Spain) to the Catholic and conservative *El Orbe Católico*, and including not only other influential newspapers, such as the independent daily *El Imparcial* and the liberal *Heraldo de Madrid*, but also magazines like *Ilustración Española* and *Blanco y Negro*, published different notes and articles on these total solar eclipses all over the country. As a consequence, astronomy and the scientific enterprises of 1900 and 1905 became a topical issue amongst Spaniards. This was noticed by foreign astronomers such as the French Théophile Moreux (1867-1954), who went to Elche to observe the total solar eclipse of 1900:

> I had time to read all the newspapers and to notice that the eclipse occupied all the columns. "Better this than a bullfight report, I thought. Decidedly Spain gets thrilled by science. This is perhaps the sign of its social

recovery." The illustrated journals acted like their colleagues and for several weeks the caricaturists had to sharpen their words and their talent on the eclipse.[23]

Similarly, Flammarion wrote after this visit to Spain:

> For fifteen days, Spain has been completely engaged with the eclipse. From the highest rank of the social scale to the humblest classes, everyone has been determined to assist in this great lesson of cosmography.[24]

Indeed, early twentieth century journalists created, on the occasion of these total solar eclipses, a public taste for science. Their stories were, on many occasions, embellished with ideas such as those of national prestige and obligation, as these lines show:

> Considering the immense publicity that this eclipse [of 1905] has afforded Burgos, the natives should be pleased to provide hospitality, for their own dignity and for that of Spain.[25]

The reports on total solar eclipses were an excellent occasion to bring the agendas of the different newspapers to light. For example, Spanish intellectual elites were very much linked to and represented by the two most influential newspapers in the country, *El Imparcial* and *El Liberal*, founded in 1867 and 1879 respectively. Spanish elites aimed for a nationwide programme, in which the State would play a prominent role, to create the infrastructure needed for a modern economy. These initiatives, which sprang from the new social and cultural framework developed in Spain during the last quarter of the century,[26] included measures to promote science and scientific education. Consequently, they were very interested in obtaining a serious commitment from the government with regard to scientific research. In order to set up a public debate, the press functioned as a resonance box to resound claims that the involvement of the government was required in the scientific development of the country, as illustrated by an article which appeared the day after the eclipse of 1900 in the Spanish newspaper *El Liberal*:

> The curiosity to see the astronomical instruments, [and] the admiration for scholars, both foreigners and locals... had as much solemnity as bustle... The people... have faith in science... There is still one thing to do: the government has to appreciate that spirit and correspond to it from its own field of action; attending and honouring the scholars who worked in the very important astronomical operations.[27]

As already mentioned, almost every Spanish journal and newspaper wrote enthusiastic articles about anything that was related to the total solar eclipses. Some of them also included suggestions and advice for amateur observers, [28] making it sometimes difficult to draw a clear line to differentiate popular scientific works from mere news reports. Rather than attempting to define this line, it is more interesting to underline how the combination of both kind of texts contributed to create in Spain during these years a new public image and understanding of science (and particularly of astronomy).

Flammarion in Spain

There is no doubt that the public image and understanding of science in the country was largely shaped by prestigious scientists and authors who wrote popular works both in books and newspapers. But how important were they? Did they really engage with the public and have large audiences at all? The total solar eclipse of 1900 provides an excellent opportunity to show how celebrated some of these authors were in Spain. For instance, on 26 May 1900, up to 25,000 of the 30,000 inhabitants of the city of Elche patiently awaited the arrival of a very special train. [29] They all wanted to welcome the man who travelled in it. As the journalist Joaquín Dicenta wrote:

> The rich and the poor, the wise and the ignorant, the workers from the city and the labourers from the fields, all walks of life that constitute the totality of Elche were represented … Who was that man …? [30]

That man was the prestigious French astronomer and popularizer Cammille Flammarion, who went to Elche to observe the total solar eclipse of 28 May 1900. He had arrived to Barcelona on 24 May 1900 after a four-day journey, leaving after lunch to Tarragona and continuing to Castellón, where despite the bad weather conditions more than a hundred people, representatives of the city, welcomed him. [31] After Castellón he stopped in Valencia, where several events were held there in his honour, including a banquet for seventy people in the Botanic Garden of the University on 25 May 1900, organized by the University and the Ateneo Científico, Literario y Artístico of the city. The Ateneo was an association founded in 1868, whose members belonged to the intellectual elites and were very interested in the social promotion of cultural initiatives, including not only literary and artistic, but also scientific works. [32]

On his train journey from Valencia to Alicante, Flammarion was welcomed at every stop he made. The interest he awoke in the general public was expressed on banners at several stations.[33] For example, at the train station of Villena there was written:

> Recognition to the one who has led us to know the Universe and who has aroused in our souls the philosophy of the infinite.[34]

Flammarion arrived to Alicante on 26 May 1900 and continued his trip to Elche, where the multitude welcomed him.[35] Here, he attended a banquet in which several local dignitaries honoured the astronomers who, like him, were in the city to observe the eclipse.[36] It was said that in total about a hundred thousand Spaniards welcomed Flammarion on his trip to Spain at the many different train stations through which he passed or stopped at.[37] His visit was recorded in almost every single Spanish newspaper and all the bibliographical sources surveyed agree on the large number of people that Flammarion attracted during his visit.

Once in Madrid, Flammarion and his wife were hosted by Ernesto de la Guardia,[38] who had promoted the Flammarion Scientific Society of Madrid, created in 1899.[39] In fact, the influence of Flammarion during the second half of the nineteenth century had been such that several scientific societies named after him were founded in Spain during these years. De la Guardia, on behalf of the Scientific Society of Madrid, awarded Flammarion a commemorative medal of his visit to Spain,[40] and he was also honoured in Madrid with a banquet organized by the Spanish newspaper *El Imparcial*, and a reception hosted by the magazine *Blanco y Negro*.[41]

Many figures from the political, cultural, and scientific sectors of Spanish society were present at these events.[42] Flammarion's party made a trip from Madrid to Toledo and then spent a couple more days in Madrid, where they visited the capital's observatory. The absence of the Director of the Observatory of Madrid, Francisco Iñiguez Iñiguez (1853-1922), meant that the First Astronomer, Vicente Ventosa y Martínez de Velasco (1837-1919), received and entertained them. He even gave them a guided tour of the museum of art, El Prado, which confirms the significance of leisure and tourism in the organization of scientific expeditions such as those of 1900 and 1905.[43] They were also received by the royal family, whose members had studied the solar event. Princess Isabel had observed the eclipse from Argamasilla, while observations from the Royal Palace of Madrid had been made by the regent María Cristina of Habsburg (who made some drawings of the phases of the eclipse, notes on the effect of the eclipse on flowers, and meteorological observations) and by her son, the future King Alfonso XIII, who took some photographs of the corona at different phases of the eclipse.[44] Five years later, the eclipse of 1905 was observed at Burgos, by Alfonso XIII, and Eugenio Montero Ríos (1832-1914), the President of the Spanish Government. Several celebrations were organized during the days immediately before and after this eclipse, including the unveiling of monuments, a young-pigeon target practice, a photographic competition, and a first-class bullfight.

Most of these Spanish "one-day observers" and amateur astronomers, who had gradually increased in number during the second half of the nineteenth century, took advantage of the well-developed state of amateur astronomy in the neighbouring country of France. Therefore, in the absence of such institutionally strong Spanish equivalents, many of them were members of the Société Astronomique de France and subscribed to its journal L'Astronomie. In fact, Ernesto de la Guardia received an award from the Société Astronomique de France in 1900 for bringing to the society fourteen new members, a significant number.[45] Amateur astronomy was gradually growing in Spain. In this sense, Flammarion's tour around Spain may be understood not only in terms of scientific research and leisure, but also of advertisement of his universal programme of popular astronomy and to legitimate his ideal of cooperation between amateur and professional scientists.

Eclipse expeditions, leisure, and business

The public perception of science was not only determined by authors of popular science books and the press. There were, of course, many other aspects which played a crucial role. A good example of the status achieved by astronomy in Spanish society on the occasion of the total solar eclipses of 1900 and 1905 is the success of the special trains sent to the path of totality by various railway companies. In this initiative, which had no precedents in the rest of Europe, the United States or elsewhere, cheap one-day return tickets to the path of totality were offered to Spaniards.[46] Some routes and fares for the eclipse of 28 May 1900 are given in Table 2.[47]

Departure	Arrival	Price
Madrid	Navalmoral de la Mata	3 pesetas (for groups of more than three people)
	Argamasilla	Luxury train: 25 pesetas
	Alcázar	Luxury train: 20 pesetas 1st class: 12 pesetas 2nd class: 8 pesetas 3rd class: 4 pesetas
All the stations from Cartagena to Beniaján	Hellín	1st class: 12 pesetas 2nd class: 8 pesetas 3rd class: 5 pesetas
All the stations from Murcia to Cieza		1st class: 10 pesetas 2nd class: 7 pesetas 3rd class: 4 pesetas
Linares, Vadollano, Vilches, Santa Elena, and Cárdenas	Manzanares	1st class: 10 pesetas 2nd class: 7 pesetas 3rd class: 4 pesetas
Valdepeñas		1st class: 4 pesetas 2nd class: 3 pesetas 3rd class: 2 pesetas

Table 2: Some train routes and fares to the path of totality on 28 May 1900

By 26 May, more than 800 tickets had been sold to travel from Madrid to Navalmoral de la Mata (Cáceres) in first, second, and third class trains. One day later, the day before the eclipse, more than 4000 tickets from Madrid to Navalmoral had already been sold.[48] Even tickets for the luxury train rapidly sold out and more had to be put on sale; the upper classes also wanted to be present at such an important astronomical (and social) event and about 600 luxury train tickets were sold for Alcázar de San Juan (Ciudad Real) and Argamasilla de Alba (Ciudad Real).[49] But all expectations were exceeded on the day of the eclipse. Some tickets were put up for sale with an extra charge of 25 per cent of their original cost.[50] In all, about 10,000 to 12,000 people left Madrid by train on the day of the eclipse for various locations in the path of totality, such as Alcázar, Argamasilla, and Navalmoral.[51] Similar initiatives were also taken by several train companies for the eclipse of 1905 to encourage these "one-day expeditions".[52]

These one-day enterprises have to be understood not only in terms of public curiosity and interest for total solar eclipses, but also of business. The commercial success of these astronomical events was related not only to railway companies, but also to opticians, who made good money out of the popular interest in eclipses. For example, the Spanish firm Casa de Oliva sold more than ten thousand "observe-eclipses" devices at a price of one peseta each on the occasion of the eclipse of 1900,[53] and more than a hundred thousand pieces of smoked glass were sold for one and two *perras gordas* (ten-cent coins) by hawkers in Madrid on the day of the eclipse.[54]

Similarly, on the days leading up to the eclipse of 28 May 1900, inns were full in cities located in the path of totality, such as Plasencia.[55] Some Spanish innkeepers had special offers for visitors during the day of the eclipse.[56] Moreover, the British journal *The Observatory* reported that the night before the eclipse of 30 August 1905 the price for bed and breakfast in Burgos was £4 (about 130 pesetas).[57] The (sometimes exploitative) business generated by these eclipses was also described in a Spanish newspaper as follows:

> A large number of *professionals* and an even larger number of learned amateurs and curious people of all kinds are arriving at Burgos to observe the solar eclipse, and some natives of Burgos, trying to make the occasion as successful and profitable as possible, aim to strip the very skin from the Spaniards and foreigners in their grip.
> They have taken from a Yankee – from what we have been told – the unbelievable quantity of 1800 pesetas for *three days'* accommodation.[58]

At the same time, eclipse expeditions sent from abroad also had an important impact on the public image of science in Spain. In some cases, like the British, amateur expeditions were a way for the wealthy to gain prestige and have to be understood in terms of tourism, scientific curiosity, and upper-class leisure. A good example of this is how travel agents developed tours in 1900 and 1905 based around observing the eclipse from the path of totality. For example, Thomas Cook & Son advertised a tour from London to Talavera, visiting Paris, Bordeaux, Biarritz, and Madrid, for £30, including tickets, accommodation, and all fees.[59] These tours were widely advertised in popular scientific journals and mainly addressed to amateur astronomers and astronomical amateur societies such as the British Astronomical Association – originally set up for those who found the subscription to the Royal Astronomical Society too high, or the papers excessively technical, or who, like women, were excluded.[60] For instance, for the eclipse of 1905 three different possible itineraries were presented by Edward Walter Maunder (1851-1928) to the British Astronomical Association: a twenty-strong party expedition organized by the Eclipse Committee, following the programme arranged by Thomas Cook, for 30 guineas;[61] a route via Barcelona and Valencia to the east coast; and a tour to Majorca, also arranged by Cook, for 33 guineas.[62]

As these initiatives show, there was a demand from the international public, mainly the upper classes, to play an active role in the observation of total solar eclipses. Companies, which obviously aimed to obtain benefits, responded successfully with new and original products and initiatives such as those here presented. Indeed, analogous to what Spanish railways companies did, some steamship companies advertised that steamers on regular services which brought them to the central line at the time of the eclipse would heave to for a few hours to give passengers the opportunity to observe the event.[63] Firms such as the Italian Steamship Company, of Genoa, also arranged excursions to the path of totality to observe the 1905 eclipse. The programme that the Italian Company advertised began with departure from Genoa on the evening of 23 August and return on 2 September 1905, including food and wine for the trip and ranging from 200 to 300 Italian lire (between about £8 and £12) according to the class of accommodation on board.[64] Similarly, the Peninsular and Oriental Company arranged to send a vessel by way of Gibraltar and thence to Marseille.[65]

Finally, it is interesting to point out how successful other initiatives, such as the projection of films of the eclipse of 1905, were in different cinemas in Spain. For example, the cinema *Napoleón*, established in Barcelona in 1896, presented to the public on 7 September 1905 five shows filmed by Spanish observers. Even when, in general terms, most cinemas were mainly addressed to lay people, this particular one aspired from its foundation to engage with the upper classes: invitations were sent to the most prominent figures of the city, tickets were relatively expensive (one peseta versus the ten to fifty cents of other cinemas by the turn of the century), etc.[66] Typically, there were daily projections and, as it was not an itinerant cinema, the programme had to be renewed almost every week. However, as recorded in contemporary Spanish press, the projections on solar eclipses lasted in many cinemas for almost a whole month.[67]

Popular science was indeed developed for a broad public of different characteristics, including not only the upper-middle, but also the working classes.[68] In fact, solar eclipses were no strangers in other forms of public science, such as the lectures of the so-called university extension movement, linked to the growth of the liberals. This movement, founded in Great Britain in the late 1860s and soon spread all over Europe, was promoted in Spain during the last years of the nineteenth century by Professors Adolfo Álvarez-Buylla (1850-1927) and Rafael Altamira (1866-1951) at the University of Oviedo – even when similar initiatives had been attempted in Spain under the freedom projected by the *Sexenio Revolucionario* (1868-1874). In particular, the lectures of the programme for the academic course 1903-1904, delivered in the mornings and the evenings to the middle and the working classes (including women), included one dealing with the total solar eclipse of 1905 – delivered by Enrique Fernández Echevarría, Professor at the Science Faculty.[69]

Conclusion

Being the only European country from which the total solar eclipses of 28 May 1900 and 30 August 1905 were observable, Spain became a focus of international interest during these years. Moreover, these astronomical events managed to engage a large and diverse audience from all classes of Spanish society. Local astronomers used these eclipses to promote the development of the practice of astronomy in Spain, helping to consolidate a compact group of people interested in astronomy and astronomical research in the country.[70] The eclipses and the events which surrounded them were enthusiastically presented in almost every Spanish journal and newspaper by professional and amateur astronomers, science popularizers, and general journalists. The press and several publishers littered the country not only with popular works to find out and observe the progress of these eclipses, but also with any news related to scientists, public figures, and any sectors of society involved in the observation of these astronomical events. As a consequence, and as a contemporary Spanish scientific publication reported in 1905, the solar eclipse awoke even more public interest than the general elections in the country, with which it coincided.[71]

Furthermore, the role of those who visited Spain from abroad was of crucial importance. In fact, these foreign travellers often had an influential position in scientific and political issues in their home countries. Their trip served their own political and scientific agendas but, at the same time, they became the agents to project in their own countries a positive image of Spain as a country engaged in modernization. This was clearly perceived by Spanish elites, who consequently hosted eminent foreign figures and contributed to the organization of their visits.

Popularization of astronomy in Spain achieved its zenith on the occasion of the total solar eclipses of 1900 and 1905. The popularization and commercialization of eclipses did not only involve astronomers, publishers, instrument makers, and other entrepreneurs who organized travelling and accommodation facilities, celebrations, and festivities to honour local and foreign scientists, but also lectures and conferences for the working classes and film projections. Indeed, the total solar eclipses were celebrated in Spain as "a national *fiesta*".[72] This aroused not only huge expectations and a curiosity for astronomy amongst Spaniards, but also helped to set up a public debate in terms of the importance of science and of national pride, prestige, and obligation – a discussion promoted and celebrated by the Spanish intellectual elites. Indeed, the popularization of astronomy in Spain on the occasion of these total solar eclipses served a triple function: professional, commercial, and political.

Notes

[1] Pyenson, L., and Sheets-Pyenson, S. (1999). *Servants of Nature: A History of Scientific Institutions, Enterprises, and Sensibilities.* New York: W. W. Norton & Company.

[2] Shapin, S. (1984). "Pump and circumstance". *Social Studies of Science* 14: 481-520; Hinton, D. A. (1979). *Popular science in Britain, 1830-1870.* unpublished PhD thesis. Bath: Bath University.

[3] Habermas, J. (1962). *Historia y crítica de la opinión pública. La transformación estructural de la vida pública.* México-Barcelona: Gustavo Gili. Note how the original word employed by Habermas in German, "Öffentlichkeit", was translated differently into other languages such as English, French, and Spanish as "public sphere", "espace public" ["public space"], and "opinión publica" ["public opinion"] respectively.

[4] This model of public sphere, closely linked to the rise of the modern state and the growth of capitalism (in the late eighteenth and nineteenth centuries), has been subsequently re-visited to include post-bourgeois models and new aspects, in order to better define and understand what constitutes *the public*. Several historians of science have, during the last two decades, applied this concept to study how the practice of science interacts within social and political aspects. See for example, Golinski, J. (1992). *Science as Public Culture: Chemistry and Enlightenment in Britain, 1760-1820.* Cambridge: Cambridge University Press, and Stewart, L. (1992). *The Rise of Public Science: Rhetoric, Technology and Natural Philosophy in Newtonian Britain, 1660-1750.* Cambridge: Cambridge University Press. However, some historians are reluctant to employ such categorization, as they identify this concept with public demonstrations for legitimization of the scientific endeavour. See for example, the introductory text of the 5th STEP meeting on *Popularisation of Science and Technology in the European Periphery* (June 2006). [http://einstein.uab.es/suab237w/PDF/STEPMeeting/InformationStepMeeting.pdf].

[5] Fusi, J. P. and Palafox, J. (1997). *España: 1808-1996. El desafío de la modernidad*. Madrid: Espasa Calpe.

[6] Walters, A. N. (1999). "English Broadsides of Solar Eclipses". *History of Science* 37: 1-43.

[7] See for instance, Buchanan, J. Y. (1905). "Letters to the Editor: Eclipse predictions". *Nature* 72: 603, and Downing, A. M. W. (1905). "Letters to the Editor: Eclipse predictions". *Nature* 72: 629.

[8] Pickering, W. H. (1899-1900). "The total eclipse of 1900, May 28". *Journal of the British Astronomical Association* 10: 333.

[9] A. A. (1905). "El eclipse de Sol de 30 de Agosto". *El Mundo Científico* 7: 544-47, p. 547. All translations from Spanish and French are by the author of this paper.

[10] For a detailed study of the figure of Comas i Solà as astronomer and populariser, see Roca Rosell, A. et al., coords. (2004). *Josep Comas i Solà: astrònom i divulgador*. Barcelona: Ajuntament de Barcelona.

[11] As defined in Spanish: "Publicar, extender, esparcir alguna cosa, diciendola à muchas personas y en muchas partes. Viene del Latino *Divulgare*, que significa esto mismo". *Nuevo Tesoro Lexicográfico de la Lengua Española* [http://buscon.rae.es/] (accessed 14 September 2007).

[12] In Spanish: "El común de la gente popular ó la plebe". See n. 11.

[13] See n. 11.

[14] I am very grateful to Josep Simon for sharing his views on this issue with me and providing me with further evidence.

[15] Govoni, P. (2002). *Un pubblico per la scienza. La divulgazione scientifica nell'Italia in formazione*. Roma: Carocci, p. 20; Bensaude-Vincent, B. and Rasmussen, A. (1997). "Introduction". In Bensaude-Vincent and Rasmussen, eds. *La science populaire dans la presse et l'édition, XIXe et XXe siècle*. Paris: CNRS Éditions, 13-30; Topham, J. R. (2008). "Rethinking the History of Science Popularization/Popular Science". In Papanelopoulou, F.; Nieto-Galan, A., and Perdiguero E., eds. *Popularising Science and Technology in the European Periphery, 1800-2000*. Aldershot: Ashgate.

[16] Govoni. *Un pubblico*, pp 25-7.

[17] This reflection is based on a paper given by Jonathan Topham at the 5th STEP meeting on *Popularisation of Science and Technology in the European Periphery* held in Mahón (Spain) in June 2006.

[18] A great example of this is the issue of 26 May 1900 of the Spanish newspaper *El Heraldo de Madrid* published on, to which contributed scientists and popularisers of science such as the Spaniards José Echegaray and Victoriano Fernández Ascarza, and the French Henri Alexandre Deslandres and Aymar de la Baume Pluvinel. Echegaray, J. (1900). "El Sol y el eclipse". *El Heraldo de Madrid* 26 May [Año 11, n° 3483]: 1; Deslandres, H. A. (1900). "Origen eléctrico de las protuberancias, de la cromosfera y de la corona solar". Ibid.: 5; Baume Pluvinel, A.

de la (1900). "Fotografías del eclipse". Ibid.; and Fernández Ascarza, V. (1900). "El análisis espectral". Ibid.: 6.

[19] Davidson, K. (2006). "Why science writers should forget about Carl Sagan and read Thomas Kuhn. On the troubled conscience of a journalist". In Doel, R. E. and Söderqvist, T., eds. *The Historiogaphy of Contemporary Science, Technology, and Medicine. Writing Recent Science.* London: Routledge.

[20] Anon. (1900). "Antes de la batalla". *El Liberal* 28 May: 1.

[21] Similarly, the historian Pere Ibarra i Ruiz wrote some articles for the Valencia-based *Las Provincias*. Ibarra, P. (1900). "Desde Elche. Multitud de sucesos y escasez de cronistas...". *Las Provincias* 23 May, and (1900). "Desde Elche. La fotografía y el eclipse...". *Las Provincias* 27 May. The director of this newspaper, Teodor Llorente, also went to Elche to observe the eclipse. See: Llorente, T. (1900). "El eclipse en Elche". *Las Provincias* 30 May. All these articles are found in Soler Selva, V. F., coord. (2000). *L'eclipsi total de sol de 1900 al Baix Vinalopó.* Elx: Ajuntament d'Elx.

[22] López de Zuazo, A. (1981). *Catálogo de periodistas españoles del siglo XX.* Madrid: Universidad Complutense.

[23] Moreux, T. (1900). "Autour de l'éclipse. Notes de voyage". *Bulletin de la Société Astronomique de France* 14: 308-24, p. 308.

[24] Flammarion, C. (1900). "L'éclipse totale de soleil". *Bulletin de la Société Astronomique de France* 14: 289-97, p. 297.

[25] Anon. (1905). "Mirando al cielo. El Eclipse de Sol". *El Universo* 29 August.

[26] Fusi, J. P. (1999). *Un siglo de España: La cultura.* Madrid: Marcial Pons.

[27] Anon. (1900). "La fiesta de la ciencia". *El Imparcial* 29 May : 1.

[28] See for instance, Anon. (1900). "Para el eclipse de mañana. Guía práctica del observador". *El Heraldo de Madrid* 27 May [año 11, n° 3484]: 1.

[29] Moreux. "Autour de l'éclipse", p. 314.

[30] Dicenta, J. (1900). "Flammarion. Instantánea". *El Liberal* 29 May: 1.

[31] Anon. (1900). "Flammarion en España". *El Imparcial* 24 May : 3.

[32] Anon. (1900). "En honor de Flammarion". *El Liberal* 26 May: 2. See also, Zelt (1900). "Flammarion en Valencia". *El Heraldo de Madrid* 24 May [año 11, n° 3481]: 1, and (1900). "Flammarion". *El Heraldo de Madrid* 25 May [año 11, n° 3482]: 2, and Anon. (1900). "Flammarion en España". *El Imparcial* 25 May: 1.

[33] See for example, Anon. (1900). "Flammarion en España". *Blanco y Negro* 2 June: 4.

[34] Moreux. "Autour de l'éclipse", p. 311. The enthusiastic reception in Villena (and later in Alicante) was also recorded in Anon. (1900). "Flammarion". *El Heraldo de Madrid* 26 May [año 11, n° 3483]: 2.

[35] Anon. (1900). "El eclipse". *El Imparcial* 27 May: 2.

[36] Anon. (1900). "El eclipse". *El Imparcial* 28 May: 1-2, p. 2. Amongst these scientists were Rear Admiral Juan Bautista Viniegra y Mendoza (director of the Spanish Observatory of San Fernando from 1891 to 1903), Josep Joaquim Lànderer (one of the most prestigious Spanish amateur astronomers), the Spanish

Father Angel Rodríguez de Prada (Director of the Vatican Observatory from 1898 to 1905), the French Comte Aymar de la Baume Pluvinel, the Dutch Nicolas Donitch, and the British Joseph Norman Lockyer.

[37] Moreux. "Autour de l'éclipse", p. 310.

[38] Ibid., p. 322. See also, Anon. (1900). "Flammarion en Madrid". *El Liberal* 1 June: 2.

[39] Oliver, J. M. (1997). *Historia de la astronomía amateur en España*. Madrid: Sirius.

[40] Anon. (1900). "Flammarion". *El Imparcial* 2 June: 2, and Anon. (1900). "Visita de Mr. y Mme. Flammarion a 'Blanco y Negro'". *Blanco y Negro* 9 June: 7-8.

[41] Anon. (1900). "Flammarion". *El Imparcial* 3 June 1900: 1-2.

[42] Anon. (1900). "En honor de Flammarion". *El Liberal* 3 June: 1. See also, Anon. (1900). "Flammarion en Madrid". *El Heraldo de Madrid* 31 May [año 11, n° 3488]: 2; Anon. (1900). "En honor de Flammarion". *El Heraldo de Madrid* 2 June [año 11, n° 3490]: 3; Anon. (1900). "Flammarion en Madrid". *El Heraldo de Madrid* 3 June [año 11, n° 3491]: 1.

[43] Moreux. "Autour de l'éclipse", pp. 323-4.

[44] Flammarion, C. (1900). "[…] Notes complémentaires". *Bulletin de la Société Astronomique de France* 14: 298-307, p. 300. See also, for example Anon. (1900). "Flammarion en Palacio". *El Heraldo de Madrid* 2 June [año 11, n° 3490] : 1.

[45] As it was reported, he received the "médaille frappe à la Monnaie de Paris pour récompenser le zèle du sociétaire qui aura amené le plus grand nombre de nouveaux adherents". Flammarion, C. (1901). "La Société Astronomique de France". *Bulletin de la Société Astronomique de France* 15: 205-17, p. 210.

[46] Anon. (1900). "El Eclipse de Sol". *El Heraldo de Madrid* 28 May [año 11, n° 3485]: 1.

[47] This information is illustrative and not complete. See for instance Anon. (1900). "Día de tregua". *El Liberal* 29 May: 1; Anon. (1900). "Trenes baratos". *El Heraldo de Madrid* 23 May [año 11, n° 3480]: 3; Anon. (1900). "El eclipse de Sol". *El Imparcial* 22 May: 1, and Anon. (1900). "El eclipse de Sol". *El Imparcial* 23 May: 1. Note that by the early twentieth century, the average salary of industrial workmen, miners, and artisans in Spain varied from 3 to 4 pesetas for a working day, while women got from 1 to 2.5 pesetas and children 1 peseta. Families spent from 75% to 90% of their incomes for necessary food. The cost of the absolute necessities of life was such that an unmarried man could not live on less than an average of 2.75 pesetas a day. Perkins, C. (1912). "The Social and Economic Problems of Modern Spain". *Political Science Quarterly* 27: 92-108.

[48] Anon. (1900). "No hay billetes". *El Heraldo de Madrid* 27 May [año 11, n° 3484]: 2.

[49] Anon. (1900). "Para el eclipse: Servicio de trenes". *El Heraldo de Madrid* 26 May [año 11; n° 3483]: 2.

[50] Anon. (1900). "El eclipse". *El Imparcial* 29 May: 1.

[51] Anon. (1900). "Día de tregua". *El Liberal* 29 May: 1; See also Anon. (1900). "Trenes baratos". *El Heraldo de Madrid* 23 May [año 11, n° 3480]: 3; Anon. (1900). "El eclipse total". *El Imparcial* 29 May: 1-2.
[52] See for example, Anon. (1905). "El eclipse en Burgos". *El Imparcial* 27 June: 2, and Anon. (1905). "Las fiestas del eclipse". *El Imparcial* 29 June: 2.
[53] Anon. (1900). "Para el eclipse de mañana. Guía práctica del observador". *El Heraldo de Madrid* 27 May [año 11, n° 3484]: 1.
[54] Anon. (1900). "Telescopios a perra gorda". *El Liberal* 29 May: 2.
[55] Anon. (1900). "El eclipse". *El Imparcial* 27 May: 2.
[56] See n. 35 and 36.
[57] Anon. (1905). "The total solar eclipse of August 30, 1905". *The Observatory* 28: 377-382. Note that one pound was equivalent, more or less, to thirty three pesetas.
[58] Anon. (1905). "Mirando al cielo. El Eclipse de Sol". *El Universo* 29 August.
[59] For this particular example, see also Anon. (1900). "[Notes:] The coming solar eclipse of May 28". *The Observatory* 23: 147.
[60] British Astronomical Association (1989). "The History of the British Astronomical Association. The First Fifty Years". *British Astronomical Association Memoirs* 42. (Chapter 1: "Formation and first session", pp. 7-10).
[61] Note that a guinea was one pound and a shilling and that there were twenty shillings to the pound. Consequently, thirty guineas were equivalent to thirty one and a half pounds.
[62] Maunder, E. W. (1904-1905). "The total solar eclipse of August 30, 1905". *Journal of the British Astronomical Association* 15: 240-2.
[63] Chambers, G. F. (1904-1905). "Proposed eclipse expedition to the Mediterranean". *Journal of the British Astronomical Association* 15: 384-5.
[64] The Company also offered another route via Marseille, including some nights of hotel accommodation, at slightly higher cost. Anon. (1905). "[Notes:] The total solar eclipse of August 30". *The Observatory* 28: 260-261.
[65] Anon. (1905). "[Notes:] The total solar eclipse". *The Observatory* 28: 328-30.
[66] Cerdán, J. (1998). "La sala Napoleón: un estudio de la exhibición cinematográfica en Barcelona 1897-1898". In *Actas del VI Congreso de la A.E.H.C.* Madrid: Academia de las Artes y las Ciencias Cinematográficas de España, pp. 49-62. [http://www.cervantesvirtual.com/servlet/SirveObras/01349420833248386977802/p0000001.htm] (accessed 21 August 2007).
[67] Soler, E. (2005). "L'eclipsi de 1905, una novetat per al cinema espanyol". In *A propòsit de Cuesta. I Congrés sobre els començaments del cine espanyol (1896-1920)*. (Valencia 24, 25 and 26 October 2005) [http://www.ivac-lafilmoteca.es/cuesta].
[68] Several works have been published on these lines. For late nineteenth and early twentieth century France, see for instance: Bensaude-Vincent, B. and Blondel, C., eds. (2002). *Des savants face à l'occulte 1870-1940*. Paris: Éditions la découverte. For the British Mechanics' Institutes, see for example: Shapin, S. and Barnes, B.

(1977). "Science, Nature and Control: Interpreting Mechanics' Institutes". *Social Studies of Science* 7: 31-74; Inkster, I. (1976). "The Social Context of an Educational Movement: A Revisionist Approach to the English Mechanics' Institutes, 1820-1850". *Oxford Review of Education* 2: 277-307, and (1985). *The Steam Intellect Societies: Essays on Culture, Education, and Industry, circa 1820-1914*. Nottingham: University of Nottingham.

[69] From: Altamira, R. (1987). *Tierras y hombres de Asturias*. In *Rafael Altamira. Alicante-México (1866-1951)*. Alicante: Generalitat Valenciana, Instituto Juan Gil Albert de la Diputación Provincial de Alicante.

[70] Fontserè, E. (1900). "Eclipse total de 28 de mayo de 1900". *El Mundo Científico* 2: 403-10, pp. 404-5. See also: Ruiz-Castell, P. (2008). *Astronomy and Astrophysics in Spain (1850-1914)*. Newcastle-upon-Tyne: Cambridge Scholars Publishing.

[71] A. A. (1905). "El eclipse de Sol de 30 de Agosto". *El Mundo Científico* 7: 544-7, p. 546.

[72] Anon. (1900). "La fiesta de la ciencia". *El Imparcial* 29 May : 1.

THE BODY OF A REAL WOMAN:
MEDICINE IN THE YOUNG WOMEN'S GUIDEBOOKS PUBLISHED IN FINLAND, 1890-1923

TIINA MÄNNISTÖ-FUNK

> It is true that girls are in lesser degree inclined to unnecessarily engage themselves in the secrets of sex life than boys and also less often go astray in that area. The menstruation appearing at the age of 12-14 offers a girl's mother the opportunity, without seeking it, to talk with her child about the natural manifestations of sex. But how many mothers are using this opportunity to educate their daughters?[1]

This quotation from Max Oker-Blom's guidebook *Mitä Martta sai oppia Tohtori-enon luona maalla* ("What Martta Learnt at the House of her Uncle, the Doctor, in the Countryside"), published in early twentieth-century Finland, shows the general claim of early Finnish young women's guidebooks: young girls need solid information about their body turning into woman's body, but their mothers are not able or willing to provide it. However, enlightened and educated citizens could help young girls and their mothers by providing easy-to-understand books and booklets about this important subject: special guidebooks for young women.

Guidebooks for young women were a kind of publication produced in different countries during the nineteenth century. The first Finnish guidebooks were translations of works by German, American and Swedish authors. These translations constituted appropriations of the international popularization of medicine to the Finnish social and cultural context, characterized by middle class values, Lutheran culture, the educational and social role of women, and the rise of the Finnish national state. During the first decades of the twentieth century, Finnish authors made their own contribution to this kind of literature, and guidebooks rapidly became a successful and consolidated genre in the Finnish publishing market.

Finnish women had a fundamental role in the promotion and circulation of these books through the network of women's, temperance, and Christian associations in which they had a prominent stand. The moral, social and national ideologies promoted by these associations interacted with the publication and circulation of the guidebooks, configuring a particular genre of medicine popularization whose language and narrative were shaped by the aim of communicating the higher and middle class social and cultural views, while preserving and promoting the authority of official medicine.

In this paper I am going to explore the nine earliest young women's guidebooks published in Finland and their ways of constructing the female body by popularizing medicine. My thesis is that the female body was the place where various ideological, political and scientific strivings of early twentieth-century Finland were culminated. In this article I am searching for the ideological and cultural connections and explanations of the popularizations of medicine, and will show how international influences, agrarian society and Finns' enthusiasm for educational associations contributed to configure Finnish popular views of medicine and understanding of female body.

My approach to the history of the body is based on the theoretical work of Judith Butler. Her theory of the performative gender stresses the idea that the gendered body is completely constructed in linguistic definitions and social practices. There is no pre-linguistic, natural body – or at least we have no access to such a thing.[2] Every real woman is a reproduction of the cultural definitions of the real woman. The popularized medical definitions of the female body found in the young women's guidebooks were a part of the textual network producing the gendered body of their time. They were not just reflecting certain general medical ideas and helping their readers to discover their bodies, but actively producing the material, bodily reality.

Almost no historical research has been done on young women's guidebooks,[3] yet they offer an interesting insight into the production of the female body and throw a new light upon the comparative study of the processes of popularization in different countries. The point of this article is not to argue that such guidebooks had a central role in producing the gender and body in late nineteenth-century and early twentieth-century Finland. Instead, I argue that these guidebooks reveal some of the most central aspects of how female body was understood and produced at the time. They bring up themes crucial to the public discussion of their time in society: sexuality, family, nation, the different roles of sexes, education, class and moral values.

In this article I will first introduce the guidebooks, their publishers, authors and distributors, and the audiences constructed by them. After providing a brief review on their style and methods of advice, I will move on to analyse their medical content including the medicalization of the female body, the uterus-centred physiology, the bodily changes of puberty described in the guidebooks and the bodily dangers that the authors of these books wanted to prevent.

Advices from doctors and educated women, Finns and foreigners

The first Finnish young women's guidebooks endorsed a middle-class motherhood and family ideal, sharpening the difference between men and women in order to create a binary middle-class world divided into the spheres of home and public world. At the same time, they promoted slightly eugenic worries about degeneration and the immoral habits of the lower classes. These were not solely Finnish concerns. Foreign guidebooks were quickly translated into Finnish and the originally Finnish guides did not differ from the foreign ones in any significant manner. Between the end of the nineteenth and the beginning of the twentieth centuries, the middle-class interest in popularizing medicine was international. Finnish translation of foreign guides and the constitution of a stable genre by publication of Finnish authors constituted an appropriation of international trends into a local setting characterized by middle-class values, Christian culture, the educational role of women and the rise of a national state. In twentieth-century Finland, young women's guidebooks became such a well-established type of literature that new titles were published more frequently than, for example, general etiquette guides.[4] The female body was thus of central interest to society.

The first guidebooks giving advice about healthy sexual behaviour were published in Finland in the 1860s, and the first ones meant specifically for women date from the 1880s.[5] At the turn of the century, the range of different kinds of guidebooks published in Finland started to rise considerably. Young women's guidebooks began to be published in Finland, the nine earliest ones differing from later guidebooks by having chosen popular medicine as their point of view. Medical advice is not their only topic, but their scope of advice mainly focuses on the physical transformation that happens in puberty and the changes this transformation triggers in the life of a young woman. The first young women's guidebook in Finnish was published in 1890. The commercial publishing house Hagelberg translated and printed the guidebook written by the German doctor Hermann Klencke, twelve years after the first edition of the German original.[6] After the turn of the century other publishing houses started to translate guidebooks. Weilin & Göös published the first Finnish editions of three different guidebooks written by the American doctor

Mary Wood-Allen in 1903, 1906 and 1907.[7] In 1905 Werner Söderström published a small guidebook written by the Swedish teacher and poet Jeanna Oterdahl.[8] The first original Finnish young women's guidebook was written by Max Oker-Blom, Professor of Hygiene at the University of Helsinki, appearing in 1907, again through the entrepreneurship of Weilin & Göös.[9] Subsequently, in 1920 and in 1922, Werner Söderström published two small books written by the Finnish temperance activist Fanny Lukka.[10] The only non-commercial publisher printing young women's guidebooks during this period was the women's temperance association Valkonauhaliitto,[11] which in 1923 published an anonymous guidebook named *Tervehdys äidiltä* (A Greeting from Mother).[12]

There are no records left about the print runs of these books. Since commercial publishing houses printed and reprinted guidebooks, they must have been profitable. Although non-commercial associations were not seeking profits, their publications were often cheap and could reach surprisingly large audiences. At the beginning of the twentieth century, educational booklets published by the temperance unions could run to 15, 000 copies,[13] which was a remarkable number considering the Finnish population was two and a half million people.[14] The guidebooks written by Lukka and the booklet published by Valkonauhaliitto were very probably promoted and circulated by the temperance and Christian associations. The guidebook by Oker-Blom was bound to achieve publicity not only because of his position as professor and public medical figure, but also through women's associations, which publicly supported his work.[15]

Five of the nine earliest young women's guidebooks in Finland were written by doctors, four by educated middle-class women. This is not surprising, as the whole culture of sexual moral education and discussion was established in Finland not only by religious and medical authorities, but also crucially by middle-class women's associations. The temperance movement was an especially strong feature in Finnish culture. Temperance unions started as working class associations, but were transformed into predominantly female middle-class associations at the beginning of the twentieth century. At the same time, middle-class women were also actively organizing themselves in smaller, specifically female associations. They were not forwarding women's equal rights with men, but rather were emphasizing a woman's role as a mother and a wife. Defining the new middle-class ideal family as woman's sphere was seen as emancipatory, because it would allow women to realize their true womanhood.[16] Women's associations argued for absolute chastity and criticized prostitution, the upper-class double standard of morality, and the morally dubious characteristics of traditional Finnish culture. The wealthy middle-class was minority in the scarcely populated, agricultural Finland, in contrast to, for example, the Middle-European countries.[17] However, the values of the middle classes were central when the Finnish citizenship and democratic system was formed in the beginning of the twentieth century.[18] Young women's guidebooks reveal that woman's body was an important issue for the birth of the new nation.

The ideals of home and motherhood promoted in discussions about sexual morals were deeply rooted in the values of the middle class. The women activists and medical authorities wanted to spread these ideas among the working classes as well, and even succeeded to some degree: the working class educators of the time, for example, adopted the typical character of a good girl from the picture gallery of the bourgeoisie.[19] The middle-class female activists emphasized the similarity of all women and wanted to secure the future of the nation by making lower class women share their own moral values.[20] The guidebooks studied in this article were mainly intended to be read by middle-class girls, although they represent also middle-class worries about the lower classes. In Klencke's and Oterdahl's guidebooks the surroundings are the grandest, upper rather than middle class; in the rest of the guides they are less luxurious, but still far from the average Finnish girl's standard of living at that time. The girls described in the guidebooks lived in homes where there were sofas, sewing tables, library rooms and doorbells. The average Finnish girl of the time, living and working in a small farmhouse, is not visible in the guidebooks at all.

The educational combination of scientific, religious and popular

The influence of hygienism is apparent in the medical argumentation of the earliest young women's guidebooks in Finland. The culture of hygienic education had been established in Europe on the basis of eighteenth-century medical ideas and developed during the nineteenth century in the direction of social hygiene. This meant that a person's health was no longer his or her private matter or characteristic, but a social issue, which affected the success of the nation. Hygienic doctrines were popularized, especially during the late nineteenth century and early twentieth century with the help of guidebooks, and by organizing large hygiene exhibitions.[21] In Finland medical professionals gained significant authority towards the end of the nineteenth century, but most of the population, especially in the countryside, were out of the reach of organised health care and doctors. Therefore it was considered important to try to educate them about the basic facts of medical science through media such as almanacs, magazines and other printed material.[22] Both doctors and middle-class women's associations used hygienic doctrines as practical tools in their battle for the health of the nation and against loose sexual morals, alcoholism and "bad mothers", as those were seen as a threat to the whole society.

Most of the guidebooks included religious ingredients in their argumentation. The strictest hygienists did not tend to mix religious and scientific arguments. However, from the end of the nineteenth century hygienism gained such popularity in Europe that even the most religious and conservative Christian public educators assimilated hygienic trends in their texts.[23] In Finland, Christian doctrines remained until the 1930s as powerful as the belief in scientific and technological progress. Therefore, both the hygienists and the middle-class female activists had to be able to write texts able to communicate in a Christian environment, created in the Finnish public arena by the middle-class associations and conservative political parties promoting the trinity of home, Lutheran religion and fatherland. Women's associations were especially eager to use Christian arguments to advance their own quests.[24] The guidebooks often called forth the authority of doctors and of God.

The guidebooks typically delivered their advice by means of a combination of scientific and popular ingredients, which is a common feature of texts belonging to the hygienic tradition. On one hand, the advice was based on medical science; on the other hand it was delivered by distancing the text from the scientific and coming close to the language of everyday life. In each of Wood-Allen's guidebooks the reader is introduced to an educated mother giving advice to her daughter. *Tytöstä naiseksi* entirely consists of a story in which Mrs. Vainio tells her daughter Lyyli about "the secrets of life".[25] In *Mitä tytön tulee tietää*, written by the same author, at the beginning the reader meets a mother and her daughter Niina, whose evening discussions are supposedly reflected in the book, although they rather resemble lectures than discussions.[26] In *Mitä nuoren naisen tulee tietää*, Wood-Allen herself also appears as a mother giving advice to her daughter, but in this case in the form of a book.[27] The guidebook of Oker-Blom tells the story of two sisters, Martta and Elsa, who spend the summer with their uncle and his wife on the countryside. The common feature of these guidebooks is that they are not just striving to fill in the gaps in sexual education caused by shy mothers. They introduce to their readers the medical view point on their bodies and try to make them act and think according to this stand.

As Roger Cooter and Stephen Pumfrey have pointed out, popularized science can take on very different meanings within a culture than those intended by its popularizers.[28] The guidebooks themselves do not give direct evidence about who the readers were, or how they read the books and reacted to them. Young women readers were possibly amused or made angry by the advice given to them. They might have received the book and left it unread, although there is a good reason to assume that the information about sex and sexual organs was scarce enough to make every text dealing with it interesting. Although it would be very valuable to have evidence about readers' responses, the analysis of the guidebook texts in itself provides a useful account of their place and function in culture and society. The concerns voiced in the guidebooks and the medical advice given in them shows an interesting part of the cultural construction of female body: the context, in which the women of the time could understand their own body, either by rejecting the advice or following it.[29] In this paper I concentrate on the analysis of the texts: having presented the authors and the context of their arguments, the audiences constructed in the guidebooks and the rhetorical strategies used in them, I will now move on to examine what the guidebooks said about female health and illness.

Know your body! (But not too well)

The main justification the guidebooks gave for their existence was young girls' dangerous lack of knowledge about their own bodies. This ignorance was supposedly caused by their mothers being too shy and prudish, and therefore neglecting their duty to explain to their daughters the functioning of the female body.[30] This in turn could lead girls to damage their bodies. Authors of the guidebooks declared their horror in the face of this ignorance. Mary Wood-Allen, for example, wrote: "I have met young female students who are either totally ignorant or indifferent about even the most basic rules of physiology. They are constantly harming themselves by breaking these rules".[31] It was especially harmful to be ignorant about one's own sexual organs. For example, the first appearance of menstruation could shock an ignorant girl badly and lead her to cause herself serious harm.[32]

Guidebooks present themselves to young women's parents as practical devices for telling their daughters about delicate matters; as the best possible compromise between decent bashfulness and the need for education.[33] Talking about sexual organs, a subject otherwise unthinkable in discussion, was problematic also in the hygienic literature until the end of the nineteenth century. This made sexual themes in public discussion, literature and theatre a taboo, as supposedly innocent women needed to be protected against such topics. In Finland this thought was clearly a part of the educated middle-class world. In the Finnish agrarian culture of the time, women were seen as sexual beings, and the border between the two sexes was more open to different interpretations.[34] The guidebooks studied in this article used two means to make the description and discussion of sexual organs acceptable. On one hand they argued for the divine function of these organs in creating new life.[35] On the other hand the sexual organs were treated from an extremely medicalized point of view, which distanced them from the everyday world.[36]

This kind of medicalization also ruled the relationship between the internal advisor and the reader of early young women's guidebooks. While providing detailed descriptions, counting rules and explaining facts, the guidebooks had to continuously keep up the undisputed authority of doctors. Indeed, the female advisers in the guidebooks always referred to the male medical authority as the ultimate centre of knowledge about body and health: the doctor was always considered to know the woman's body better than she did, and he had the power to take decisions in order to make it function properly. Earlier popularizations of medicine had often aimed to provide their readers with means of curing themselves, but towards the end of the eighteenth century, enforcing the authority of medical profession became an aim in its own right. Guidebooks therefore started to discourage their readers from trying to become their own physicians and avoided teaching any concrete medical knowledge besides hygiene, preventive medicine and first aid.[37] The guides studied here are perfect specimens of this type of popularization of medicine. They permit the reader to glimpse the hard core of medical science through cosy stories, domestic examples and simulated friendly discussions. When the imaginary Mrs. Vainio is explaining about ovaries to her daughter Lyyli in the guidebook *Tytöstä naiseksi*, she reminds Lyyli that she has seen the ovaries of a chicken, when the kitchen maid was preparing meals for the family. Lyyli responds to this: "But mother, do not say that women also have eggs in their bodies. Is it really so? I do not like the idea". At that Mrs. Vainio tells her soothingly that the eggs women have in their bodies are microscopically small and can be called by their Latin name ovum, if calling them eggs creates discomfort.[38] The medical authority is also emphasized directly: if the author has medical profession or experience, this is made as clear as possible, and the authors frequently refer to the sayings and deeds of real or fictional doctors. They construct an image of woman, who is doubly subordinate from the medical viewpoint: firstly to the medical expert, the doctor, but even more importantly to the functions of her body. The authors want to persuade their readers to take up the only possible role left for them in the medical system: that of an observer.

Learning about the womb, observing the menstruation

In modern Western medicine, the problematic nature of the female body generally revolves around menstruation.[39] In the guidebooks studied in this paper, this emphasis is rather logical because of their hygienic view of the female body: the womb is a woman's leading organ, and therefore also rules over her health and sickness.[40] The guidebooks emphasize that regular, scant and painless menstruation is the best proof that the young woman's health is also generally good.[41] The limits of normality are very tight: for example Klencke urges that girls with painful or irregular menstruations (that is, those which last longer than four days or do not come regularly every 29 days), should seek a doctor's help immediately. Seeing menstruation as a bodily crisis causing numerous instabilities both in body and in mind – a female irregularity in comparison to the male system – was the prevailing opinion until the 1930s, when the hormonal process behind the bleeding started to gain publicity.[42] Setting the strictly regular menstruation as goal appears as an attempt to medically control the uncontrollable, and an effective way of making most women feel the need of medical advice or doctor's help. All guidebooks stress that menstruation should not be seen as a sickness, and congratulate themselves for approaching the subject rationally. However, they are of the opinion that the reckless behaviour of girls is often the reason for their painful or irregular menstruation, and present long lists of precautions. During the menstruation, three things are to be avoided: exhaustion, catching a cold and being overwhelmed by emotions, which are all equally seen as causing bodily symptoms. The reasons for such caution is, according to the guidebooks, that the nervous system of a woman is more sensitive (and the womb heavier) during the menstruation.[43]

The authors of guidebooks assume that girls are aware of even the minutest irregularities in their menstrual cycle and point out that only an ignorant girl, reckless of true womanhood, does not pay attention to her menstrual health. Klencke's advice to change the sanitary towel every day is based not on cleanliness, but given because daily fresh towels help the girl to observe the amount and quality of her menstrual blood.[44] However, ceaseless observation is not enough: after the beginning of menstruation, girls need to adjust their whole life to the rhythm of their menstrual cycle. This thinking was typical also in the health advice given to married

Finnish women. Women were not thought to be able to control their body actively like men, but needed to direct their energy inwards, to the womb, and give up their individuality in the face of their menstruation.[45] It is interesting that although guidebooks claimed to give practical advice, Klencke was the only author mentioning sanitary towels, which might be seen as the one subject concerning menstruation, in which girls most acutely and concretely needed advice. In Finland, women commonly used self-made sanitary towels until well after the Second World War. Especially in the early decades of the twentieth century, advice about making and using them was very hard to find, because the menstruation was usually a difficult and shameful subject even between mother and daughter.[46] We can argue that, by leaving this subject aside, the guidebooks were not so keen in helping young women to deal with the bodily changes of puberty as they were in teaching the medical view of female body suitable for the ideological purposes of the middle classes.

The authoritative and omniscient way in which the guidebooks discussed about menstruation is especially interesting, because the reasons and functions of menstruation remained a partial mystery to medical science well into the 1920s, when the connection between hormones and menstruation was discovered.[47] At the beginning of the twentieth century, medical authorities widely assumed that menstrual blood was foetus nutrition, which was removed from woman's body every month if she did not become pregnant.[48] Beside the scientific uncertainties, a lot of traditional beliefs surrounded menstruation, connecting it to impurity and strong sexual desire, comparable to animals' heat.[49] Of the guidebook authors, for example, Klencke writes that the special sour smell of sweat appearing during menstruation can make milk, beer and fruit turn sour.[50]

Because menstruation was connected to sexuality, it belonged to subjects which had to be concealed from other people. For example, in Oker-Blom's guidebook Martta decides with her aunt that they will not tell anybody else about the beginning of Martta's menstruation, because "it is after all so nasty that one has something like that".[51] In such comments, the girl's body turning into a woman's body appears at the same time as the most private and the most public thing, something to be hidden, but also something to be exposed, discussed and paid attention to.

The wakening of the sleeping femininity

The guidebooks do not mention the word sexuality, but still their main messages are tightly concentrated around this subject. The modern idea of sexuality was formed during the last decades of the nineteenth century on the basis of, and side by side with, hygienic thought. At the beginning of the twentieth century the science of sexuality was born and sexuality started to be seen as an important feature.[52] This modern sexuality was clearly defined as normatively heterosexual and based on the functions of sexual organs. In practice, this was thought to mean that as the sexual organs started to function, girls became automatically interested in boys.[53]

The guidebooks I have studied energetically promote the common hygienic idea that the difference between male and female genitals is logically manifested throughout the rest of the physical, mental and emotional system, as the thoughts and emotions have their basis on material foundations. This means that the whole of a woman's being is shaped by her reproductive function. The womb is her leading organ and makes her passive and emotional, whereas a man's system is mastered by his brain, which makes him rational, creative and able to control himself and others.[54] Guidebooks promoted the idea that a girl is actually a woman from her birth on, but her womanhood is sleeping until her womb starts to function. Lukka for example writes: "A girl will once become a woman, and this sleeping woman in the soul of a girl makes her much more emotional and sensitive than a boy".[55] The word "puberty" is not used in the guidebooks, although it was a widely used term at the time. In the story-telling style of the guidebooks it might have appeared too clinical. Instead, the authors use such phrases and euphemisms as "time of changes", "time of development", "developing into a woman", "physical turning point" or "wakening of the reproductive organs". Emotionality is described as one typical feature of this time of changes. Wakening of the womb turns a girl into a daydreamer.

Doctors and other public educators in early twentieth-century Finland also taught that puberty made men and women mentally different. When the sexual organs started to function, will developed in young men and emotion in young women. According to the guidebook authors, young women felt in a purely romantic way. They did not even realize that they had a sexual drive because this drive was naturally weak and virtually sleeping until marriage. Young men, instead, had to struggle with such a forceful sexual drive that they could hardly control it, even with their stronger will and more powerful mind. The activists of Finnish women's movement used this notion as a positive asset by claiming that women were to function as the supporting pillars of the nation's morals, and also lift men to a higher level of chastity. The prerequisite of this was that women themselves would have strong morals in sexual matters. Therefore, it was crucial to provide sexual education to girls and young women.[56]

Every woman's responsibility for simultaneously keeping up her own and every young man's chastity is one of the strong themes in the guidebooks. Young women must not lead men directly or indirectly into situations where the men might not be able to control themselves.[57] Thus, women were supposed to enforce double control: the control over their own sexuality, but also that of men. The guidebooks were effectively normalizing and materializing this ideological thesis by linking it to the physiology of female body. The wish to control women's sexuality and make them control that of men is the deep current which runs under the advice given by the authors. It gives their voice its urgent and threatening edge even when they are appearing as the warm-hearted, caring advisers. The catastrophes that the authors hint at, and from which they want to protect their readers, are ultimately sexual in nature. There were two main catastrophes looming over a young woman's life: she could lose her double control over sexuality and thus end up becoming an unmarried mother without any hope of a normal, happy future or she could risk her chances of getting married and having children by harming her sexual health, high morals and attractiveness by ignoring the needs of her body.

Endangering the womb, endangering the future

In the guidebooks, developing into a woman meant developing into a human being organized around the womb and resembling the womb. The guidebooks were persuading their readers to let their womb define their life. Young women were meant to prepare themselves for maternity and should concentrate on fulfilling this mission as perfectly as possible.[58] Poor health was seen as a danger to girls who used too much of their energy by studying hard during their puberty. This meant robbing the energy the womb needed in order to develop properly. According to the guidebooks, such use of energy could cause anaemia and other consuming sicknesses, which make the normal development of a woman's body impossible.[59] This worry of future mothers exhausting themselves by straining their brain explains why in Finland, from the year 1920 on, middle school for girls lasted one year longer than the same education for boys, as girls were thought to need a lighter syllabus.[60]

In the hygienic tradition, masturbation was also thought to drain the energies of the body and cause serious physical and mental sicknesses by damaging the nervous system. The campaign against masturbation had arrived in Nordic countries already in the eighteenth century, although masturbation was initially considered to be exclusively a male problem.[61] The guidebooks severely warn about this vice, which can easily ruin health and morals. Wood-Allen warned that masturbation also alters the brain by making the part reacting to sexual stimuli permanently larger.[62] At the time when the guidebooks were published, masturbation had already lost some of the pathological symptoms linked to it in the earlier medical theory. It was getting more difficult to claim that masturbation directly caused diseases. Rather, symptoms connected to it were more markedly representing the offence the masturbator made against the civilization process of middle class culture.[63] The guidebooks mention bodily ills caused by masturbation, but these appear as signs of the damage done to the high morals, in the same way as they see reading bad novels as a possible cause of unnaturally early and quick puberty.[64]

Masturbation and other vices were also dangerous because children might inherit such bad habits from their parents. The international eugenic ideas arrived in Finland in the beginning of the twentieth century. Racial ideologies did not gain much popularity, but the general worries about degeneration were common, as bad lifestyle and especially alcoholism were thought to seriously diminish the power of the whole nation.[65] Guidebooks urged young women to build solid health for the benefit of the next generation by eating moderately, sleeping enough and exercising through light gymnastics and other not-too-exhausting sports.[66] They were emphasizing the idea of balance by comparing the body to a fine-tuned machine.[67] These kinds of hygienic notions of body culture gained importance in Finland at the turn of the nineteenth and twentieth centuries. Public discussion began to try to make people responsible for their own health by promoting healthy diet and rest. Little by little physical exercise also began to be promoted. However, girls had to beware of dangerous exhaustion and therefore do lighter exercise than boys.[68] Beside exhaustion, guidebooks also mention numerous other things which can quickly and dangerously damage the balance of a young woman's body. These particularly included dancing, drinking alcohol, smoking and reading morally questionable books. Acceptable pastimes for a good girl were relatively few: light physical exercise, observing nature and doing chores at home.[69]

Taking part in housekeeping is, according to the guidebooks, an essential part of preparing a young girl for her ultimate calling as a mother and a wife. The ideal they present is a middle-class woman, who does not have to work to provide for her family, but who joyfully busies herself with the endless household tasks and so creates a good, happy and healthy home for her children and husband.[70] This housewife ideology of the middle-class was reflected in the school programme for Finnish girls. Especially after the Finnish civil war of 1918,[71] the upper classes were worried about lower-class women's suitability for motherhood, since they were not devoted to the middle-class values of home, religion and fatherland. The middle-class ideas were promoted for whole generations of Finnish girls through the household education of the elementary school system established in 1921. Cleanliness was especially emphasized in this education, as the clean home was thought to be directly linked to the morally respectable family life.[72] Factory work done by working-class women was seen as morally dubious, as it could not be understood as an extension of motherly work at home. Agricultural work did not raise as many worries, as it took place inside the sphere of home. Basic school education again gave girls better preconditions to bring up their future children. Too much and too high education was thought to be a serious risk for women's health. It was not natural to them to strain their brain, as their natural calling was that of creating new life in their body.[73]

Conclusion: Bodies on the border

The popularization of medicine in the Finnish young women's guidebooks was part of a bigger project: the building of the independent Finnish nation. Although the actual Finnish middle-class was thin, its values were central in this building work. Such phenomena as loose sexual morals, alcoholism and unfit mothers were seen as threats to the young nation. Hygienic doctrines were used to fight these threats. The middle-class women's movement was active in public debate, promoting the idea of women as the motherly guardians of the nation's morals. The young women's guidebooks show the medical roots of their thinking.

Guidebooks presented the hygienistic view of the womb as the leading organ in woman's body: it was seen as the most important issue in women's health care, but also as the most formative factor in a woman's being and life. Publication of guidebooks was justified by the supposed ignorance young women had about their body and its needs. This ignorance was thought to possibly make young women weak in morals and health: they might end up as unmarried mothers or harm their health and sexual organs so seriously that they could not become mothers at all. Both of these scenarios were catastrophes not only for the victims, but to the young and small nation as whole. Keeping the female body healthy and in balance appears in the guidebooks as a very problematic task. Young women are taught to observe their bodily functions ceaselessly. This observation is also the only active relation that the guidebooks suggest women should have to their body. Although giving advice and explaining medical knowledge, guidebook authors did not want to make their readers experts on their own bodies. Instead, they medicalized the female body and subordinated it to the authority of doctors. According to the guidebooks, women's physiology did not only make them more prone to medical problems, but also more emotional and passive than men. Even more importantly, it defined women's aims in life. The whole being of a woman was shaped by her reproductive function; and so should her life be. The guidebooks were persuading their readers to concentrate on the aim of becoming healthy, married mothers of healthy Finnish citizens. For this purpose women needed to control their sexuality and the sexuality of men, use their energy in household tasks, observe their body and trust in the medical knowledge of doctors. This ideology became also institutionalized in education and health care.

The female bodies produced in young women's guidebooks, and the guidebooks themselves, are on the border, or crossing borders, in several ways. The bodies are on the border of childhood and adulthood, and this is exactly why they attract the special attention of medical advisers. These advisers see their readers as wavering between right, happiness-ensuring decisions and catastrophically wrong decisions. The border between private and public was not in force when the health and future of the nation was seen at stake: the most private bodily functions and habits of young women were not only their concern, but the concern of the whole society. The other border, which the guidebooks crossed with ease, was the one between science and ideology, in particular, between medicine and morals. The popularization of medicine was directly serving the spreading and normalizing of the middle-class worldview based on the importance of home, religion and fatherland. The medical views the guidebooks presented did not provide help in the everyday bodily problems their readers might encounter during their puberty and did not aim to spread the cutting-edge medical knowledge. Instead, they were promoting the hygienic views of the body in which the organs and their functions were also the basis of mental functions. This meant that especially for women the border of mind and body could not exist. The guidebooks studied in this article are examples of a double popularization of medicine: the idea of women as subordinate to their womb was based on the popularized medical views of hygienism and this idea led to the need of providing young women with sufficient medical information about their bodies. In this way they are also challenging the border of pure science and popularized science. These two do not appear as clearly differentiated entities, but rather as the two ends of a culturally constructed continuum, on which different actors, with or without scientific education, are creating and appropriating mixtures of scientific and popular, medical and ideological knowledge.

Notes

[1] Oker-Blom, M. (1907). *Mitä Martta sai oppia Tohtori-enon luona maalla. Kirjanen tyttöjen vanhemmille.* Helsinki: Weilin & Göös, pp. 8-9. All translations from Finnish are by the author of this paper.

[2] Butler, J. (1993). *Bodies that Matter. On discursive Limits of "Sex".* New York and London: Routledge, for example pp. 2-13, and (1997). *Excitable Speech. A Politics of the Performative.* New York and London: Routledge, p. 5.

[3] Susanne Pellatz's *Körperbilder in Mädchenratgebern* (1999) concentrates on the German young women's guidebooks in the 17th and 18th centuries, but she uses the methodology of the psychoanalytical study of literature, which makes her analysis somewhat ahistorical.

[4] Männistö, T. (2003). *Haluathan tulla todelliseksi naiseksi. Naisruumiin tuottaminen Suomessa ilmestyneissä nuoren naisen oppaissa 1890-1972.* Turku: Turun yliopisto, historian laitoksen julkaisuja 63, p. 26; Räisänen, A. (1995). *Onnellisen avioliiton ehdot. Sukupuolijärjestelmän muodostumisprosessi suomalaisissa avioliitto- ja seksuaalivalistusoppaissa 1865-1920.* Helsinki: Suomen Historiallinen Seura, p. 22.

[5] Räisänen. *Onnellisen*, pp. 23-4.

[6] Klencke, H. (1890). *Neitsyt. Ruumiin ja sielun terveysoppi täysikasvuisen neidon itsehoidossa ja itsekasvatuksessa.* Tampere: Hagelberg.

[7] Wood-Allen, M. (1903). *Tytöstä naiseksi.* Helsinki: Weilin & Göös, (1906). *Mitä tytön tulee tietää.* Helsinki: Weilin & Göös, and (1907). *Mitä nuoren naisen tulee tietää.* Helsinki: Weilin & Göös.

[8] Oterdahl, J. (1905). *Tytöille.* Porvoo: Werner Söderström.

[9] Oker-Blom. *Martta.*

[10] Lukka, F. (1920). *Nuori nainen.* Porvoo: Werner Söderström; Lukka, F. (1922). *Kukkien puhjetessa.* Helsinki: Werner Söderström.

[11] Valkonauhaliitto was the Finnish branch of the WCTO (Woman's Christian Temperance Union).

[12] Anon. (1923). *Tervehdys äidiltä.* Helsinki: Suomen valkonauhaliitto.

[13] Räisänen. *Onnellisen*, p. 22.

[14] The level of literacy in Finland was high. Already in the 1860s, 95% of all Finns over age 14, could read. By the early 20th century this percentage had risen to almost 100%, and especially the literacy of children had been developed by establishing a more extensive school system. See for example: Kaarninen, M. (2007). "Koulu ja kansa". In *Suomalaisen arjen historia 2.* Helsinki: Weilin & Göös, pp. 231-249.

[15] Rajainen, M. (1973). *Naisliike ja sukupuolimoraali. Keskustelua ja toimintaa 1800-luvulla ja nykyisen vuosisadan alkupuolella noin vuoteen 1918 saakka.* Helsinki: Suomen Kirkkohistoriallinen Seura, p. 116.

[16] Sulkunen, I. (1987). "Naisten järjestäytyminen ja kaksijakoinen kansalaisuus". In Alapuro R. et al. eds., *Kansa liikkeessä.* Helsinki: Kirjayhtymä, pp. 157-75.

[17] In 1910 still 70 percent of the whole population earned its living from agriculture and forest economy.

[18] Finland was an autonomic part of Russia until it gained independence 1917. In 1906 Finland had reformed its system of government and granted all adult citizens the right to vote. Almost immediately after gaining independence Finland experienced a traumatic civil war.

[19] Salmi-Nikander, K. (2002). "Pahan tytön viimeiset sanat. Työläisnuorison sukupuolikeskustelua 1900-luvun alkupuolella". In Aaltonen, S. and Honkatukia, P., eds. *Tulkintoja tytöistä.* Helsinki: Suomalaisen Kirjallisuuden Seura, pp. 185-206, on p. 194.

[20] Markkola, P. (2002). *Synti ja siveys. Naiset, uskonto ja sosiaalinen työ Suomessa 1860-1920.* Helsinki: Suomalaisen Kirjallisuuden Seura, pp. 361-70.

[21] Sarasin, P. (2001). *Reizbare Maschinen. Eine Geschichte des Körpers 1765-1914.* Frankfurt am Main: Suhrkamp; Mäkinen, K. (1992). "Ihmeellinen kone – haave uudesta ihmisestä". In Onnela, T., ed., *Vampyyrinainen ja Kenkkuinniemen sauna. Suomalainen kaksikymmenluku ja modernin mahdollisuus.* Helsinki: Suomalainen Kirjallisuuden Seura, pp. 18-37; Räisänen. *Onnellisen,* pp. 61-2.

[22] Piela, U. (2006). "Luonto ja muuttuvat maailmat kansanlääkinnässä". In Helsti, H.; Stark, L., and Tuomaala, S., eds., *Modernisaatio ja kansan kokemus Suomessa 1860-1960.* Helsinki: Suomalaisen Kirjallisuuden Seura: pp. 277-331.

[23] Sarasin. *Reizbare,* pp. 121-2.

[24] Markkola, P. (2000). "The Calling of Women. Gender, Religion and Social Reform in Finland, 1860-1920". In Markkola, ed. *Gender and Vocation. Women, Religion and Social Change in the Nordic Countries 1830-1940.* Helsinki: Suomalainen Kirjallisuuden Seura. Studia Historica 64, pp. 113-145.

[25] Wood-Allen. *Tytöstä,* especially pp. 3-8.

[26] Wood-Allen. *Mitä tytön,* especially pp. 9-13.

[27] Wood-Allen. *Mitä nuoren naisen,* especially pp. 3-9.

[28] Cooter, R. and Pumfrey, S. (1994). "Separate Spheres and Public Places: Reflections on the History of Science Popularization and Science in Popular Culture". *History of Science* 32: 237-67, p. 249.

[29] See also Jordanova, L. (1999). *Nature Displayed. Gender, Science and Medicine 1760-1820.* London: Longman, p. 105.

[30] Klencke. *Neitsyt,* p. 45; Oker-Blom, *Martta,* pp. 6-7; Wood-Allen. *Mitä nuoren naisen,* pp. 5-7.

[31] Wood-Allen. *Mitä nuoren naisen,* p. 21.

[32] Lukka. *Kukkien puhjetessa,* p. 14; Wood-Allen. *Mitä nuoren naisen,* p. 84.

[33] See also Brumberg, J. (1997). *The Body Project. An Intimate History of American Girls*. New York: Random House, pp. 35-7.

[34] Löfström, J. (1999). *Sukupuoliero agraarikulttuurissa*. Helsinki: Suomalaisen Kirjallisuuden Seura, pp. 173-95.

[35] *Tervehdys äidiltä*, pp. 13-4.

[36] Wood-Allen. *Tytöstä*, pp. 11-5.

[37] Ramsey, M. (1992). "The Popularization of Medicine in France, 1650-1900". In Porter, R., ed. *The Popularization of Medicine, 1650-1850*. London: Routledge, pp. 97-133, on pp. 121-2.

[38] Wood-Allen. *Tytöstä*, p. 13.

[39] Helén, I. (1997). *Äidin elämän politiikka. Naissukupuolisuus, valta ja itsesuhde.* Helsinki: Gaudeamus, p. 198.

[40] Sarasin. *Reizbare*, p. 371.

[41] For example Klencke. *Neitsyt*, p. 52; Oterdahl. *Tytöille*, p. 6; Wood-Allen. *Mitä nuoren naisen*, p. 103.

[42] Johannisson, K. (1994). *Den mörka kontinenten. Kvinnan, medicinen och fin-de-siècle*. Stockholm: Nordstedts, pp. 114-9.

[43] For example Lukka. *Kukkien puhjetessa*, pp. 14-5; Klencke. *Neitsyt*, p. 52; Wood-Allen. *Mitä tytön*, p. 88, and *Mitä nuoren naisen*, p. 106.

[44] Klencke. *Neitsyt*, p. 56; Wood-Allen. *Mitä nuoren naisen*, pp. 85-6.

[45] Räisänen. *Onnellisen*, pp. 175, 192-3.

[46] Sohlman, A. (2007). *Linnunkiikut ja kallen kaulukset. Kotitekoiset kuukautissiteet 1900-luvun alkupuolen Suomessa.* Pro gradu. Helsingin yliopisto, kotitalous- ja käsityötieteiden laitos, pp. 54-5, 64-5.

[47] Brumberg. *The body project*, pp. 7-8.

[48] Räisänen. *Onnellisen*, pp. 175-6.

[49] Brumberg. *The Body Project*, p. 7; Delaney, J; Lupton, M., and Toth, E. (1988). *The Curse. A Cultural History of Menstruation*. Revised edition. Urbana and Chicago: University of Illinois Press, pp. 7-17; Räisänen. *Onnellisen*, p. 176.

[50] Klencke. *Neitsyt*, p. 55.

[51] Oker-Blom. *Martta*, p. 68.

[52] Duden, B. (1987). *Geschichte unter der Haut*. Stuttgart: Klett-Cotta, pp. 44-5; Ekenstam, C. (1993) *Kroppens idéhistoria*. Hedemora: Gidlunds Bokförlag, pp. 248-9, 287-9.

[53] Aapola, S. (2001). "Liian varhaisen ruumiillisen kehityksen ongelma". In Puuronen, A. and Välimaa, R. eds., *Nuori ruumis*. Helsinki: Gaudeamus, pp. 30-44, on pp. 36-7.

[54] Brumberg. *The Body Project*, pp. 8-9; Räisänen. *Onnellisen*, pp. 136-41; Sarasin. *Reizbare*, pp. 64-8, 90-1.

[55] Lukka. *Kukkien puhjetessa*, p. 7.

[56] Rajainen. *Naisliike*, pp. 104-14.

[57] Oterdahl. *Tytöill*, pp. 11-20; *Tervehdys äidiltä*, p. 5; Wood-Allen. *Mitä nuoren naisen*, pp. 120-4.

[58] For example Klencke. *Neitsyt*, p. 212; Lukka. *Kukkien puhjetessa*, pp. 3-4; Wood-Allen. *Mitä nuoren naisen*, p. 84.
[59] Klencke. *Neitsyt*, p. 46; Wood-Allen. *Mitä nuoren naisen*, p. 103.
[60] Kaarninen, M. (1995). *Nykyajan tytöt. Koulutus, luokka ja sukupuoli 1920- ja 1930-luvun Suomessa.* Bibliotheca Historica 5. Helsinki: Suomen Historiallinen Seura, pp. 160-4.
[61] Ekenstam. *Kroppens idéhistoria*, pp. 147-8; Sarasin. *Reizbare*, pp. 64-8, 403-17.
[62] Wood-Allen. *Mitä nuoren naisen*, pp. 114-6.
[63] Laqueur, T.W. (2004). *Solitary Sex. A Cultural History of Masturbation.* New York: Zone Books, p. 370.
[64] For example Wood-Allen. *Mitä nuoren naisen*, p. 91.
[65] Helén. *Äidin*, p. 184; Hietala, M. (1985). "Rotuhygienia". In Kemiläinen, A. ed., *Mongoleja vai germaaneja? Rotuteorioiden suomalaiset.* Historiallinen Arkisto 86. Helsinki: Suomen Historiallinen Seura, pp. 147-57; Räisänen. *Onnellisen*, pp. 64-6; Sarasin. *Reizbare*, pp. 403-46; Tiihonen, K. (2000). "Puhtaan nuoruuden ihanne. Sukupuolikasvatusta nuorison opaskirjoissa 1920- ja 1930-luvuilla". In Immonen, K.; Hapuli, R.; Leskelä, M., and Vehkalahti K. eds., *Modernin lumo ja pelko. Kymmenen kirjoitusta 1800-1900-lukujen vaihteen sukupuolisuudesta.* Helsinki: Suomalaisen Kirjallisuuden Seura: pp. 169-208; on p. 177.
[66] Klencke. *Neitsyt*, pp. 63-5; Wood-Allen. *Mitä nuoren naisen*, pp. 52-4, 136-44.
[67] Wood-Allen. *Mitä nuoren naisen*, p. 16; Lukka. *Nuori nainen*, pp. 10-1.
[68] Mäkinen. "Ihmeellinen", pp. 30-5. In Finland Tahko Pihkala published in 1909 a guidebook for men's physical exercise called "Everyman's machine science" ("Joka miehen koneoppi eli liikunnon merkityksestä").
[69] Klencke. *Neitsyt*, pp. 57, 192; Lukka. *Nuori nainen*, p. 18, and *Kukkien puhjetessa*, p. 14; Oterdahl. *Tytöille*, pp. 18, 20, and *Tervehdys äidiltä*, pp. 4-5; Wood-Allen. *Mitä tytön*, pp. 53-60, 70-5, 89, and *Mitä nuoren naisen*, pp. 145-7.
[70] For example Klencke. *Neitsyt*, pp. 50, 66, 191, 202-3; Lukka. *Nuori nainen*, pp. 16-17; *Tervehdys äidiltä*, pp. 9-10; Wood-Allen. *Mitä tytön*, pp. 70-4.
[71] The Civil War in Finland was fought from January to May 1918 between the Reds and the Whites. The red side consisted of Social Democrats, Communists, workers and non-independent farmers. On the white side fought the conservatives, middle and upper-middle class men and farm owners. The Reds were defeated and Finnish society traumatically divided into two groups.
[72] Kaarninen. *Nykyajan tytöt*, pp. 41-6.
[73] Ibid., pp. 89-95, 166-7, 231-45, 195; Markkola. "The Calling of Women", pp. 113-25; Räisänen. *Onnellisen*, pp. 116-20, 179-80.

LEAVING SUSPICION BEHIND:
SPANISH PUBLIC DISCOURSES ON PRIVATE FUNDING OF HUMAN GENETICS

MATIANA GONZÁLEZ-SILVA[*]

During the last three decades of the twentieth century the international science system changed dramatically. Some scholars have argued that the main changes have been the "privatization" of research and the increasing influence of "extra-scientific" pressures on science;[1] others have stated instead that what has been lost is the balance between private and public science that had prevailed since the nineteenth century, turning knowledge into a commodity.[2] Whatever the differences in the interpretation of the main features of contemporary technoscience, there is broad agreement that there have been during recent years major transformations in the relationships between science, law, economy and social agents such as the media.[3]

This paper aims at deepening the understanding of the transformations of the contemporary science system, exploring what role the media played in one of the main changes it has experienced: the increasingly high participation of private funding in scientific research. In particular, it analyses the Spanish public discourses on private funding of human genetics, as they were published in the leading newspaper in the country, *El País*, between 1976 and 2006. The paper pays special attention to genetic diagnoses, as these technologies are intimately interwoven with scientific research directed by the search for economic profit.[4]

The last twenty years have witnessed a boom in studies of science in the media; the presuppositions of these, however, vary significantly. A considerable amount of scholarly work still assumes a model of communication in which journalists act only as mediators between two allegedly "separated" realms – science and the public – in an unidirectional process in which lay people play the role of mere passive receptors.[5] This approach has increasingly been criticized on the grounds that it uses a simplistic, inaccurate and unproblematic concept of popularization. The new historiography of science popularization – in which this paper is inscribed – assumes instead that the public plays a

fundamental role in the local construction of scientific systems. Studies
have focused on the relationship between actors that participate in the
various processes in which science is communicated,[6] the political agendas
associated to certain models of scientific popularization,[7] the implicit idea
of science conveyed by the media,[8] or the way in which different
audiences appropriate scientific knowledge made publicly available
through certain communication channels.[9]

In this paper, I analyze scientific news published in a daily newspaper,
presupposing the important role the media play both as a reflection of and
as an influence on the general context in which science is carried out. By
the end of the twentieth century, the media had become a major source of
images of science, which were very influential in public attitudes towards
it. On the other hand, the media were the target of a wide number of actors
interested in using them for achieving their particular objectives. Studying
texts published in a newspaper therefore allows us to trace the public
discussions and controversies that surrounded science at a given historical
moment, which actors participated in them and what their respective
arguments were.

The temporal, geographical and political focus of this case is
particularly interesting for its relevance to unveil the relations between the
state, the media, the economy and the public. The life of *El País* coincides
with major political and economical changes in Spain. The launch of this
newspaper in 1976 occurred during a period of political transition towards
democracy, after four decades of military dictatorship.[10] The following
decades thus witnessed important cultural, social and political changes in
the relationship between Spanish citizens and the State, as well as
economical transformations leading to the full integration of Spain in the
international capitalist economy. *El País* had a fundamental role in all
these processes, becoming the most read general newspaper in the
country.[11]

During these decades, important changes leading to the
institutionalization and professionalization of the biomedical sciences also
took place. An incipient community of geneticists consolidated, mainly
devoted to the study of the genetic aetiology of different diseases and to
the development and implementation of genetic diagnosis tools and
programmes.[12] On the other hand, politicians began to conceive science as
a driving force for the Spanish economy. The first National Plan of
Research and Development was launched for the period 1988-1991.[13]
Concomitantly, private funding of science appeared in certain sectors of
Spanish scientific practice. Particularly during the nineties, new companies
based in the most recent scientific and technological developments

appeared, many of which were founded by biomedical scientists and were dedicated to the performance of genetic tests.

The rapid development of both Spanish genetics and the biomedical industry offers the opportunity of showing the changing relationship established between scientists, policy makers and the media. In the context of the definition of local scientific and economic policies, the assessment of private funding of scientific research in *El País* is far from trivial. It reflects local attitudes that contributed to shape the context in which these policies were designed, and thus influenced the kind of science that was promoted at a local level.[14] It also contributed to the acceptance or rejection of practical applications of new biomedical knowledge, such as genetic tests, and had a central role in the shaping of Spanish public opinion on human genetics.

Money and science: a changing relationship

Coverage of the economic aspects of scientific research was one of the most difficult and ambiguous issues in *El País'* journalistic approach to science. While in some areas of research these aspects were completely obscured, fostering the idea of science as an independent endeavour completely alien to external influences, in other cases journalists tried to explain the existing links between commercial interests and researchers. From a more subjective point of view, now the newspaper supported the potential benefits of private research, now it expressed its worries of the contamination this could mean for the values traditionally linked to science. This was the result of the changing context in which the process of scientific communication took place and of the transformations within *El País'* editorial team.

Until the early nineties, *El País* presented the involvement of scientists in private ventures as typical of the US science organization. It regarded it as clearly antagonistic to other traditions, such as the French one, which was depicted as more "altruistic" and morally superior. Early reports mentioning financial aspects of research informed the reader, for instance, about the French decision to make available all data regarding their first map of the human genome, "while in the United States they are trying to patent genetic sequences whose function is still unknown".[15] Other reports explained that the French were trying to prevent misuse of new genetic knowledge, which could derive from the "entrance of commercial interests where scientific and human criteria should be paramount".[16]

In general, *El País'* journalists distrusted private funding of research as potentially damaging the quality of science, or even prompting deliberate bias. In 1990, when reporting on the results of a study on the genetic factors of lung cancer, it was pointed out from the very headline that the public firm Tabacalera – a leader in the Spanish tobacco sector – had funded a study whose results somehow exonerated smoking from causing the disease.[17] Moreover, on the occasion of a controversy about the right of a mother to choose the sex of her future baby, an editorial stated that public powers should respect individual decisions, but that they should prevent any action aimed at achieving "unmentionable objectives of social domination or economic benefit".[18] According to this position, *El País* fostered an image of science as an endeavour completely alien to commerce.

This attitude was shared by local geneticists, particularly those who expressed their opinions on scientific areas in which they were not participating. When the first controversies on the patenting of human genes arose in the United States, the Spanish geneticists' greatest concern was how patents would affect the "philosophy" of scientific progress,[19] while their foreign counterparts focused instead in more practical consequences such as the patents' impact on the cooperation among scientists. [20] Science was ideally isolated from any "external" contamination in editorials too; in 1991, *El País* stated that patents were a "practical" matter to be solved, but not a moral or scientific issue that could ever affect the very "scientific" core of research.[21]

As early as in 1989, the molecular biologist Pere Puigdomènech noticed the adverse climate towards the search of economic profit through science which was prevailing in Spain. He wrote an opinion article deploring that "for some people, the worse evils of our world" would be derived from scientists, perceived as arrogant, and sold out to the interests of the most perfidious multinational companies.[22]

In the years that followed, *El País'* attitude towards private funding of research began to change, coinciding with the emergence of political and economical pressure groups linked to genetics research in Spain. Initially, this change was more implicit in the model of scientific journalism than explicitly stated. In reports on the discovery of new genes related to disease, journalists completely obscured any non technical information – including economic information – according to a model of scientific popularization traditionally linked to science promotion and centred on scientific "truths".[23] *El País* did not publish information on local scientific policies either – as explicitly stated in the journalistic programme of its science editor[24] – thus contributing to an image of scientists as neutral professionals only committed to the search for knowledge.

However, a different model was used when treating branches of human genetics perceived mainly as foreign to the Spanish science system, such as the Human Genome Project (HGP). In this case, *El País* was keener to offer a more complex image of research, which included its sociological and economic aspects. While in countries leading the research on the human genome, reports tended to stress the geneticists' medical promises and the technical details of different scientific breakthroughs,[25] *El País'* journalists were attentive to the expectations of profitability of the new knowledge, the stock performance of biomedical firms and how the emergence of new actors was affecting the general organization of scientific research, among other aspects.[26]

The advancement of the HGP led to more positive attitudes towards the commercial interests in science in *El País*. Reports began to depict patents as the only way in which private investment in science could be guaranteed, accepting that this source of funding was crucial for the better future of genetics and medicine. With the passing of time, commercial interests among scientists became less a source of scandal and more a given reality, and the newspaper reported it in an increasingly more "neutral" manner.[27] But, mentions of the increasingly powerful economic interests in human genetics presented this as an issue completely alien to Spanish science, the achievements of which were presented instead as pure knowledge detached from any practical details.

Despite the increasingly benevolent way in which the private sector participation in science was treated, distrust sporadically reappeared. Criticizing the search for economic profit through science was an important rhetorical weapon of the HGP against Celera Genomics, and it was reflected as such in *El País*. It also appeared in extreme situations, such as the sale of the Iceland medical records to the private firm DeCode Genetics, or when a crucial gene in the fight against the HIV was patented. In those cases, *El País* tried to maintain the good image of science, stating whatever seemed to lead to its defence. This implied that the fostering of a positive image of Celera Genomics when it was seen as a promoter of the advancement of science was followed by reviling reports when its actions were interpreted as stock market strategies,[28] or the citation of "economic reasons" for the failures of scientists.[29]

The technical framing of local controversies

Even if most reports on local genetics in *El País* did not mention the economic interests of Spanish researchers, these became visible during the second half of the nineties. In contrast to early approaches to the subject in relation to foreign science, Spanish cases were treated very indulgently. This can be seen in some explicit statements, but mainly in the technical framing of controversies that arose on the occasion of the launching of genetic diagnosis programmes. Application of these tools was quite a problematic issue in Spain. In this context, researchers with particular interests in this area modified their public stances in order to make them appear more reliable and beneficial, while journalists supported them in this particular aim.

Early diagnosis of Alzheimer's disease was with no doubt the most problematic genetic test ever made available in Spain. In 1992, *El País* informed on the creation of a local "gene bank" for the study of this disease, which was dependent on the public Instituto para Enfermedades del Sistema Nervioso Central.[30] Its director, Ramón Cacabelos, explained on that occasion that "among the causes" of the disease, "there are genetic, racial, sexual, infectious and psycho-social factors". According to Cacabelos, only "when the disease appears before the patient is 65 years old, concrete genetic alterations, located in chromosome 21, seem to exist".[31]

In 1995, *El País* informed that a genetic test for Alzheimer's disease had become available at an international level.[32] Only three weeks later, Cacabelos reappeared in the newspaper, now as the director of a private institute he had created, devoted to the study of this disease. In this context, his discourse completely changed. He stated that the main risk for developing Alzheimer's disease was to have relatives who had suffered from it. Alzheimer's diseased was depicted as a genetic ailment that could be diagnosed and prevented thirty years before the appearance of its first symptoms.[33]

In 1997, Cacabelos began to promote a programme for the "genetic prevention" of Alzheimer's disease, arguing that early diagnosis and preventive treatment could retard the appearance of the dementia "between 6 and 18 months", with the consequent saving of money for the national social security system. "Alzheimer's disease is a purely genetic illness. Even 30 per cent of the cases thought to be sporadic (because they appeared in people with no previous cases in their families), are due to the combination, in one child, of the father's and the mother's genes", Cacabelos said.[34] Such an extreme genetization brought on reactions from other members of the Spanish scientific community. In a letter to the editor, María Asunción Morán, lecturer at the Faculty of Medicine at the Universidad Autónoma de Madrid, attacked every statement Cacabelos had made: "It is not true that 98 per cent of the cases can be diagnosed", she said, "there are false positives and false negatives, which could not justify the beginning of treatment when people are 20 years old".[35] Some weeks later, two members of the Sociedad Española de Neurología,[36] Félix Bermejo and Teodoro del Ser, stated that Alzheimer's disease was "the result of multiple factors, genetic and environmental" and that Cacabelos' opinions had "no scientific basis".[37]

The controversy around the predictive capacity of genetic tests for Alzheimer's disease reappeared in 2002, when the public Hospital Clinic, in Barcelona, launched a programme to determine which patients were suffering from a genetic case of the disease. The hospital was planning to test relatives of ill people, even as they recognized that there was no treatment they could offer them. Trying to persuade people to let their genes be analysed, they argued that the diagnosis would allow people to better plan their lives, and that "those that adhere to this programme will be the first to participate in the clinical trials for drugs intended to retard or to block the disease".[38] Once again, other scientists were swift to react, disagreeing with the causation link that could be established between certain genetic mutations and the disease.[39] Tests for Alzheimer's disease were by that time very common in Spain, as stated in a 2003 report. However, in the words of its author, it was "not very clear" why so many of these tests were performed, given that the disease had "a multi-causal origin".[40]

These examples show that the more controversial aspects of the tests were their predictive capacity and thus the degree of genetic causation of Alzheimer's disease. Scientists responsible for genetic test programmes tended to stress this causal link, in their bid to establish new tools. Whether for business reasons – in the case of private institutes – or for maintaining their research protocols – in the case of public ones – geneticists needed to consolidate new diagnoses and to guarantee a steady flow of patients to their laboratories. By circumscribing debates to the tests' reliability, geneticists thought they were more likely to acquit themselves well. They tried therefore to avoid any non-technical discussion which could have engendered suspicion regarding the unselfishness and benevolence of their actions.

Surprisingly enough, detractors of genetic diagnoses did not resort to other arguments against new tools and their promoters, either. In the Spanish genetic tests controversy, the only danger that was constantly mentioned was genetic discrimination by insurance companies or employers, which in fact was not too serious, given the solid Spanish public health system. On the contrary, there was almost no mention of the psychological cost of "knowing" that one would develop certain disease in the future,[41] nor were the conflicts of interests of certain Spanish geneticists pointed out. Cacabelos' private research institute, for instance, was not only developing genetic tests, but also studying the drugs that would be prescribed to patients carrying mutated genes. He was therefore "financially" interested in stressing the role of genes in Alzheimer's disease, something that not even his more violent detractors ever mentioned.

El País contributed to divert discussions away from problematic aspects of genetic tests, focusing reports in the pertinence of genetic explanations of disease and framing controversies in a technical way. This was certainly a way of implicitly backing geneticists, together with other actions, such as stressing positive aspects of diagnoses and giving test promoters a privileged space in the newspaper's pages. In 2002, the head of the detection programme for genetic cases of Alzheimer's diseases at the Hospital Clinic was interviewed several times within a couple of days, while his detractors had to resort to the marginal space of the letters to the editor. On that occasion, *El País* defended the programme in an editorial,[42] and in the days that followed it published a long story significantly headed "Reasons for the genetic prediction of risks", in which the positive assessment of genetic tests was very much reinforced.[43]

The intention of deactivating potential conflictive issues regarding genetic tests can be noticed in other reports that also reflect the positive attitude of *El País* towards private funding of research in Spain. A 2001 report on the development of a genetic test for Familiar Hypercholesterolemia made clear that the foundation promoting it was funded by the pharmaceutical firm producing the drug that would be prescribed to the affected patients, but the reporter did not mention any possible conflict arising from this circumstance.[44] Moreover, when in 2004 this foundation began to market the corresponding biochip for detecting the mutation, the origin of funding was completely left out. The report praised on the contrary the fact that patients carrying the mutated genes would benefit from reduced prices for the drug they needed, thanks to an agreement between the pharmaceutical firm and the Spanish public health system. The reporter went further, stating which Spanish regions were already offering diagnoses and which were not, contributing to the expansion of the firms' market through the generation of public demand.[45]

It is quite clear that Spanish geneticists saw in *El País* a means to gain legitimacy for the new diagnostic tools they were offering, and that they found in the newspaper an ally in their objectives. Genetic diagnoses were removed from any suspicion regarding the negative consequences of the private funding of research, while every novelty in the genetic diagnosis realm was praised.

The biomedical sector of the economy

While reports on genetic tests obscured the origin of the funding of scientific institutions, the emergence of Spanish biomedical firms was covered for its own sake by *El País*.[46] Alzheimer's disease test promoter, Ramón Cacabelos, was once again considered to be a pioneer in this field. As stated in the 1995 report on his new institute, "Cacabelos has achieved something that in Spain is still a very rare accomplishment: to establish an advanced research medical institute on a completely private basis".[47]

Private genetic laboratories gained a common presence both in Spain and in *El País* by the turn of the century, coinciding with the end of the HGP, but they were far from being problematized. Journalists reproduced their founders' reassuring statements,[48] praised their enterprising capacity and took private participation in research as proof of the high quality of local science.[49] Other texts adopted a more economically oriented approach, defining genetics as "a scientific challenge that the pharmaceutical industry expects to transform in a flourishing market",[50] while others depicted the new "scientist-businessmen" as young, brave biologists who found in industry what they could not achieve in the academic world.[51] Only when the conflict of interests was too evident, as when Cacabelos was appointed professor in a private university in a chair funded by two firms he headed, *El País* cast a shadow on the innocuousness of the new financial state of research.[52]

Many of the reports on the new biomedical sector of the Spanish economy were published in sections not traditionally involved in the coverage of scientific news, such as the regional and technology supplements and *El País'* economic pages. They also coincided with the emergent conception of genetics as a new driving force for the Spanish economy. The lack of Spanish participation in the HGP brought fierce criticisms against the lack of public support to genomics, which in turn led to the foundation of the public Fundación Genoma, the aim of which was to promote the "transformation of genetic knowledge into technology and therapeutic tools, such as new drugs and products for diagnosis",[53] and to turn genomics into a new "sector" of the market economy.[54] Although the newspaper's editorial position was in those days extremely critical towards almost every governmental action related to the advancement of genomics, the commercial approach to it was never criticized. Other public policies aimed at promoting biomedicine as a main pillar for the Spanish economy were also praised.[55]

The visibility of private enterprises devoted to genetics both at a national and at an international level fostered the publication of different columns on the subject, the positions of which contrast with the general positive disposition of *El País'* reporters towards the search of economic profit through science. Like in the newspaper's early years, columnists distrusted scientists' participation in private ventures and linked commercial interests with the misuse of science and bias in results. "Commercialization of life" was perceived as a major risk to modern biology, and one which should be prevented with appropriate regulations. "Of course I don't doubt the efficacy of genetic therapy. (…) But as drugs are part of the marketplace, so now are scientists, I fear 'distortions'", columnist Manuel Lloris wrote in 2000.[56] After the publication of the human genome sequences, reader Ignacio Alonso Marín regretted that science had always been used for "constructing weapons" and "winning money" and called for strong regulations on the study of the human genome.[57] On his side, columnist Manuel Vicent congratulated himself because, when new biomedical technologies "would become profitable", he would be already dead.[58]

Criticisms, however, were articulated in very general terms. With some very punctual exceptions,[59] *El País* did not house a real debate on what the role of private funding in Spanish genetics should be, nor on the consequences, advantages or disadvantages of promoting it. The necessity for increasing private participation was taken for granted on almost every occasion, while biomedical firms and genetic tests expanded through the country with apparently no major opposition.

Conclusion

Economic interests in science became increasingly more common in Spain during the late 1990s, but this new situation did not prompt in *El País* a growing problematization of the effects it could bring to the organization of national research. Its reporters generally obscured funding of scientific centres; when they occasionally discussed financial features, they usually linked private funding to innovation and to the advancement of science. This approach contrasts with *El País'* early years, when the search of economic profit through science was approached with much scepticism. The change in the editorial attitude coincided with the growth of private interests in Spanish science, the promotional campaigns of local geneticists and the proximity of the newspaper's journalists to this increasingly powerful professional group.[60]

Geneticists involved in the launch of genetic diagnostic tests in Spain took advantage of *El País* to institutionalize these new tools. They found in the newspaper an ally that praised the novelties and privileged them in tone and space, compared to the treatment ascribed to critical statements against new tests. By framing controversies in a technical way, journalists deactivated possible resistance that could have derived from the commercial orientation of research institutes; in so doing, they favoured the expansion of genetic diagnoses for adults in Spain.

The same applies to the business approach to human genetics and new Spanish biomedical firms. Reporters did not mention any potential conflict of interest arisen from scientists' involvement in private ventures, and supported governmental initiatives aimed at increasing private funding of scientific research. This positive depiction of the role of science in the economic development of the country can be explained by the coverage of biomedical ventures by reporters outside *El País'* "society" section in which scientific subjects traditionally appeared. Business and technology journalists were certainly less influenced by the ideal image of science as necessarily detached from any interest alien to the search of knowledge, and thus more likely to praise original and technology-rich new initiatives. On the other hand, by the turn of the twenty-first century, journalists in charge of scientific news were also convinced that the lack of private participation in science had been one of the major problems of Spanish research.[61]

A study of *El País* shows an unequal distribution of the capacity of reaching the public – and thus of influencing it – among different Spanish actors. Thanks to their economic power, their proximity to *El País* journalists and their public relations campaigns, geneticists were a privileged professional group, to the promotion of whose interests this newspaper contributed. Such a supporting attitude, however, did not prevent *El País* from publishing fierce criticisms of commercial interests within science, mainly written by its regular columnists. However, this does not necessarily mean that *El País* was a public forum in which this issue was widely discussed. Criticisms of genetic tests were relegated to marginal spaces of the newspaper, or referred to very general issues, leaving the Spanish scientific organization untouched. Neither criticism represented a real menace to the emerging Spanish biomedical sector.[62]

The lack of real debate in *El País* sends us back to the general organization of the current international science system, as recently analysed by Dominique Pestre. According to the French historian, transformations of the "social body" were crucial for the consolidation of a new "financial market-driven" scientific regime during the last thirty years of the twentieth century. These transformations included new decision-making processes in which democratic procedures "occupy a subaltern position with respect to the *de facto* situations created by financial and economic actors".[63] *El País* participated in this new situation, since "debates" in its pages never challenged the growing commercial interests within it. On the contrary, this newspaper contributed to the effect of such economic forces, discouraging the problematization of private funding of science and obscuring references to legal and regulatory aspects of Spanish human genetics.[64] In that sense, during the nineties *El País* acted both as a reflection of and an influence on the context in which it existed. It reported on the general trend towards the increase of private participation in research, helping its local consolidation through the beneficial image it conveyed. But, contrary to the fundamental role *El País* played in the development of Spanish democracy,[65] this newspaper did not assume the role of a public arena in which discussions on the desirable balance between private and public science really took place.

Notes

[*] A preliminary version of this paper was prepared for the conference *Science for Sale? Public Communication of Science in a Corporate World*, held at Cornell University (Ithaca, New York). 15-17 April, 2005. This Research was funded by the Mexican *Consejo Nacional de Ciencia y Tecnología* (CONACYT). I also wish to thank the Department of History of Science at the Institució Milà i Fontanals-CSIC, in Barcelona, for its support.

[1] Nowotny, H.; Scott, P, and Gibbons, M. (2001). *Re-thinking Science. Knowledge and the Public in an Age of Uncertainty*, Londres: Polity Press.

[2] Pestre, D. (2005). *Ciencia, dinero y política*. Buenos Aires: Ediciones Buena Visión.

[3] As Dominique Pestre has put it, "interests in science have extended, and new institutions have proliferated: venture capital, pension funds, Nasdaq, and *start-ups* are now part and parcel of the science business", while patent laws and the definition of intellectual property have been radically reformulated. Pestre, D. (2005). "The Technosciences between Markets, Social Worries and the Political: How to Imagine a Better Future?". In Nowotny, et al., eds. *The Public Nature of Science under Assault: Politics, Markets, Science and the Law*. Berlin: Springer, pp. 29-52.

[4] Cassier and Gaudillière exemplify the "mode of research" of the scientist-entrepreneur, with private companies that control the patents of key genes, do research, and sell genetic tests. Cassier, M. and Gaudillière, J. P. (2000). "Recherche, médecine et marché: la génétique du cancer du sein". *Sciences Sociales et Santé* 18: 29-50.

[5] Dornan, C. (1999). "Some Problems in Contextualizing the Issue of Science in the Media". In Scanlon, E.; Whitelegg E., and Yates S., eds. *Communicating Science*. London: Routledge, pp. 179-205, on p.182.

[6] See, for instance, Nelkin, D. (1987). *Selling Science. How the Press Covers Science and Technology*. New York: W. H. Freeman and Company.

[7] Hilgartner, S. (1990). "The Dominant View of Popularization: Conceptual Problems, Political Uses". *Social Studies of Science* 20: 519-39.

[8] Malone, R.; Boyd E., and Bero, L. A. (2000). "Science in the News: Journalist's Constructions of Passive Smoking as a Social Problem". *Social Studies of Science* 30 (5): 713-35; Altimore, M. (1982). "The Social Construction of a Scientific Controversy: Comments on Press Coverage of the Recombinant DNA Debate". *Science, Technology, & Human Values*, 7: 24-31.

[9] Secord, J. (2003). *Victorian Sensation. The Extraordinary Publication, Reception and Secret Authorship of Vestiges of the Natural History of Creation*. Chicago: The University of Chicago Press.

[10] Dictator Francisco Franco died on November 20th 1975, after having led Spain since his triumph in the Spanish Civil War, in 1939.

[11] *El País* acted as socializer of local conflicts, introducing discussion on important subjects that had been completely veiled during the dictatorship. A general history of this newspaper can be found in Seoane, M. C. and Sueiro, S. (2002). *Una historia de* El País *y del grupo Prisa. De una aventura incierta a una gran industria cultural*. Barcelona: Plaza y Janés. See also Imbert, G. (1988). *Le discours du journal* El País. Paris: Editions du Centre National de la Recherche Scientifique.

[12] By contrast, there was almost no local participation in the cartography and sequencing of the human genome.

[13] The political interest in science also increased media attention to it. *El País* scientific supplement, *Futuro*, was launched in 1985, leading to the professionalization of scientific journalism within the newspaper.

[14] Even if this is an issue very difficult to assess, Nelkin has linked media coverage with generous funding of certain branches of research in the detriment of others. Nelkin. *Selling Science*.

[15] "Investigadores franceses logran el primer mapa del genoma humano", "Casi un cuento de hadas", "El mapa del metro". *El País* [hereafter *EP*] 16-12-1993. All translations from Spanish are by the author of this paper.

[16] "Francia tendrá en 1992 un código bioético para proteger el patrimonio genético". *EP* 20-12-1991. "Investigadores franceses logran el primer mapa del genoma humano". *EP* 16-12-1993.

[17] "Tabacalera financia la investigación de mutaciones genéticas inducidas por el humo del tabaco". *EP* 03-10-1990.

[18] "Varón o hembra". *EP* 11-08-1990.

[19] "La patente de genes humanos indigna a los científicos". *EP* 24-10-1991.

[20] "EE UU quiere patentar genes humanos antes de saber para qué sirven". *EP* 23-10-1991.

[21] "Alarmismo ético". *EP* 05-11-1991.

[22] Puigdomènech, P.. "Tribuna: Ciencia y castas". *EP* 04-10-1989.

[23] González-Silva, M (2005). "Del factor sociológico al factor genético. Genes y enfermedad en las páginas de *El País* (1976-2002)". *Dynamis* 25: 487-512.

[24] Ruiz de Elvira, M (1990). "Las fuentes de la noticia en ciencia". *Arbor*, 534-535: 93-102.

[25] Van Dijck, J. (1998). *Imagenation. Popular Images of Genetics*. New York: New York University Press.

[26] González-Silva, M. "With or without scientists. Reporting on human genetics in the Spanish newspaper *El País* (1976-2006)". In Papanelopoulou, F.; Nieto-Galan, A., and Perdiguero E. eds. (2008). *Popularising Science and Technology in the European Periphery, 1800-2000*. Aldershot: Ashgate.

[27] "Las multinacionales se adueñan de los genes". *EP* 13-12-1998. "La carrera para descifrar el genoma humano ya tiene tres contendientes". *EP* 02-09-1998.

[28] "Celera Genomics pierde un 22% en la Bolsa tras descifrar el genoma". *EP* 28-06-2000.

[29] This was the case when a patient participating in a clinical trial of genetic therapy died in 1999. The event gave rise in the United States to a big controversy about transparency in biomedical research, in which the scientists' economical interests came to play a leading role. In the US media, the controversy prompted a peak in the number of reports mentioning the scientists' conflicts of interest. McComas, C. and Simone, L. (2003). "Media Coverage of Conflicts of Interest in Science". *Science Communication* 24: 395-419.

[30] Institute for the Research of the Central Nervous System.

[31] "Primer banco de genes en España para estudiar el mal de Alzheimer". *EP* 23-03-1992.

[32] "Propensos al Alzheimer". *EP* 04-12-1995.

[33] "El Alzheimer ya se puede diagnosticar y prevenir con treinta años de antelación". *EP* 24-12-1995.

[34] "4.000 familias españolas con casos de Alzheimer tienen ya su 'ficha' genética". *EP* 02-07-1997.

[35] "Sobre la enfermedad de Alzheimer". *EP* 09-07-1997.

[36] Spanish Society of Neurology

[37] Félix Bermejo y Teodoro del Ser. "Tribuna: ¿Se puede tratar la predisposición a padecer Alzheimer?". *EP* 04-09-1997.

[38] "Diagnóstico genético del Alzheimer". *EP* 19-02-2002.

[39] "Aprendices de brujo". *EP* 20-06-2002.

[40] "La desordenada revolución de los tests genéticos". *EP* 15-07-2003.

[41] "El cáncer de mama y ovario en la misma familia, síntoma de una alteración genética". *EP* 26-10-1992.

[42] "Saber o no saber". *EP* 09-06-2002.

[43] "Razones para la predicción genética de riesgos". *EP* 08-06-2002.

[44] "80.000 españoles tienen un alto riesgo de infarto precoz debido a un gen defectuoso". *EP* 02-07-2001.

[45] "Un 'biochip' para evitar 80.000 infartos". *EP* 25-01-2004.

[46] Most of these firms were dedicated to the performance of genetic tests.

[47] "El Alzheimer ya se puede diagnosticar y prevenir con treinta años de antelación". *EP* 24-12-1995.

[48] "You win more money developing a treatment for cancer, than making a virus able to kill half of the population", the head of a new laboratory once stated. "Laureano Simón: 'Los criterios éticos en biotecnología están garantizados'". *EP* 03-01-2001.

[49] According to a report on oncological genetic therapies, "all the clinical trials performed in Spain in the last three and a half years are entirely funded by the pharmaceutical industry. This industrial giant has committed itself to the Spanish medical oncology, because of its great prestige in Europe". "Primeros pasos de la terapia genética del cáncer". *EP* 25-07-2000.

[50] "Exploradores del mapa genético". *EP* 16-07-2001. "Genética para América Latina". *EP* 27-09-2004.

[51] The image was that science suffered from permanent lack of resources. "Jóvenes científicos españoles se convierten en empresarios de servicios biotécnicos". *EP* 24-10-2002.

[52] The report stressed from the very title how abnormal this situation was. "La universidad privada Camilo José Cela otorga una cátedra al director de la empresa que lo financia". *EP* 05-11-2003.

[53] "De los genes a la industria". *EP* 26-03-2001.

[54] "José Luis Jorcano: 'Es difícil saber quién es el Bill Gates de la biotecnología". *EP* 06-04-2003.

[55] For *El País* coverage of discussions on genomics related scientific policies, see González-Silva, M. (2007). "Funding through the press: Genomics as a new political issue in the pages of *El País*". In Herran, N. et al. eds. *Synergia: Primer Encuentro de Jóvenes Investigadores en Historia de la Ciencia*. Madrid: CSIC, pp. 77-94.

[56] Lloris, M. "Tribuna: La vida en sus manos". *EP* 30-06-2000.

[57] "El genoma y su futura utilización". *EP* 22-02-2001.

[58] "Muy sencillo". *EP* 11-08-2002.

[59] In 2005, the historian of medicine, Josep L. Barona, stated that public science should not subordinate itself to the interests of the industry, at the risk of losing "science as culture, knowledge and social good, first objective of any public policy". Barona, J. L. "Tribuna: Ciencia y tecnología. Entre la industria y las políticas públicas". *EP* 27-06-2005.

[60] During the late nineties *El País* hired journalists very close to the Spanish biomedical community, with which they shared values and a geneticized view of biology. Javier Sampedro was specialized in genetics of *Drosophila*, while Xavier Pujol had been the director of the official journal of the Spanish Society of Biochemistry and Molecular Biology.

[61] In 2002, *El País* science editor, Malen Ruiz de Elvira, regretted the scarcity of research carried out by the private industry, stressing that "science and technology

are the basis of the economic development of a country". Ruiz de Elvira, M.
(2002). "Visión periodística de una normalización frustrada". *Quark. Ciencia,
medicina, comunicación y cultura* 22-23: 48-50.

[62] Criticisms to the search of economic profit through science could have generated
precisely the effect contrary to the one that columnists were hoping, as they
reinforced the idealized image of science as intrinsically alien to commercial
interests.

[63] Pestre. *Ciencia dinero y política.*

[64] According to a survey carried out by the Spanish Human Genetics Association,
in 2002 there was no specific regulations regarding genetic tests, nor an official
accreditation system or a standard protocol for the referral of patients.
[http://www.aegh.org/docs/EncIPTSfinal.pdf] (accessed 27 May 2004).

[65] See n.11.

PART IV

SCIENCE AND NATION

SCIENCE AND NATION

NÉSTOR HERRAN

The concept of nation is a central element in the discourse of economic, social and political history. As far as scientists and scientific institutions are embedded in society, it is therefore worth considering if this concept is also a valid category in the historical analysis of science. In a limited sense, most historians agree that, as far as the modern period is concerned, the nation is a pertinent framework for the historical analysis of science. According to David Edgerton, there is a strong case for treating science nationally "since it has been constituted by national education systems, by national funding (not least for that ultimate purpose of the nation-state – waging war), and by national industrial and economic policies".[1] Indeed, in late nineteenth-century and twentieth-century science, creation and consolidation of national states went together with the establishment of national research systems. Interestingly, this is the same period in which the word "scientist" was adopted in several European languages and science adopted its modern features.[2]

The restriction of historical enquiry to national settings has been traditionally driven either by scientists' desire to improve their status in a national setting, state-based financing, or because of historians' nationalistic biases.[3] Focusing on the recovery or exaltation of scientists and scientific institutions in the framework of contemporary nation-states, these narratives include everything from chronological or thematic reconstructions of a nation's scientific history to the wide array of dictionaries of scientific biographies or books of a more restricted disciplinary approach focusing on a single nation.[4] It is also important to note the vigour of this literature in countries of the so-called "scientific periphery", which in the most extreme variety of this analytical model would include everything outside the "big three" (Germany, Great Britain and France) in the nineteenth century or the "big four" (the three plus the USA) in the twentieth century.[5] This is the case, for example, of histories of science in Latin America, aimed at vindicating the existence of science in every nation of this continent and, more or less explicitly, to evaluate the faithfulness of local scientific developments to mainstream science.[6]

   The Latin American case highlights that most important issue affecting
national histories of science: the tension between the claims of universal
validity of science and the local character of knowledge production. This
tension underlies the main issues raised by our contributors, that is, the
existence of national styles of science, the construction of nation by means
of science and technology, and the role of scientific internationalism in its
various forms. This introduction aims to provide a larger framework to
these topics by setting their background and analysing the issues at stake.

   Built upon the notion of literary and artistic style, the origin of national
styles of science can be traced back to the late nineteenth century.
However, it is generally associated with the work of John Theodore Merz
and Pierre Duhem, who established typologies of scientists by defining
national stereotypes. According to Duhem, for example, French deep and
narrow minds were better suited for mathematical abstraction, while broad
and shallow British minds favoured model-building.[7] Merz, in his study of
nineteenth-century European scientific thought, also characterized typical
English, French and German scientists according to stereotypes such as
"thoroughness", "idiosyncrasy" or "pedagogical aim".[8]

   Despite the often parochial and even racist overtones, the concept of
national style of science stuck in the history of science under more
sophisticated reformulations. Despite Marc Bloch's early caution that
comparative history was not restricted to comparisons between different
nations, focusing on national differences has also been commonly assumed
as a prerequisite for comparative analysis.[9] Andrew Jamison, building on
Henry Brookman's characterization of style as one of the central aspects
of nationality in science, identifies the existence of styles with dominant
schools of thought in a country, which would shape how scientists think
about the results of their research.[10] More recently, Jonathan Harwood has
related the notion of national style to an aggregate of values, norms,
assumptions, research traditions and funding patterns. In his particular
case study about the history of twentieth-century genetics, he claims that
this aggregate shaped the range of questions that German geneticists took
to be central to their discipline.[11] Most historians, nonetheless, have
proved comfortable with loose definitions of national styles, using the
category to explain differential successes of research in local settings.[12]

   As well as supporters, the notion of national styles has harsh critics.
This is the case, for example, of Nathan Reingold, who argues against the
use of "cultural monuments, essences, determinisms and idealized national
styles", but at the same time asserts the adequacy of the study of science in
national context.[13] Lewis Pyenson argues similarly, contending the
usefulness of national styles by pointing out fundamental inconsistencies

of this concept. For instance, nations can hardly be considered homogeneous entities, and scientific practices bear more relation to local institutional settings than to a transcendent, national spirit. Pyenson, however, goes much farther than Reingold in his distaste for national styles, and extends his criticism to the whole field of national histories of science, claiming that the nation is not an appropriate analytical category in the discipline. Thus, "the nation, notwithstanding its decisive power in matters of war and taxation, is only an encumbrance". In his opinion, historians of science should limit themselves to the study of a science "which progresses on a local level and which judges its progress by transnational, usually disciplinary, norms".[14]

Aligning with these criticisms, Faidra Papaneloupoulou's paper in this part provides new elements for the deconstruction of the notion of national styles. Her analysis of thermodynamics in mid-nineteenth century France shows that communities of physicists and engineers working in this field could hardly been fitted under a single "national style". In fact, their approaches to the study of thermal phenomena show a great diversity, more related to professional and disciplinary allegiances than to geographic boundaries. Papaneloupoulou's deconstruction of the notion of national style does not imply a full rejection of "styles" as analytical category, as she argues that Kostas Gavroglu's idea of "limited styles" (a reformulation of Alistair Crombie's and Ian Hacking's idea of "styles of scientific reasoning") provide with a valid framework for understanding the intellectual perspectives and interactions inside French thermodynamics.[15]

Criticisms of the notion of national styles set out a reconsideration of the hierarchy of nation and science as analytical concepts, and invite considering the causal axis the other way round. Building on Benedict Anderson, it is useful to consider nations as socially constructed entities, emerging in the context of the nineteenth-century European industrial revolution.[16] Indeed, the spread of communication technologies and industrialization could be considered as the force behind the creation of nationalism and nationalist ideologies.[17] If we relate this perspective with the reasonable assumption that science is intrinsically tied to techno-industrial development, it is possible to consider science as an important factor in the shaping of national identities. This change of focus can lead, in our opinion, to renewed and important insights about the shaping of both science and modern societies.

A good departure point for this analysis is, as Ludmilla Jordanova has pointed out, studying the construction of national communities of scientists by means of commemoratory practices such as medals, memorials, portraits, collective biographies and "histories of their fields which emphasized nationalities of presiding geniuses". [18] The first instances of this programme had been explored by means of studies of universal exhibitions or science policy systems. This is the case, for example, of Auerbach's analysis of the functions of the Great Exhibition of 1851 in building British national identity; or Morris Low's volume on the role of science, technology and medicine in the construction of Japanese national identity.[19] However, Jordanova's emphasis on the need to think about the same role of the history of science in promoting nationalistic agendas has scarcely been taken into account. [20]

The construction of the nation by means of science and technology is a topic addressed by Jorge Lossio's paper. His study of the emergence of the concept of high-altitude diseases shows the ways in which Peruvian doctors studying these diseases specifically focused on the importance of geographical conditions, rather than socio-economic ones. Lossio argues that this geographical focus not only strengthened their political discourse regarding the particular national character, but also consolidated racial stereotypes about Andean natives by relating their difference no only to language, clothing and lifestyles, but also to physiological and medical disparities. His analysis resonates with some topics of colonial science studies. As stated by Palladino and Worboys, knowledge created in colonized areas shaped science in the imperial metropolis. This process could be seen as equivalent of the shaping of the metropolis national identity by means of its colonial expansion.[21]

The importance of science and technology for the imperialist endeavour indicates a strong correlation of these institutions with the emergence of discourses of "technological nationalism". This term involves all kind of statements assuming that the power and supremacy of a nation is crucially dependent on the development of particular technologies.[22] As shown by David Edgerton and others, technological nationalism is strongly related to discourses about national decline, characterized by emphasis on comparison of industrial and technological progress with other countries and calls for increased state investment in research and development. [23] These discourses had traditionally been important rhetorical tools in scientists' hands, mediating their relationship with the state in the struggle for status, resources and improved scientific education. [24] Historical scrutiny of these discourses shows that, despite their popularity and political opportunity, they can hardly be considered as a good assessment of a country's scientific and technological status, but more as an indicator of scientists' ability to spread their concerns in the public sphere.[25] Building on this literature, Faidra Papanelopoulou's paper in this volume provides with a new instance for the deconstruction of declinist discourses. According to her, the backward image of French thermodynamics in the nineteenth-century is the result of bold generalizations about the state of French science in general produced by declinism and its historiography, which does not take into account disparities between different scientific or regional developments.

Criticism of the nation as a central unit of comparison and emphasis on transnational interactions can also foster the deconstruction of declinism and a better understanding of the role of the nation in the history of science. In this sense, the rich literature on scientific internationalism could be a good starting point. The idea has the advantage of not only being a category in sociological and historical analysis, but also a concept used by historical actors in different historical periods, from the Enlightened *République des Lettres* to the international projects of contemporary Big Science.[26] On the other hand, scientific internationalism has also been a central element of sociology of science. Robert Merton considered universalism as one of the main ingredients of the scientific ethos, providing ground for the description of science as an intrinsically internationalist enterprise. According to this perspective, the internationalist nature of scientists could and should bring nations together through international cooperation.

Historians and sociologists of science have stressed that scientific internationalism is widespread by raising many contemporary and past examples of international institutions or collaborative enterprises (UNESCO, for example), international congresses, and excellence-building schemes in science such as the Nobel prizes, which are explicitly conceived as international.[27] This is the case, for example, of Elisabeth Crawford's analysis of the awarding of Nobel Prizes during the first decades of the twentieth century. According to Crawford, Swedish academics fostered internationalism by preferring candidates who had received nominations from scientists of different countries to their own.[28] The atomic scientists' movement for the international control of the atomic weapons race, which articulated events such as the Pugwash conferences, has also been used as a good example of scientists' prevailing internationalist attitudes.[29]

Even though the spread-of-internationalism discourse is difficult to contest, other studies have shown the lack of correspondence between scientific practice and scientists' political attitudes. Brigitte Schroeder-Gudehus has argued, for example, that most scientists favoured the ideology of scientific internationalism for pragmatic reasons, as it was considered indispensable for the progress of their careers. According to her, there is no paradox in the parallel growth of nationalism and internationalism among scientists, as they are part of the same struggle for the rational organization of scientific research. This process, involving division of labour, large-scale research structures and improved communication, explains both their nationalistic rhetoric – aimed at national support – and their appeal to internationalism, which was essential as a way to share scarce resources and to ease international communication. Indeed, the increased number of international scientific conferences, associations or exchanges since the mid nineteenth century is not exclusive of science, but parallel to similar interactions in other professional areas.[30]

Despite the apparent undermining of the Mertonian paradigm, these perspectives could be potentially productive for setting the scientific ethos in its historical context, and bounding its emergence – as well as science universality – to particular communication practices. In this approach, scientific internationalism should be taken into account as a powerful ideological force aimed at increasing communication over national barriers. This force should be analysed in parallel with other phenomena with similar aims, such as corporate internationalism. In fact, transnational companies had played – and play – an important role in shaping science and promoting international communication and standardization. This important role has yet not been sufficiently explored by historians of science, who have privileged the study of foundations and state-based institutions.[31] Jorge Lossio's paper in this volume is a happy exception to this tendency. Focusing on the Cerro del Pasco mining company's activities in Peru, it sets out interesting questions on how its policy shaped Peruvian medicine and medical research, and how its transnational character boosted the international recognition of Peruvian doctors working on hypoxia.

The emphasis on these unexplored facets of internationalism, as well of previous contributions focusing on national styles or the role of science and technology in national construction, intend to provide a fresh look at central questions of the debates about science and national identities, from a perspective that – as discussed in the general introduction to this volume – approaches current debates by challenging essentialist approaches and stressing historically-situated perspectives. In this respect, we hope that the contributions to this part could accordingly enrich our current perspectives of science and the nation, reflexively building a less parochial and more "international" perspective.

Notes

[1] Edgerton, D. (1997). "Science in the United Kingdom. A Study in the Nationalization of Science". In Krige, J. and Pestre, D., eds. *Science in the Twentieth Century*. Amsterdam: Harwood Academic, p. 759-76.

[2] Cunningham, A. and Williams, P. (1993). "De-centring the 'big picture': The Origins of Modern Science and the Modern Origins of Science". *British Journal for the History of Science* 26: 407-32.

[3] See Homburg, E. (2008). "Boundaries and audiences of national histories of science: insights from the history of science and technology of the Netherlands". *Nuncius* [forthcoming].

[4] A good example of the first kind is Van Berkel, K.; Van Helden, A., and Palm, L. eds. (1999). *A History of Science in The Netherlands*. Leiden: Brill. For a review of Dutch national history of science, see Kragh, H. (2000). "National Histories of Science: The Dutch Case". *Minerva* 42: 235-9. This article shows how the Dutch achievement has inspired other massive projects in other European nations, such as the History of Danish Science project. As an example of national dictionaries of scientists, see Lightman, B., ed. (2004). *The Dictionary of Nineteenth-Century British Scientists*. Bristol: Thoemmes; López Piñero, J. M. et al. (1982). *Diccionario histórico de la ciencia moderna en España*. Barcelona: Península.

[5] See part V in this volume, which discusses centre/periphery issues in history of science.

[6] Recent examples of this literature are Guimaraes, M. and Montoyama, S. (1979). *Historia das ciências no Brasil*. São Paulo: Epu/Edusp; Gortari, E. (1980). *La ciencia en la historia de México*. Ciudad de México: Grijalbo; Vessuri, H. M. C. (1984). *Ciencia académica en la Venezuela moderna: Historia reciente y perspectivas de las disciplinas científicas*. Caracas: Fondo Editorial Acta; Babini, J. (1986). *Historia de la ciencia en la Argentina*. Buenos Aires: Solar; Saldaña, J. J. (1992). *Los orígenes de la ciencia nacional*. Ciudad de México: SLHCT-UNAM, and Quevedo, E. et al. (1993). *Historia social de la ciencia en Colombia*. Bogotá: Colciencias.

[7] Duhem, P. (1906). *La théorie physique, son objet, sa structure*. Paris: Marcel Rivière.

[8] Merz, J. T. (1903). *A History of European Scientific Thought in the Nineteenth Century*. Edinburgh: W. Blackwood. This book is interesting as a precursor of the "Big three historiography", for his restriction to English, French and German sources.

[9] Bloch, M. (1928). "Pour une histoire comparée des sociétés européennes". *Revue de synthèse historique* 46: 15-50; Sewell, W. H. (1967). "Marc Bloch and the Logic of Comparative History". *History and Theory* 6 (2): 208-18; Sewell, W. H. and Thrupp, S. L. (1980). "Marc Bloch and Comparative History: Comments". *American Historical Review* 85: 847-53. On comparative history, see Pyenson, L. (2002). "Comparative History of Science". *History of Science* 60: 1-33. Paradigmatic examples of comparative history of science are Forman, P.; Heilbron,

J. L., and Weart, S. (1975). "Physics circa 1900: Personnel, funding, and productivity of the academic establishments". *Historical Studies in the Physical Sciences* 5: 1-185; Landes, D. S. (1969). *The Unbound Prometheus: Technological Change and Industrial Development in Western Europe from 1750 to the Present.* Cambridge: Cambridge University Press, and Glick, T., ed. (1987). *The Comparative Reception of Relativity.* Boston: Reidel.

[10] Brookman, F. H. (1979). *The Making of a Science Policy.* Amsterdam: Vrije Universiteit; Jamison, A. (1982). *National Configurations of Scientific Knowledge.* Lund: University of Lund.

[11] Harwood, J. (1993). *Styles of Scientific Thought: The German Genetics Community, 1900-1933.* Chicago: Univ. of Chicago Press.

[12] See, for example, Marjorie Malley's study of Curie and Rutherford laboratories, which tries to relate Rutherford's success in discovering atomic transmutation to particular national styles. Malley. M. (1979). "The Discovery of Atomic Transmutation: Scientific Styles and Philosophies in France and Britain". *Isis* 70: 213-23.

[13] Reingold, N. (1991). "The Peculiarities of the Americans, or are there National Styles in the Sciences?". *Science in Context* 4: 347-66.

[14] Pyenson, L. (2002). "An End to National Science: Extension of Local Knowledge". *History of Science* 40: 251-90, p. 2.

[15] Gavroglu, K. (1990). "Differences in Style as a Way of Probing the Context of Discovery". *Philosophica* 45: 53-75; Crombie, A. C. (1994). *Styles of Scientific Thinking in the European tradition. The History of Argument and Explanation especially in the Mathematical and Biomedical Sciences and Arts,* 3 vols. London: Duckworth; Hacking, I. (1994). "Styles of Scientific Reasoning: A New Analytical Tool for Historians and Philosophers of Science". In Gavroglu K., et al. eds. *Trends in the Historiography of Science.* London: Dordrecht, pp. 31-46

[16] Anderson, B. (1991). *Imagined Communities: Reflections on the Origin and Spread of Nationalism.* Revised Edition. London and New York: Verso. See also Hobsbawm, E. and Ranger, T., eds. (1992). *The Invention of Tradition.* Cambridge University Press.

[17] Ibid.

[18] Jordanova, L. (1996). "Science and national identity". In Chartier, R. and Corsi, P., eds. *Sciences et langues en Europe.* Paris: École des Hautes Études en Sciences Sociales.

[19] Auerbach, J. A. (1999). *The Great Exhibition of 1851: A Nation on Display.* New Haven: Yale Univ. Press; Low, M., ed. (2005). *Building a Modern Japan: Science, Technology, and Medicine in the Meiji Era and Beyond.* New York: Palgrave-Macmillan.

[20] Here, Spanish historiography of science provides an interesting case, as offering examples of both the potential use of histories of national science to reinforce a centralist view of Spain or to sustain regional nationalisms. López Piñero, J. M. and Navarro Brótons, V., eds. (1995). *Història de la ciència al País Valencià.*

València: Edicions Alfons El Magnànim; García Ballester, L., coord. (2002) *Historia de la ciencia y de la técnica en la Corona de Castilla.* Valladolid: Junta de Castilla y León; Vennet, J. and Pares, R. (2005). *La ciència en la història dels Països Catalans.* València: Universitat de València.
[21] Palladino P. and Worboys, M. (1993). "Science and Imperialism". *Isis* 84 (1): 91-102. The point on the shaping of the imperial metropolis by its expansion has been raised by Headrick, D. (1988). *The Tentacles of Progress: Technology Transfer in the Age of Imperialism.* New York: Oxford University Press.
[22] On technological nationalism, see Low, M. (2003). "Displaying the Future: Techno-nationalism and the Rise of the Consumer in Post-war Japan". *History and Technology* 19: 197-209; Charland, M. (1986). "Technological Nationalism". *Canadian Journal of Political and Social Theory* X (1-2): 196-212; Hecht, G. (1998). *The Radiance of France: Nuclear Power and National identity after World War II.* Cambridge, Mass.: MIT Press; Edgerton, D. (2005). "Science and the Nation: Towards New Histories of Twentieth-Century Britain". *Historical Research* 78: 96-112. The construction of national identities by means of technology has been addressed, for example, by Bud, R. (1998). "Penicillin and the New Elizabethans". *British Journal for the History of Science* 31: 305-33; Fritzsche, P. (1992). *A Nation of Fliers: German Aviation and the Popular Imagination.* Cambridge, Mass.: Harvard University Press.
[23] Edgerton, D. (2006). *The Shock of the Old: Technology and Global History science 1900.* London: Profile Books. On technological nationalism, see also Nye, D. (1994). *American technological sublime.* Cambridge, MA: MIT Press, and Charland, M. (1986). "Technological nationalism". *Canadian Journal of Political and Social Theory* 10: 196–220.
[24] On the introduction of sciencific matters in educational curricula, see Stichweh, R. (1992). *Zur Entstehung des modernen Systems wissenschaftlicher Disziplinen: Physik in Deutschland.* Frankfurt: Suhrkamp. The relationship between the constitution of national states and education systems are addressed by Archer, M. S. (1979). *Social Origins of Educational Systems.* London: Sage Publications Ltd, Green, A. (1990). *Education and State Formation. The Rise of Education Systems in England, France and the USA.* Basingstoke and London: Macmillan; Vaughan, M. and Archer, M. S. (1971). *Social conflict and educational change in England and France, 1789-1848.* Cambridge: At the University Press.
[25] See, for example, Gooday, G. (2000). "Lies, Damned Lies and Declinism: Lyon Playfair, the Paris 1867 Exhibition and the Contested Rhetoric of scientific Education and Industrial Performance". In Inkster I. G. C. et al., eds. *The Golden Age. Essays in British Social and Economical History, 1850-1870.* Aldershot: Ashgate, pp. 105-20. The French case has been addressed in Fox, R. and Weisz, G. (1980). "Introduction: The Institutional Basis of French Science in the Nineteenth Century". In Fox and Weisz, eds. *The Organization of Science and Technology in France, 1808-1914.* Cambridge: Cambridge University Press, pp. 1-28. On Spanish declinism, see Nieto-Galan, A. (1999). "The Images of Science in Modern

Spain". In Gavroglu, K., ed. *The Sciences in the European Periphery During the Enlightenment*. Dordrecht: Kluwer Academic Publishers, pp. 73-94.

[26] Daston, L. (1991). "The Ideal and Reality of the Republic of Letters in the Enlightenment". *Science in Context* 4 (2): 367-86; Avramov, I. (1999). "An Apprenticeship in Scientific Communication: The Early Correspondence of Henry Oldenburg (1656-63)". *Notes and Records of the Royal Society* 53 (2): 187-201; Thackray, A., ed. (1992) *Science after '40*. Osiris 7.

[27] Alter, P. (1980). "The Royal Society and the International Association of Academies". *Notes & Records of the Royal Society London* 34: 241-64; Salomon, J. (1971). "The Internationale of Science". *Science Studies* 1: 23-42.

[28] See Crawford, Elisabeth (1990). "The Universe of International Science, 1880-1939". In Frängsmyr, T., ed., *Solomon's House Revisited: The Organization and Institutionalization of Science*, Canton, Mass.: Science History Publications, 251-69. Crawford, E.; Shinn, T., and Sörlin, S. (1992). "The Nationalization and Denationalization of the Sciences: An Introductory Essay". In Crawford; Shinn, and Sörlin, eds. *Denationalizing Science: The Contexts of International Scientific Practice*. Dordrecht: Kluwer, 1-42. Crawford, E. (1992). *Nationalism and Internationalism in Science, 1880-1939*. Cambridge: Cambridge University Press; Elzinga, A. and Landström, C., eds. (1996). *Internationalism and Science*. London: Taylor Graham.

[29] On the atomic scientists movement, see for example Smith, A. K. (1971). *A Peril and a Hope: The Scientists' Movement in America, 1945-47*. Cambridge, Mass.: MIT Press; Boyer, P. (1985). *By the Bomb's Early Light: American Thought and Culture at the Dawn of the Atomic Age*. New York: Pantheon; Jones, G. (1988). *Science, Politics and the Cold War*. London: Routledge; Wittner, L. S. (1993). *The Struggle Against the Bomb Volume 1: One World or None: A History of the World Nuclear Disarmament Movement Through 1953*. Stanford: Stanford University Press.

[30] Schroeder-Gudehus, B. (1978). *Les scientifiques et la paix: La communauté scientifique internationale au cours des années 20*. Montréal: Montréal University Press, and (1982). "Division of Labour and the Common Good: The International Association of Academies, 1899-1914". In Bernhard, C. G.; Crawford, E., and Sörbom, P., eds., *Science, Technology and Society in the Time of Alfred Nobel*. Oxford: Pergamon, pp. 3-20.

[31] On corporate internationalism, see Crawford; Shinn, and Sörlin. *Denationalizing science*. Classical accounts of the relationship between the corporation and science are Noble, D. F. (1977). *America by Design: Science, Technology, and the Rise of Corporate Capitalism*. New York: Knopf, and Dennis, M. A. (1987). "Accounting for Research: New Histories of Corporate Laboratories and the Social History of American Science". *Social Studies of Science* 17: 479-518. Among the most recent contributions, see Marchand, R. and Smith, M. L. (1997). "Corporate Science on Display". In Walters, R. G., ed. *Scientific Authority and 20th-century America*. Baltimore: Johns Hopkins Univ. Press, pp.148-82, and McGrath, P. J. (2002)

Scientists, Business, and the State, 1890-1960. Chapel Hill, NC: University of North Carolina Press. Steven J. Harris and Kapil Raj have also stressed the importance of considering transnational corporations in the early modern period. Harris, S. J. (1998). "Long-Distance Corporations, Big Sciences, and the Geography of Knowledge". *Configurations* 6 (2): 269-304; Raj, K. (2007). *Relocating Modern Science: Circulation and the Construction of Scientific Knowledge in South Asia and Europe, 1650-1900*. Basingstoke: Palgrave Macmillan.

THE EMERGENCE OF THERMODYNAMICS IN MID-NINETEENTH-CENTURY FRANCE: A MATTER OF NATIONAL STYLE?

FAIDRA PAPANELOPOULOU[*]

The development of "energy" physics in the nineteenth century is commonly considered by British historians as a key phenomenon defining Britishness in science. Important work has been produced in this field setting the emergence of energy physics within British culture, and especially within nineteenth-century British engineering.

By contrast, little has been done to explore the development of this subject in France. This is due, on the one hand, to the fact that developments in this field were affected by French scientists' rhetoric of decline, aimed at increasing their research resources by comparing France with other countries, particularly Germany. This historical discourse has often been too closely followed by historians, obstructing the historical study of French science in this period. On the other hand, the geographical spaces, the contexts of practice and the actors leading the development of thermodynamics in France do not fit in the conceptualization of the "national" commonly used by historians of French science.

Both in the French and in the British case historians have commonly made implicit use of the concept of "national styles". In this framework, science has been considered as a practice contained within the nation, international comparisons are scarce, mainly represented by a few comparative works undertaken by sociologists in the 1960s, and little attention has been given to the study of international communication. In addition, scholars work too often under the assumption that the national units framing their claims are homogeneous. However, most often they are not.

In this paper, I aim to de-construct the concept of "nation" commonly applied by historians to French science. I show that the notion of "national styles" tends to obscure the cultural diversity and complexity of the country. This is fundamental in order to write the history of French thermodynamics, as it allows integration of the study of practices and practitioners which have traditionally been left out due to a restrictive and inaccurate concept of French (national) science. Conversely, I advocate that the use of the concept of "styles of scientific reasoning" allows for a more flexible and accurate treatment of the diversity of actors and contexts involved in the development of thermodynamics in France. I can thus enrich the picture of scientific practice in France, as well as the current historiography on French science.

Although I do not argue for microanalysis *per se*, I believe that comparative history in an international perspective is in need of richer methodological tools than scholarly articulations of national stereotypes. The analysis of this case could hopefully offer heuristic tools to tackle other national contexts in order to proceed to accurate international comparisons devoid of artificial and restrictive national boundaries.

The historiography of thermodynamics

The history of thermodynamics is a privileged field where the historian can examine an intriguing confluence of scientific and technological practices. In 1957 Thomas Kuhn brought to light a significant number of actors, whom he considered to be the "twelve discoverers" of the principle of energy conservation.[1] Despite his goal-directed historiography, which gave rise to debates concerning the choice of characters, and a use of the concept of "discovery" evoking a truth hidden in nature,[2] Kuhn's identification of three trigger factors that made possible the "simultaneous discovery" of the concept of energy (availability of conversion processes, concern with engines, and *Naturphilosophie*) was a first attempt to integrate thermodynamics within a wider intellectual and cultural context.

More recently, Crosbie Smith has further explored the biographical approach by considering scientific work not as the product of isolated individuals but as crucially contingent upon the cultural resources of the period in which it was produced. His work shed new light on the emergence of energy physics within the cultural and economic circumstances of a group of north British scientists and engineers, highlighting as important factors such as the increasing use of industrial machines, the existing social and institutional networks, and political and religious ideologies.[3] Drawing on recent sociological work, he built his narrative through the use of the notion of credibility, analysed as believability, personal power based on the confidence of others, and business trust.[4] These were the crucial factors that enabled the acceptance and further consolidation of the new energy doctrine, considered, like the term "energy" itself, as an intrinsically British construct.

Despite Smith's re-evaluation of the history of thermodynamics in Britain, little has been done in this respect for other nations,[5] and even less concerning comparative studies and the study of the role of international communication in this field. In fact, various aspects of what was finally consolidated as the theory of thermodynamics were developed outside Great Britain, reaching British natural philosophers through various vehicles of communication.

Although it was Sadi Carnot (1796-1832) who stated, in 1824, what became known as the second law of thermodynamics, French physicists, chemists and engineers came to terms with the theory much later. This relative "delay", combined with the fact that it was the British scientists James and William Thomson who revived Carnot's theory in the 1840s, is often used as an illustrative example of the decline of French science in the mid-nineteenth century. Correspondingly, there have been very few studies of the French participation in the early creative phases of the development of thermodynamics, if we exclude scholarly work on Sadi Carnot.[6]

In this paper I would like to redress this historiographical state of play, using two case studies related to the emergence of the mechanical theory of heat. I am particularly interested in exploring the issue of decline, and the related notion of "national styles" in science, which I believe have obscured aspects of the scientific activity taking place in mid-nineteenth century France. If we were to raise questions about the internal features of scientific practice and their influence within a specific cultural, social and political framework, the views concerning decline fall short of capturing a rather complex and lively picture.

The issue of "decline" is a much-discussed theme in the history of nineteenth-century science. In the case of France, the image of a "science in decline" first emerged from detailed reports on the state of French science by Louis Pasteur, Claude Bernard and Adolphe Wurtz in the 1870s, after the French defeat in the Franco-Prussian war. Their rhetoric was intended to persuade the French government to increase its funding of higher education and research infrastructure. Interestingly, this strategy coincides with that followed by British scientists through the 1830s polemics around the decline of science in England, and the debates led by international comparisons after the 1850s and 1860s International Exhibitions.[7]

However, it was the "decline of French science" that had a prominent role in the 1960s in leading the analysis of social scientists, such as Joseph Ben-David, and historians such as John Herivel.[8] The notion of "decline" was, therefore, transformed from a rhetorical tool used by historical actors in order to obtain the support of the State to a methodological one employed by scholars who wished to examine in a comparative way the infrastructure of scientific education and research across nations. Declinist literature inspired comparative studies that aimed to explore the institutional organization of science and the educational system in various nations.[9] It also led to revisionist accounts, which redressed the question of decline by drawing attention away from the stultifying environment of Paris to scientific activity in the provinces.[10] Despite the careful distinction between different scientific fields, which prevents bold generalizations about the state of French science in general, the literature still revolves around the issue of decline. For instance, in his recent book on Adolphe Wurtz, Alan J. Rocke takes the revisionist literature into account but concludes, nonetheless, that in chemistry France exhibited a dramatic decline, vis-à-vis Germany, in both quantity and quality of published research between 1820 and 1880.[11]

Obviously, the use of the notion of decline, currently and in the past, is related to the use of a national unit of analysis. In this sense, the historical and historiographical claims about decline are also related to the historiographical category of "national styles". Discussion of national styles first appeared in the early twentieth century, in the works of John Theodor Merz and Pierre Duhem, in terms of national stereotypes.[12] The notion of national styles has served the function of bringing together elements that are deemed to characterize the entire scientific production of a nation, opening the way to comparative analysis. Apart from engaging organizational and social settings, research traditions and schools and state intervention, this notion endorses certain values and norms of scientific practice that imply continuity with the past. However, problems begin at just this stage. The overarching category of national styles, as much as it may seem convenient for comparative analysis, is inadequate to capture the diversity of scientific practice and the importance of the local context in which it emerges. Even in countries like France, which are characterised by rigid institutional structures and extreme centralization, the notion of national styles has proved inadequate to embrace scientific activity.[13]

In French historiography the notion of national style is often implied by the term "science officielle", initially employed in a pejorative way by peripheral and amateur scientists as well as science popularizers. It referred to scientific research that originated in scientific institutions and joined the "approved" corpus of French science in a national and international context. It basically meant, nevertheless, science produced in Paris. The shortcomings of using such a biased unit of analysis are discussed extensively in the relatively recent literature that has directed attention to the scientific activity of the French provinces. In this paper, although I will not particularly focus on the Paris-Provinces debate, I will seek to unveil some of the various ways in which French physicists, chemists and engineers – in Paris, and in the provinces – dealt with the mechanical theory of heat. By shifting attention from potential French contributions to the mechanical theory of heat to two case studies of the appropriation of the theory in France during the mid-nineteenth century, this paper aims at exploring new historiographical perspectives on the emergence and development of energy physics.

My approach in the study of these two cases will be led by the use of the concept of "styles of scientific reasoning", which allow a more flexible and more diversified approach than the concept of "national styles".

The notion of styles is usually linked to the work of Alistair Crombie and Ian Hacking,[14] but my use of the concept is closer to Kostas Gavroglu's notion of "limited" styles, a notion that refers to the discourse and practices adopted by different groups of scientists, or even by different individuals.[15] The notion of styles can account for the work of particular groups of individuals who, due to institutional links, educational backgrounds and cultural affinities, developed practices which shared characteristics leading to the resolution of theoretical and practical problems. It has thus a relatively restricted scope and refers to specific dynamic units of analysis which are, nevertheless, well situated into time and space.

This perspective may bear similarities to the Kuhnian paradigms defined as exemplars of actual scientific practice or models from which spring particular coherent traditions of scientific research.[16] However, the concept of styles has a more restricted scope. It defines answerable questions and soluble problems through a set of constraints imposed by institutional and cultural affiliations, and delineates the criteria that determine what counts among these as acceptable or not. Styles of scientific practice do not necessarily lead to new phenomena, the development of new research methods and techniques and new explanatory themata. If a paradigm represents a consolidated framework within which scientific research takes place, styles, on the other hand, offer a more flexible reading of the diversity of scientific research, especially when dealing with novel quests. Retrospectively one may claim that it is usually during the early creative phases of the development of a theory that one observes a proliferation of styles, considered to reflect the simultaneous development of idiosyncratic notions and methods within a specific cultural milieu.

By focusing on two different groups of people, arbitrarily but deliberately naming them "exact measurers" and "die-hard Laplacians", my aim is to show how attention to local context and the identification of particular styles of scientific practice brings to the fore a richer picture of the landscape of French thermodynamics.[17]

Exact measurers and the mechanical equivalent of heat

In mid-nineteenth century France, the professional training for scientific liberal professions was assigned to the Faculties of Medicine and Pharmacy, the École Normale Supérieure and the Faculties of Science. The function of the École Normale Supérieure, and, to a lesser extent, of the Faculties of Science, was largely to prepare secondary education teachers for service in the *lycées*. During the first half of the century, only a small percentage of *normaliens* pursued scientific research. Among these, only a few began their career as research assistants. Most of them followed a typical *normalien* career with their assignment upon graduation to a *lycée*. The *lycées* were ranked according to their size and proximity to Paris, and graduates who were placed highest in the *agrégation* – a public examination instituted for the provision of teaching staff in secondary education and other levels of state education – were more likely to move faster to Paris. A post in Paris offered greater opportunities for promotion to institutions of higher education or research.[18]

Scientific research was undertaken in specialized institutions, such as the Collège de France, the École Pratique des Hautes Études and the Muséum d'Histoire Naturelle. These institutions, apart from accommodating research that was often too esoteric for teaching institutions, also offered lectures for the learned public. Local societies and academies also accommodated scientific research, conducted usually by peripheral and amateur scientists. However, disciplines that required specialized training or laboratory facilities, such as mathematics and the physical sciences, were hardly represented.

Charles-Cléophas Person (1801-1884), Pierre-Antoine Favre (1813-1880) and Victor Regnault (1810-1878) were some of the first experimental physicists and chemists to have explicitly discussed the interconnection between heat and mechanical phenomena, and to redress their research within the new framework of the mechanical theory of heat. All of them worked almost exclusively on experimental physics. In fact, all French research on the mechanical theory of heat between the late 1840s and the 1860s, which appeared in the *Comptes Rendus* and the *Annales de Chimie et de Physique*, involving calorimetric researches, was a practice requiring laboratory facilities and training in precision measurement. Such training was not adequately offered at the École Polytechnique, the École Normale, or the Faculties of Science. The teaching of physics in all schools involved hardly any experimentation and was confined to the description of apparatus and experimental techniques, which were demonstrated to the students. Laboratory experience was usually gained after graduation, with appointments to research assistantships and the completion of a doctoral thesis.

Person and Favre were both graduates of the École de Médicine. They both held the *agrégation* and a doctorate in science. Person first taught physics in the lycées of Nancy, Metz and Rouen before becoming professor of physics at the Faculty of Science at Besançon. Favre started his career in science as a research assistant and only later obtained teaching positions. Like most young scientists of the time, he moved from one laboratory to another and accordingly undertook various research projects.[19] From the chemistry laboratory at the École Centrale des Arts et Manufactures and the chair of chemistry at the Institut Agronomique of Versailles in 1851, Favre moved in 1854 in the same capacity to the Faculty of Science in Besançon and a year later to that of Marseille, where he remained for the rest of his life.

Regnault, on the other hand, held an intermediate position between science and engineering. He was a graduate of the École Polytechnique and the École des Mines but was exempted from his engineering duties and performed research related to both engineering and scientific matters. From a very early age, and following the Parisian practice of holding multiple professorships – known as *cumul* – he obtained posts in many prestigious scientific institutions in Paris. By 1840 he had accumulated the chairs of chemistry at the École Polytechnique, at the École Centrale des Arts et Manufactures, at the Académie des Sciences, and a year later he obtained the chair of experimental physics at the Collège de France, which he held until 1871.

Person's first publications dealt with the phenomena of heat within the tradition of calorimetric research. In a series of articles dating from 1842 he determined the latent heats of vaporization and fusion of various substances, trying to formulate empirical laws that governed the phenomena observed. It was only in 1854 that Person published his last article in the *Comptes Rendus* suggesting the calculation of the mechanical equivalent of heat based on the specific heats of gases.[20] Person's note on the mechanical equivalent of heat gives no information on whether his previous work on the latent heats of vaporization and fusion had led him to similar results. He never made any priority claim nor did he argue that he had already considered the possibility of heat being converted into work and vice versa. However, an examination of his memoirs before the 1854 note on the mechanical equivalent of heat demonstrates that he was aware of the interconnectedness of heat and mechanical work, and probably not only in relation to the specific case of the heats of fusion and vaporization.

Most striking is his conclusion in a memoir on the coefficient of elasticity of metals and their latent heat of fusion, inserted as a note in 1848 in the *Comptes Rendus* and later published in full in the *Annales de Chimie et Physique*.[21] Person's conclusion was based on the idea that the heat that disappeared during the fusion of a metal was proportional to the mechanical work consumed due to changes in its molecular structure. His discussion of the similarities present in the relation observed and the operation of engines and the laws of gases, indicates that he had considered the conversion of heat and work beyond the case of fusion of metals.

Person's case is an interesting one, for it suggests that French physicists did not ignore the interconnectedness between heat and mechanical work, but remained content with the formulation of empirical laws whose validity was restricted to specific cases. Although he came close to articulating the concept of the mechanical equivalent of heat, he did not reach his conclusions from a thermodynamic point of view. The empirical relations established, which were verified by and large through experimentation, were not adequate for the foundation of a general theory of heat.

Similarly, Favre's first publications on thermochemistry, which were conducted in collaboration with Johann Silbermann, were exclusively based on experimentation and included no theoretical discussions of their results. They carried out the first large-scale calorimetric determination of heat involved in chemical reactions, especially the heat of combustion of a large number of simple and combined mineral and organic bodies. Their undertaking superseded in accuracy previous results and data, and aimed to link heat with modifications in the molecular state of the bodies and chemical affinity.[22] They constructed tables of "calorific equivalents", defined as the quantities of heat evolved by equivalent weights in combinations, and correlated them with the energy of chemical affinities.[23] Favre and Silbermann attempted to define affinity by providing a growing number of data, though without attempting to give a functional definition of the concept. They did not attempt to place their work under a theoretical framework, and were later reproached for their neglect of the mechanical theory of heat in their work.[24]

Favre's career took another direction when, under the guidance and encouragement of Jean-Baptiste Dumas, he submitted, in 1853, two doctoral theses on physics and chemistry, respectively. The former, based on the examination of the heat evolved in hydroelectric currents, was one of the first French theses, if not the first, to discuss the latest research on the phenomena of heat.[25] Although he stated his sympathy for the mechanical theory of heat, it was only in 1858 that he was finally converted to the new theory, after conducting a series of experiments for the determination of the mechanical equivalent of heat. The gradual acceptance of the new theory is evident in the language he employed in his papers; the terms "sensible" and "latent" heat, reminiscent of the caloric theory of heat, were gradually replaced with the terms "production" and "destruction" of heat. The adoption of a theoretical concept was achieved only after what he considered to be a rigorous verification through experiment. When Favre came across results inconsistent with the principle of energy conservation, he tried to justify them by locating the causes of perturbation in the specific characteristics of the experimental set-ups used.[26]

Lastly, the same characteristics are met in Victor Regnault's work on experimental physics. His work was characterized by a laborious accumulation of data, necessary for the qualification of physical phenomena, but deprived of any theoretical speculation or hypothetical reasoning.[27] Although he was held in high esteem by the scientific community and, more widely, by Parisian society, he was often reproached for his contempt for the newly developed mechanical theory of heat. Despite his open support to the new theory in 1853, in one of his lectures at the Collège de France, Regnault was said to have expressed his conviction that the principles of thermodynamics, and especially the concept of the mechanical equivalent of heat, were still hypotheses to be confirmed by experiment.[28] According to his close friend Jean-Baptiste Dumas, he was not interested in the hasty relations built upon the belief in the unity of forces.[29] The conversion of light into heat, magnetism into electricity and of these four phenomena into one another seemed to be of no interest to him. Having discredited so many physical laws concerning the phenomena of heat by his own experimental data, he felt no confidence in the new laws arising from the mechanical theory of heat.

Despite his caution in undertaking any work related to the mechanical theory of heat, Regnault was attracted by one aspect of the mechanical theory of heat: the measurement of the mechanical equivalent of heat. Long after the first estimations of the mechanical equivalent of heat by foreign and French scientists, Regnault made his own determination of the value. The importance Regnault attached to his research concerning the mechanical equivalent of heat is evident from the large scale of his experiments.[30] He decided to conduct a new series of experiments for the determination of the value through the measurement of specific heats of gases at constant pressure and volume. However, despite his expertise in the field, he was never satisfied with the results he obtained, which gave a slightly bigger value than the one found by other means of experimentation.[31]

Person's work on the latent heat of vapours and the fusion of metals, Favre's work on thermochemistry and electrochemistry, and Regnault's work on calorimetry are interesting examples of the ways in which French experimentalists began to discuss their experimental data in terms of "equivalence" and "convertibility". However, their earliest attempts to identify heat with mechanical phenomena rested on purely experimental foundations and were not associated with any general theory of heat. We can then associate Person, Favre and Regnault to an identifiable scientific style, characterized by their attachment to the strong tradition of French experimental physics, the importance of instruments of precision measurement, and the prevailing attitude towards the quantification of physical phenomena and against theoretical speculation.[32] Paradoxically, the "style of scientific reasoning" characterizing the work of these French physicists is closer to the "national style" that, according to Duhem, defines British physics, than to the one he uses to characterize French science.[33] Only a few isolated French practitioners, those I identifty as "die-hard Laplacians" had a profile more closely related to what Duhem considered as the French methodology in physics.

Die-hard Laplacians and the nature of heat

During the late eighteenth century and the early nineteenth century, the work undertaken by Pierre-Simon Laplace was intended to account for all physical phenomena on a celestial, terrestrial, and molecular scale in terms of central forces between particles. Although his system was modelled upon Newton's gravitational force, Laplace's central forces were either attractive or repulsive and acted upon imponderable and ponderable matter. The imponderable fluids (heat, electricity, light and magnetism) were a necessary component of Laplacian physics, and were thought to consist of particles that were mutually repulsive but in all cases attractive by ponderable matter. Furthermore, the forces between ponderable and imponderable matter were thought to be active only over insensibly small distances. Although Newtonian in essence, Laplacian physics differed from all previous accounts in that it offered a far more comprehensive approach of the physical phenomena in terms of molecular forces.

The Laplacian programme for the unification of celestial, terrestrial, and molecular physics flourished during the first decade of the nineteenth century due to a strong network of physicists and chemists, such as Jean-Baptiste Biot, Siméon-Denis Poisson, Louis-Joseph Gay-Lussac, Louis-Jacques Thénard and Étienne-Louis Malus, whose activities were connected through the Société d'Arcueil and the collaborative nature of their research activities.[34] Although by 1812 the Societé d'Arcueil begun to lose its impetus, the Laplacian programme was sustained due to administrative centralization and the curriculum of elementary and higher education in which it featured as a standard doctrine. It was not until the mid 1820s that Laplacian physics collapsed, and only few of its previous adherents continued to pursue its programmatic claims.[35]

However, theoretical speculation, and notably in Laplacian terms, was not absent in the work of some of the actors of thermodynamics in mid-nineteenth-century France. Although those may be considered among the few "isolated die-hards" who continued to pursue the Laplacian programme, none of them had strong or even any connections with the Laplacian network of the early nineteenth century.[36]

Marc Seguin (1786-1875), a self-educated engineer and entrepreneur, was one of the first to combine a microscopic approach to the mechanical theory of heat, which was compatible with his early ideas about the molecular structure of matter and the identification of heat as internal molecular motion.[37] His re-elaboration of his early views on the nature of matter were prompted by the publication of James Prescott Joule's calculation of the mechanical equivalent of heat in the *Comptes Rendus* of the Académie des Sciences.[38]

The interconnectedness of heat and work was considered by Seguin among the wide range of phenomena produced by the law of universal gravitation. His insistence on linking macroscopic with microscopic phenomena through the concept of universal gravitation has been described by some historians as "extreme Newtonianism", but within the French context it could have another connotation.[39] The use of the laws of planetary motion as a model for explaining molecular action was part of the Laplacian programme. Despite the decline of Laplacian physics during the 1840s some aspects of it survived. Although the name of Laplace is never cited, Seguin frequently referred to his disciple Biot as a source of inspiration.

Seguin was not the only one to depict the interconnectedness of heat and mechanical work in molecular terms. The military engineer Charles

Laboulaye (1813-1886) also made a considerable attempt to express the interconnectedness of heat and mechanical work in molecular terms.[40] Laboulaye believed that a clear understanding of the molecular constitution of matter could provide the key to all problems related to the properties of bodies, and therefore their behaviour on a macroscopic scale.[41] His main aim was to show that the mechanical theory of heat did not need to bring any changes to the belief that the mode of action of heat was intimately linked with the constitution of the bodies. Replacing Laplace's repulsive force[42] with what he called *impulsion primitive* of the molecules, Laboulaye proceeded to analyse molecular motion in a way similar to the analysis of planetary orbits. The use of the laws of planetary movement as a model for explaining molecular motion indicates once more that the Laplacian programme still appealed to a small group of French engineers and scientists who were not satisfied with the macroscopic approach of thermodynamics. Indeed, Laboulaye cited Biot in order to legitimize the transposition of the law of gravitation from planetary to molecular movement.[43]

Both Seguin and Laboulaye suggested a molecular model that explained the phenomenon of heat, reminiscent of Laplacian physics in its use of particles and central forces. However, they provided no rigorous mathematical analysis and did not attempt to provide any laws governing the attractive force.[44] The combination of both an "astronomical view of nature" and a rigorous mathematical analysis was to be found in the work of Athanase Dupré (1808-1878).

Dupré's preoccupation with the molecular constitution of bodies and the explanation of physical phenomena on the microscopic level was evident from the very beginning of his scientific career, when he was still a student at the École Normale. In 1828 he published a book on the explanation of the phenomena of heat, light and electricity, where he elaborated his opposition to the theories of imponderable fluids.[45] Dupré considered the ultimate particles of matter to be impressed with a certain quantity of motion, which accounted for the phenomena usually attributed to the action of the imponderable fluids. Furthermore, he suggested that all particles were bound together with the force of universal attraction, which influenced the kind of motion of the material particles. Heat was identified as the rotational movement of atoms, whose intensity depended on the shape of the atoms, whereas the temperature of a body was linked to the velocity of the rotational motion.[46] Although Dupré strongly opposed the existence of imponderable fluids that could be considered as one of the doctrines of the Laplacian school, his insistence in treating physical phenomena in terms of short range forces acting on the ultimate particles of matter can be considered as a re-elaboration of Laplacian physics.[47] His work between 1865 and 1869 is an example of a microscopic approach to thermodynamics, which echoed elements of the Laplacian tradition.

The tradition of molecular mechanics, based on Laplace's ambition to reduce the explanation of all natural phenomena to the action of intermolecular forces, contributed to the existence of a microscopic interpretation of the mechanical theory of heat. The Laplacian influence is evident in the reluctance of some French physicists to accept the kinetic theory of gases, and the survival of the Laplacian model for the constitution of gases based on the action of short-range forces. Such was the case of Athanase Dupré, who, although accepting the plausibility of the kinetic theory of gases, built his theory on the assumption of short-range forces in all physical states of bodies. Marc Seguin's ideas were also derived from the Laplacian programme since he attempted to advance a model of the molecular structure of bodies, based only on the action of a gravitation-like force. Closer to molecular mechanics was Charles Laboulaye's conviction that a clear understanding of the molecular constitution of bodies could provide the key to all problems related to the properties of bodies and therefore to their behaviour on a macroscopic scale.

Conclusion

The two brief case studies related to the emergence, development and integration of thermodynamics in mid-nineteenth-century France bring out the diversity of the approaches to the mechanical theory of heat that could not have been captured with the overarching category of national styles.

If the Laplacian school of the early nineteenth century was characterized by the union of experimental chemistry and mathematical physics, the mid-nineteenth century witnessed their separation. Experimental chemistry and physics, underpinned by the emergence of an empirical epistemology, became one of the legitimizing tools of the new theory. However, the difficulties encountered due to the increasing complexity of experimentation created a cautious perception of the value of theoretical reasoning. Person, Favre and Regnault did not attempt to make any hypothesis or draw any conclusions that went beyond the data carefully compiled from their experiments. The emphasis on precision and exact measurement also led to the design and construction of complex apparatus, which played a major role in the acceptance and validity of the data obtained.

On the other hand, the tradition of molecular mechanics, based on Laplace's ambition to reduce the explanation of all natural phenomena to the action of intermolecular forces, contributed to the existence of a microscopic interpretation of thermodynamics. Although one cannot deny the decreasing influence of Laplacian physics, elements of it continued to exist and were manifest in the way in which Seguin, Laboulaye and Dupré dealt with the emerging mechanical theory of heat.

There was not one single approach to the appropriation of the mechanical theory of heat in France, but rather a variety of them, characterized by different styles of scientific reasoning and practice. Among the reasons why certain styles may have become more assertive than others are the role of national policies promoting scientific education and research, the privileges gained by certain groups of scientists or engineers working for the State, and the power claims secured by the occupation of chairs in various prestigious institutions. Throughout Regnault's career, for example, governmental support for his work was exceptionally generous, extending from the provision of financial aid for the construction of his apparatus to the provision of experimental space that went beyond the restricted space of his laboratory. In turn, the unsurpassed precision obtained by his elaborate scientific instruments and his experimental skills contributed in resolving various problems of practical character that preoccupied French society.

Using "styles of scientific reasoning" as a methodological tool to unveil diversity renders the category of decline obsolete. The notion of decline seems to make sense only when taking as a unit of analysis the nation, in order to then proceed to comparative analysis. However, using the nation as a unit of analysis has in fact often implied considering only a small part of the scientific production taking place in a country, that most often is the science that is being produced either with the support of the State, or within institutional settings. Work conducted in industry, in private laboratories, companies, provincial institutions and academies has hardly been considered, particularly in the case of France, a country with historiographical and cultural features favouring this neglect.

The focus in national decline in mid-nineteenth-century French science has disregarded the study of a number of complex processes of appropriation, which involved a variety of analytical methods, experimental corroborations, engineering/practical applications and cultural affinities. Only by acknowledging and identifying the diversity extant in the work of different groups of scientists and engineers within the same nation will we be able to get away from inaccurate generalizations induced by the employment of overarching categories, and we will be able to enrich the "big picture" with what we have been able to extract from microanalysis and the examination of scientific activity within specific localities.

The advantages of the identification of different approaches in a discipline or a subject are many. Clashes between different approaches could lead to a better understanding of scientific controversies, the establishment of scientific hierarchies and even the configuration of educational curricula in one country. On the other hand, affinities between similar approaches in different countries can help to understand patterns of international communication. The need to think globally and comparatively about science requires methods that do not privilege certain aspects, groups of individuals or institutions in the scientific production of a country, but recognize and acknowledge diversity as the driving force of scientific activity.

Notes

[*] I am grateful to Kostas Gavroglu, Theodore Arabatzis and the editors of this volume for valuable suggestions and advice.

[1] Kuhn, T. S. (1959). "Energy Conservation as an Example of Simultaneous Discovery". In Clagget, M., ed. *Critical Problems in the History of Science*. Madison: University of Wisconsin Press, pp. 321-56.

[2] Elkana, Y. (1970). "The Conservation of Energy: A Case of Simultaneous Discovery?". *Archives internationales d'histoire des sciences* 90: 31-60; Smith, C. (1998). *The Science of Energy. A Cultural History of Energy Physics in Victorian Britain*. Chicago: The University of Chicago Press, pp.10-4.

[3] Smith. *The Science of Energy*, p. ix.

[4] See Latour, B. and Woolgar, S. (1986). *Laboratory Life: The Construction of Scientific Facts*. Princeton: Princeton University Press.

[5] For the Spanish case, see Stefan Pohl (2007). "La 'circulación' de la energía: Una historia cultural de la termodinámica en la España de la segunda mitad del siglo XIX". unpublished PhD Dissertation. Barcelona: Universitat Autònoma de Barcelona.

[6] These few works treat French thermodynamics from the perspective of individual contributions. See for example Locqueneux, R. (1990). "Charles Combes (1801-1872). Les principes de la 'théorie mécanique de la chaleur' fondés sur des principes métaphysiques". *Archives internationales d' histoire des sciences* 124: 11-29, and Cotte, M. (2002). "Les apports de Marc Seguin à la naissance de la thermodynamique". In Kragh, H.; Vanpaemel, G., and Marage, P., eds. *History of modern physics: proceedings of the XXth international congress of history of science. Actes du colloque de liege, juillet 1997*. Turnhout: Brepols, pp. 125-32.

[7] Gooday, G. (2000). "Lies, Damned Lies and Declinism: Lyon Playfair, the Paris 1867 Exhibition and the Contested Rhetoric of scientific Education and Industrial Performance". In Inkster I. G. C. et al., eds. *The Golden Age. Essays in British Social and Economical History, 1850-1870.* Aldershot: Ashgate, pp. 105-20.

[8] Ben-David, J. (1960). "Scientific Productivity and Academic Organisation in Nineteenth-Century Medicine". *American Sociological Review* 25: 828-43; Herivel, J. W. (1966). "Aspects of French Theoretical Physics in the Nineteenth Century". *British Journal for the History of Science* 3: 109-32.

[9] Fox, R. and Guagnini, A., eds. (1993). *Education, Technology and Industrial Performance in Europe, 1850-1939.* Cambridge: Cambridge University Press.

[10] See for example Fox, R. (1980). "The *Savant* confronts his Peers: Scientific Societies in France 1815-1814". In Fox, R. and Weisz, G. *The Organisation of Science and Technology in France 1808-1914.* Cambridge: Cambridge University Press, pp. 151-81; Fox, R. (1984). "Presidential Address: Science, Industry and the Social Order in Mulhouse, 1798-1871". *British Journal for the History of Science* 17: 127-68; Paul, H. W. (1980). "Apollo Courts the Vulcans. The Applied Science instituted in Nineteenth-Century French Faculties". In Fox and Weisz. *The Organisation of Science*, pp. 155-82.; Nye, M. J. (1986). *Science in the Provinces: Scientific Communities and Provincial Leadership in France, 1860-1930.* Berkeley: University of California Press.

[11] Rocke, A. J. (2001). *Nationalising Science: Adolphe Wurtz and the Battle for French chemistry.* Cambridge, Mass.: MIT Press, pp. 397-419.

[12] Duhem, P. (1906). *La théorie physique. Son objet, sa structure.* Paris: Chevalier-Rivière.; Merz, J. T. (1903). *A History of European Thought in the Nineteenth Century.* Edinburgh: W. Blackwood & Sons.

[13] Gert Schubring sees in the examination of institutional structures of schools of higher education materializations of underlying cultural values that can be employed to explore national differences. Although he does not use the notion of 'national styles' his unit of analysis is characterised by a common culture, language and nation and is therefore not far away of the *problematique* of national styles. Schubring, G. (1996). "Changing Cultural and Epistemological Views on Mathematics and different Institutional Contexts in 19th-Century Europe". In Goldstein et al., eds. *Mathematical Europe. Myth, History, Identity.* Paris: Editions de la Maison des Sciences de l'Homme, pp. 363-88.

[14] Crombie, A.C. (1994). *Styles of Scientific Thinking in the European Tradition. The History of Argument and Explanation especially in the Mathematical and Biomedical Sciences and Arts.* 3 vols. London: Duckworth; Hacking, I. (1994). "Styles of Scientific Reasoning: A New Analytical Tool for Historians and Philosophers of Science". In Gavroglu K., et al., eds. *Trends in the Historiography of Science.* Dordrecht: Kluwer Academic, pp. 31-46.

[15] Gavroglu, K. (1990). "Differences in Style as a Way of Probing the Context of Discovery". *Philosophica* 45: 53-75, and (1994). "Types of Discourse and the

Reading of the History of the Physical sciences". In Gavroglu et al., eds. *Trends in the Historiography of Science*, pp. 65-86.

[16] Kuhn, T. (1996). *The Structure of Scientific Revolutions.* 3rd ed. Chicago: University of Chicago Press, p. 10.

[17] The argument could be strengthened by examining the different styles of reasoning of various groups of French engineers who worked on steam engines, but lack of space prevents me from doing so. An example of a particular style of reasoning can be found in Papanelopoulou, F. (2006). "Gustave-Adolphe Hirn (1815-1890). Engineering thermodynamics in mid-19th-century". *British Journal for the History of Science* 39: 231-54.

[18] See Zwerling, C. (1980). "The emergence of the École Normale Supérieure". In Fox and Weisz. *The Organisation of Science*, 31-60

[19] Leblanc, F. (1880). *Notice nécrologique sur P. A. Favre.* Paris: Impr. de Tremblay.

[20] Person, C.C. (1854). "Note sur l'équivalent mécanique de la chaleur" *Comptes Rendus* 39: 1131-3

[21] Person, C.C. (1848). "Relation entre le coefficient d'élasticité des métaux et leur chaleur latente de fusion. Chaleur latente du cadmium et de l'argent" *Comptes Rendus* 27: 258-62 ; Ibid. *Annales de Chimie et Physique* 24: 265-76

[22] Favre, P. A. and Silbermann, J. T. (1853). *Recherches sur les quantités de chaleur dégagées dans les actions chimiques et moléculaires.* Paris: Bachelier.

[23] Levere, T. A. (1971). *Affinity and Matter: Elements of Chemical Philosophy 1800-1865.* Oxford: Clarendon Press, p. 204.

[24] Hermann von Helmholtz criticised the work of Favre and Silbermann for this reason. See Kragh, H. (1993). "Between Physics and Chemistry. Helmholtz's Route to a Theory of Chemical Thermodynamics". In Cahan, D., ed. *Hermann von Helmholtz and the Foundations of Nineteenth-Century Science.* Berkeley: University of California Press, p. 408

[25] Published as Favre, P. A. (1853). "Recherches thermo-chimiques sur les combinaisons formées en proportions multiples". *Journal de Pharmacie* 24: 241-50, 311-46, 412-23.

[26] See for example Favre, P. A. (1868). "Recherches thermiques sur la pile" *Comptes Rendus* 67: 1012-7.

[27] Fox, R. (1971). *The Caloric Theory of Heat from Lavoisier to Regnault.* Oxford: Oxford University Press, pp. 281-314. See also Chang, H. (2004). *Inventing Temperature. Measurement and Scientific Progress.* Oxford: MIT University Press, pp. 74-102.

[28] Mentioned in Laboulaye, Ch. (1863). "De la constitution moléculaire des corps, compatible avec la théorie mécanique de la chaleur" *Annales du consérvatoire des arts et métiers* 4: 66-8.

[29] Dumas, J. B. (1881). "Eloge de Victor Regnault". *Annales de Mines* 20: 237.

[30] Regnault's experiments on the mechanical equivalent of heat were not conducted in laboratory conditions. They were conducted first in the open air at the

military camp of Satory, near Versailles, and later in the sewers of Villemonble (4886m), the drain along the boulevard de Sebastopol (951m) and that of the boulevard St. Michel (1590m).

[31] Regnault, H. V. (1868). "Mémoire sur la vitesse de la propagation des ondes dans les milieux gazeux". *Mémoires de l'Académie des Sciences* 37: 3-575, and (1870). "Mémoire sur la détente des gaz". *Mémoires de l'Académie des Sciences* 37: 579-968.

[32] In Wise, M. N., ed. (1995). *The Values of Precision*. Princeton: Princeton University Press, the various contributions of the volume discuss the issue of precision in the British and German context. These studies demonstrate how styles of practice cut across national boundaries.

[33] Duhem. *La théorie physique*, pp. 99-167.

[34] Crosland, M. (1967). *The Society of Arcueil. A view of French science at the time of Napoleon I*. London: Heinmann.

[35] For an account of the political, institutional and intellectual reasons of the fall of Laplacian physics see Robert Fox's seminal paper: Fox, R. (1974). "The Rise and Fall of Laplacian Physics". *Historical Studies in the Physical Sciences* 4: 89-113

[36] In fact, the Laplacian programme had appealed to a group of engineers, notably V. Poncelet, proponents of the 'physical mechanics'. See Chatzis, K. (1998). "Jean-Victor Poncelet (1788-1867) ou le Newton de la mécanique appliquée. Quelques réflexions à l'occasion de son cours inédit à la Sorbonne", *SABIX* 19: 72. An indicative example of Poncelet's reasoning related to steam engines is his microscopic approach to the production of work through the expansion of gases. See Poncelet, J. V. (1841). *Introduction à la mécanique industrielle, physique ou expérimentale*. Paris-Metz : Gauthier-Villars, pp. 202-204

[37] Seguin was heavily influenced by his uncle Joseph Montgolfier, who believed in the convertibility of heat and work. See Seguin, M. (1824). "Letter to Sir J.F.W. Herschel. Observations on the effects of heat and motion". *Edinburgh Philosophical Journal* 10: 280-83

[38] Joule, J. (1847). "Expériences sur l'identité entre le calorique et la force mécanique. Détermination de l'équivalent par la chaleur dégagée pendant la friction du mercure". *Comptes Rendus* 25: 309-11; Seguin, M. (1847). "Note à l'appui émise par M. Joule, sur l'identité du mouvement et du calorique". *Comptes Rendus* 25: 422

[39] The characterisation of Seguin as an extreme Newtonian is found in Redondi, P. (1980). *L'accueil des idées de Sadi Carnot et la technologie française de 1820 à 1860*. Paris: J. Vrin, n. 49, p. 180.

[40] Through his position as director of the periodical *Annales du Conservatoire des Arts et Métiers*, Laboulaye was probably in close relations with V. Poncelet and A. Morin, and promoted their microscopic approach to mechanics.

[41] Laboulaye, Ch. (1863). "De la constitution moléculaire des corps, compatible avec la théorie mécanique de la chaleur". *Annales du consérvatoire des arts et métiers* 4: 64-112.

[42] Laplace had introduced a repulsive force between the particles of a liquid that was supposed to be caused by the presence of heat. The effect of heat was to reduce the attractive force between the particles of a liquid. See Fox, R. (2000). "Short-Range Forces". In Gillispie, C. C. *Pierre Simon Laplace. A Life in Exact Science*. Princeton: Princeton University Press, p. 207

[43] Laboulaye, Ch., "De la constitution moléculaire des corps", p.82

[44] An attempt to calculate the mechanical equivalent of heat through mathematical analysis of the forces acting between the particles of a gas had been made by Theodore d'Estocquois, professor at the Faculté des Sciences of Besançon. However, his attempt was highly speculative as he considered the structure of gases to be similar to that of liquids, and applied to the former the laws of fluid mechanics and equilibrium state. Estocquois, T. de (1854). *Note sur l'équivalent mécanique de la chaleur*. Besançon : impr. de Outhenin-Chalandre fils.

[45] Dupré, A. (1828). *Sur l'explication des phénomènes de la chaleur, de la lumière et de l'électricité*. Paris: Papinot.

[46] Ibid., pp. 7-12

[47] As J. Rowlinson has noted, Dupré made extensive use of Laplace's work on the phenomena of capillarity and cohesion. See Rowlinson, J. S. (2001). *Cohesion: A Scientific History of Intermolecular Forces*. Cambridge: Cambridge University Press, p. 161.

NATION, DISEASE AND HEALTH:
MEDICAL RESEARCH IN THE PERUVIAN ANDES AND THE EMERGENCE OF "HIGH-ALTITUDE DISEASES"

JORGE LOSSIO

In this study I trace the intersections of nation, science and corporate R&D policies through a history of the origins of the concept of "high-altitude pathology". This term originated in the first half of the twentieth century to describe a group of diseases produced by exposure to hypoxia, for example chronic mountain sickness, high-altitude pulmonary oedema, and high-altitude cerebral oedema. Their particularity was that they were cured by descending to sea-level.[1]

Observations on the clinical and pathological effects of altitude exposure were produced by several doctors in the most diverse settings. In this study I will pay particular attention to the work produced at the Chulec Hospital, a private hospital established in 1921 in the mining compounds of the North American Cerro de Pasco Corporation in the central high Andes. Due to the facilities offered for medical research and its proximity to Lima (just three hours away by car), the Chulec Hospital became an internationally renowned centre for the study of high-altitude related disorders and the site of many innovations in altitude medicine. My primary focus, therefore, is on the work carried out in Peru and amongst the Andean population, although I set this in the context of developing international medical and research networks. Although working in a "distant" and "isolated" Andean community, Peruvian doctors were engaged in broader international debates over the effects of altitude in health and developed links with North American and European scientists and institutions, such as the US Air Force and the Wellcome Medical Research Centre.

By tracing the origin of "high-altitude diseases" I also intend to cover a gap in medical history, since there is almost nothing written on the history of high-altitude medicine. In his *History of High-Altitude Medicine and Physiology*, John West has discussed the idea of "high-altitude pathology"; however, he assumes the category is given by nature rather than constructed by doctors. In his *Isis* article, "Andean Biology in Peru",

Marcos Cueto analysed the case of the Institute of Andean Biology, uncovering the institutional, social and political factors that allowed scientists from peripheral countries to produce internationally relevant scientific knowledge. Marcos Cueto's pioneering work on the social history of medicine in Latin America informs this research.[2] However, my work focuses on the Chulec Hospital as a key centre in the production of high-altitude scientific knowledge. The case of the Chulec Hospital is also important because it illustrates the intertwined relations between private corporations' medical research and development policies, local political factors and the professional interests of scientific communities.

In this study I also assess the importance of "locality" in twentieth century medical practice. Medical historiography tends to emphasize the universalistic (a-spatial) approach in medicine that emerged in the last third of the nineteenth century, when germ theories of disease and ontological conceptions replaced neo-Hippocratic disease models. Within this new paradigm, the focus of medical practitioners shifted towards the individual, towards disease entities, and towards the essential causes of disease. Although I do not intend to challenge this overall characterization, I do think that by overemphasizing biomedical reductionism, medical historians have ignored the persistence of geographical approaches to health. This study demonstrates that neo-Hippocratic approaches in medicine persisted concretely in Peru until at least the 1960s, and that, in certain specific contexts, the relationship between the environment and human health remained important to medical practice.

This study also contributes to the emerging literature on the history of science and medicine in Latin America. Contrary to earlier very descriptive, hagiographic and institutional histories of science and medicine in the region, predominant until the 1980s, this emergent literature has focused instead on local attitudes towards clinical medicine, governmental and popular responses to epidemics, the transference of scientific ideas, the implementation of international health initiatives, the role of international scientific agencies in shaping regional institutions, and the discriminatory components of medical work and public health policies.[3] However, there are still many gaps in this literature and this study addresses neglected topics in Latin American historiography such as the role of science in the forging of national identities and the relation between the interests of foreign corporations and those of local scientists in the production of scientific knowledge.

My argument is that the "discovery" of several high altitude diseases was produced firstly, by the development of the mining compounds of Cerro de Pasco in the central high Andes, which encouraged the migration of thousands of workers and their families from moderate to high altitudes; secondly, by the establishment of modern hospitals and medical research facilities in these mining compounds, which gave medical investigators access to laboratories, the most sophisticated medical technology, accommodation and hundreds of patients for research; and, thirdly, by the emerging interest in altitude scientific research amongst the Peruvian medical community. Contrary to the case with tropical diseases in the late nineteenth and early twentieth centuries, which were the product of the colonial and public health "gaze" of colonial doctors working in India or Africa, the diagnosis of high-altitude disorders was heavily dependent on laboratories and medical technologies. It was the combination of greater medical surveillance of the high Andean population and the access to the most sophisticated medical technologies provided by the North-American mining Cerro de Pasco Corporation that led to the construction of high-altitude related ailments. This case study will therefore allow me to explore the role of private international corporations in shaping scientific research in a "peripheral" country like Peru, where funds and facilities for research were otherwise limited.

Discoveries on high altitude diseases also reinforced the notion that, in order to properly practice medicine in high-altitude locations, such as the Andes, it was necessary to adjust medical practices, therapeutics and surgical procedures to the conditions imposed by hypoxia. Most important, physicians had to be able to promptly identify the symptoms of high altitude diseases, and be able to distinguish them from common respiratory and circulatory problems, as patients with chronic mountain sickness or high-altitude pulmonary oedema were liable to permanent disability or death if they did not descend to sea-level rapidly. [4] Peruvian doctors claimed that their country had a unique medical environment. At a meeting at the World Health Organization in 1968, physician Alberto Hurtado from the Cayetano Heredia University stated that: "Medical problems encountered at high-altitudes are unique. Thus there are diseases which are altitude specific and the rarer problems at sea-level are sometimes common at altitude. There also exist some unique therapy problems particularly in surgery and in the use of drugs". [5]

According to physician Carlos Monge – the leading altitude medical researcher in Peru during my period of study – the need to understand the health effects of altitude constituted a patriotic duty, as the highlands was the most populated region in the country. The interest in high-altitude scientific medical research in Peru began in the early 1920s, when Monge observed a patient from the highlands suffering of severe polycythemia, an increased count of red blood cells. The patient, an engineer from the high

Andean mines, presented a purple colour of the skin and face, nose bleeding and a sensation of losing consciousness during sudden efforts. In the same year, an account of the illness was presented to the National Academy of Medicine as an interesting case but nothing more. Two years later, after leading the First High-Altitude Peruvian Scientific Expedition to the Andes, Monge claimed that his 1925 case had in fact been a new condition-*Andean disease*, so called because the malady had not been observed before in any other region of the world.[6]

Monge claimed that populations living at high-altitudes developed unique physiological, anatomical and chemical peculiarities due to the effects of hypoxia; in fact, Andean people belonged to a unique "high-altitude race", and, it followed that "...a new biology implies a new pathology". In 1931 Monge established the Institute of Andean Biology and Pathology, and began to actively promote high-altitude medical research.[7] In reference to high altitude diseases, Monge wrote in 1942: "Peruvian medical problems must be observed by Peruvians and be codified for Peruvians".[8] Hence, this study intends to show how medical research was partly encouraged by nationalistic impetus and how science shaped and helped to construct national identities, as in the notion of "high-altitude diseases" as "Peruvian diseases".[9]

The rise of Cerro de Pasco, the highest city in the world

Cerro de Pasco is a mining city located in the Peruvian central Andes at 4,380 metres above sea level with a current population of approximately 60,000. Mining activity in the district can be traced back to pre-Columbian times. In fact the name Cerro de Pasco derives from the Quechua word *Yauricocha*, which means "a site of minerals".[10] The rise of Cerro de Pasco as a modern industrial centre began in the early twentieth century when the Cerro de Pasco Corporation – a United States based financing group – took possession of most of the lands and mines of central Peru and initiated what was to be the greatest mining project in South American history.[11] Huge investments in mining infrastructure, such as the construction of smelters and refineries, the installation of the most sophisticated machinery, and the completion of railroads that connected these mining districts with the main port of Lima, placed Peru amongst the leading producers of copper and silver in the world.[12]

Although the labour force was initially recruited from the Pasco population, in order to meet the demands of the growing mining operations, the managers were forced to recruit workers from neighbouring peasant communities, usually from lesser altitudes. For the company there was no doubt whatsoever about the need to rely upon the native Andeans, rather than importing a specialized work force. This was because of the proximity of many Indian communities to the mineral areas, the familiarity of the Andean population with the mining processes, the need of this population to supplement their peasant activities with the cash wages that were paid in the mines and most important, the conviction of the unique capacity of the Andeans to resist the effects of high-altitude.[13]

The recruitment process included the use of promotional posters that announced the promise of advanced payments, housing, electricity, and free medical attention and hospital services. The Corporation agents also resorted to more questionable methods, such as the monopolization of land tenure in the areas surrounding the mining districts in order to force peasants to become miners. In many cases entire peasant communities from the central Andes were forced to migrate to the mining towns as a result of the destruction of their lands, pastures and livestock as a consequence of the environmental pollution produced by the mining activity.[14]

Within a few decades, between the 1920s and the 1950s, this impetus had encouraged an important population movement and the establishment of a permanent workforce in Cerro de Pasco. In addition to the Andean natives, the Pasco Corporation also attracted a smaller group of North American managers, office staff and engineers who settled with their families in Cerro de Pasco.[15] By 1908 the Pasco Corporation had a workforce of 4,100 employees, a number that almost doubled to 8,100 by 1923. By 1956, the number of employees already reached 15,100. By the end of the Second World War the Pasco Corporation had become the biggest single employer in Peru. Its workforce was very young; in 1925 half of the workers were minors, between 14 and 18 years old.[16] Initially, the workforce was mostly composed by Quechua young males of very recent peasant origin from less elevated altitudes. According to Arregui, Velarde and Valcárcel: "it was a population out of place".[17]

In addition to the social disarticulation produced by their transformation from peasants to miners, and by the migration from moderate to high altitudes, the workforce had to confront extremely hard working and living conditions. In a book entitled *The Conduct of the Pasco Mining Company* (1913), social activist Dora Mayer condemned the working conditions and complained that the Corporation disobeyed almost every safety regulation demanded by law, such as the use of helmets or the installation of proper illumination inside the mines, which resulted in many accidents and deaths.[18] Mayer also noticed an extremely elevated rate of alcoholism among the miners, and related this to their abusive treatment, the poor salaries, and the "unbearable climate".[19]

Housing conditions were no better than the working conditions. In the 1940s, the medical missionary J. K. Clothier visited Cerro de Pasco, and prepared a report on the conditions he found. In 1943 he wrote: "Eight hundred families were living in barrack-like rows of houses, each barrack comprising twenty four homes, twelve by fifteen feet. Many were supplied with beds and some were furnished in a pathetic imitation of European amenities. There was little or no privacy and children and parents usually occupied the same bed. One latrine sufficed for twenty-four families and a row of cement laundry tubs ran down the middle of each area between the barracks. Two water taps served the needs of the twenty four families".[20]

Although conditions remained poor, the Corporation managers made some concession to workers' welfare in order to secure a stable and experienced work force. Working conditions were also improved in response to the increasing criticism from government officials and workers' uprisings. In addition to moderate wage increases, the Corporation developed a welfare programme that included the establishment of "a large and furnished school" for the children of the workers, the construction of recreational clubs that offered "billiards, books and refreshments", and outdoor film screenings (some of which addressed health topics such as the importance of hygiene). The Corporation also hired social workers to visit the miners' houses, control sanitation and cleanliness, and more generally, introduce "Euro-western values of domestic life".[21]

It was within this context that one of the most modern hospitals in Peru, the Chulec Hospital, was established. According to the North American physician L. Harris, the establishment of the Hospital was also a way of improving the reputation of these mining districts, and thus attract workers from distant communities, who were reluctant to move to the mines. The establishment of the Hospital also responded to the aim of the Corporation to control every aspect of life in these mining districts, as even an autonomous police force was established by the Pasco Corporation. Harris observed: "the physician of the company delivers the children; afterwards they attend the school of the company; listen to the preacher of the company; are employed by the company; and are buried by the undertaker of the company".[22] This "captive population" was the research material that allowed the Chulec Hospital to become an international leading centre for medical research in high altitude diseases.

The Chulec Hospital

The Hospital was established in 1921 to provide medical assistance to the Corporation workers, staff members and families (a population of around 70,000). Located in the mining town of La Oroya (3,782 metres above sea level), it was the central hospital of the Corporation, which also had two satellite hospitals, the Esperanza Hospital in Cerro de Pasco and the Puquiococha Hospital in Morococha.[23] The Hospital was established with three North-American doctors, and a staff of North-American and Peruvian nurses.[24] In the following years a dental clinic and a ward for infants were also established. By the end of my period of study, 1950, the number of doctors had increased to sixteen.[25] In addition to providing medical assistance, the Hospital became a centre of medical training, particularly for nurses and for medical students from Lima who were interested in high-altitude related disorders, and a centre for the promotion of sanitary education, mostly in areas related to infant care, tuberculosis, and sexually transmitted diseases.[26] Although the Chulec was a private hospital, its staff took charge of important public health duties in the remote districts of the central Andes, something which illustrates the weakness of the national public health services and the decision of the Peruvian state to rely upon private bodies for medical research and training.

A central figure in the history of this hospital was Chief Surgeon Harold Crane. He graduated from the University of Michigan, and arrived in Peru in 1920. He found Cerro de Pasco to be an extremely violent city, with high levels of alcoholism and prostitution; indeed, he later observed: "my patients came just about as often from the saloons as from the mines".[27] He remained as Chief Surgeon of the Hospital for thirty years, until his retirement in 1950. During this time he constantly pushed for the modernization of the Hospital and himself became an internationally recognised expert in mountain sickness. He also became a close friend and regular correspondent of leading Peruvian, North-American and European altitude physiologists, who visited and took advantage of the installations of the Hospital. Most important for this study, Harold Crane was the first to describe a new form of mountain sickness, later recognised as a distinct disease – High-Altitude Pulmonary Oedema (HAPE).[28] In 1927 he wrote an article in which he referred to some patients who combined some of the common symptoms of mountain sickness with coughing, sputum containing blood and congestion in the lungs. He also observed how quickly patients recovered upon descending to sea level.[29]

Ten years later the Peruvian physiologist Alberto Hurtado (1901-1983) described this new malady in a book entitled *Physiological and Pathological Aspects of Life at High-Altitudes* (1937). Hurtado studied medicine at Harvard University. After graduating in 1924 he returned to Peru where he became interested in high-altitude physiology, and joined the Institute of Andean Biology directed by Monge. Hurtado stated: "There is undoubtedly a type of mountain sickness which is quite rare and infrequent and is characterized by the presence of intense congestion and oedema of the lung".[30] He presented the case of a 58-year-old man "of indigenous race" who developed altitude pulmonary oedema after returning to the highlands from a three day stay in Lima. Hurtado described the case as follows: "When he reached the altitude of Casapalca (13,600 feet) he felt a strong discomfort principally on coughing, and noted that his sputum contained a large amount of black blood. Soon there were added to his symptoms headache, intense dyspnoea and some mental dis-coordination. [A chest radiograph taken the day after] showed evidence of reduced transparency of the lungs".[31]

Hurtado was impressed by the research conditions he found in Cerro de Pasco. In 1936, in a letter sent to Allan Gregg, a Rockefeller Foundation official, he wrote: "The Cerro de Pasco Copper Corporation has installed an excellent laboratory for the study of industrial diseases, particularly pneumoconiosis, and the physiological and pathological aspects of life at high-altitudes. The laboratory contains all the necessary material to make haematological, circulatory and respiratory studies, as well as other investigations that are fundamental from the point of view of altitude. As you know this company has thousands of men working at high-altitudes and the conditions for work are splendid".[32]

It is interesting to notice the constant support the Cerro de Pasco Copper Corporation gave to medical research, despite the fact that some of this research was related with working conditions in the mines and led to increased intervention of public health officers in mining sanitary legislation.[33] The Corporation not only expected direct benefits from this research, but also used this support to improve its public image in Peru in the context of increasingly adverse reactions to the presence of American-owned companies in Latin America by anti-capitalist and anti-Imperialist socialist movements that emerged in the region during the 1930s. The Cerro de Pasco Corporation became a main target of these socialist movements. The Peruvian governments of the 1930s and 1940s, however, being mostly right-wing, defended and protected the presence of foreign companies, including the Pasco Corporation.[34] By praising the medical assistance and research facilities provided by the Chulec Hospital, the medical community acted in appeasement of anti-Imperialist discourses and local resistances to foreign corporations. Also it should be mentioned that Hurtado benefited personally from the collaboration, and despite his constant criticisms of the working conditions of the miners, this cooperation continued even after he persuaded the government into reinforcing occupational mining legislation. For instance, in 1946 he was granted "a furnished room and board for seven men for one week each month for a period of one year, and also the use of the laboratory of the Chulec Hospital for the same period".[35]

In 1939 Alberto Hurtado was named General Health Director of the Ministry of Health, almost a vice-minister. A year later he visited the mining camps of Cerro de Pasco to investigate health conditions and wrote a report to the government based on the health records of the companies. His report had two important consequences: companies were forced to perform periodical medical tests on workers to detect the presence of pneumoconiosis, and, if the case demanded it, to grant compensations. Mining constituted one of the main economic sectors of the country, and to public health officers pneumoconiosis was as much a hazard for the miner's health as a menace to the economic development of the country. Hurtado also persuaded the government to create a Department of Industrial Hygiene, within the Ministry of Health, in order to address occupational health problems.[36]

Hurtado, however, did fear that his work in the Ministry of Health had damaged his relationship with the Pasco Corporation. In 1949 he wrote:

> The economic situation of the University is almost disastrous and my efforts to get a fair and decent pay in Pan American Grace Airways have not been successful. After a very thorough consideration I am thinking in the possibility of doing some work for them. You know of my long controversy with the Cerro de Pasco Copper Corporation in regards to the medical care of their workmen. For this reason I did not talk personally to Mr. Russell [Director of the Corporation]. I am writing to you, knowing your close friendship with Mr. Russell. Will you be willing to make a preliminary inquiry with him? I am 47 years old, healthy and quite adapted to high altitudes.[37]

Hurtado's case illustrates the reliance Peruvian scientists had upon the economic resources and equipment of the Cerro de Pasco Corporation for doing high-altitude scientific research.

In the following years several other cases of high-altitude pulmonary oedema were reported in the Chulec Hospital. Leoncio Lizarraga, a native from the high Andes and medical resident in Chulec, wrote the first dissertation on the condition in 1954, entitled *Mountain Sickness: Acute*

Pulmonary Oedema. In this study Lizarraga stated that pulmonary oedema of high altitude may occur either in acclimatized dwellers returning to a high altitude after a stay at sea level, or in lowland individuals during their first ascent. It was portrayed as a serious condition that could lead to death. The symptoms were relieved by oxygen administration or by descending to sea-level.[38] To Lizarraga a unique expertise was required to practice medicine properly in high altitude regions, given the distinctiveness of their medical problems. In 1954 he wrote: "Climatic aggression of altitude has created a human race that offers biologic characteristics different from those of the man of sea level...high-altitudes confront us with new and ignored clinical pictures".[39] After HAPE, the second most important "high-altitude disease" was that of chronic mountain sickness or Monge's disease.

It was in the mining districts of Cerro de Pasco during the 1950s that important epidemiological findings were produced by Carlos Monge's son, Carlos Monge Cassinelli, also a well known altitude physiologist, on the extent of chronic mountain sickness.[40] Contrary to the common view at that time, that chronic mountain sickness only affected a few individuals who had lost their acclimatization to high-altitudes; Monge Cassinelli's studies demonstrated that chronic mountain sickness affected around 15% of the population of Cerro de Pasco. He also found that "most of the population" was suffering from at least some of the symptoms associated with chronic mountain sickness and that this disease became more common with increased elevation.

Furthermore, Monge Cassinelli argued that chronic mountain sickness was a natural concomitant of permanent stays at high-altitudes and increasing age. He also noted an extensive number of native adults with clubbed or drumstick fingers and cyanosis. Clubbed fingers, he argued, demonstrated that the oxygen supply to these tissues was failing at some point; it did not affect only susceptible individuals, but was a constitutional characteristic of the adult population of Cerro de Pasco. Monge Cassinelli attributed the extent of chronic mountain sickness to the effects of rapid industrialization, urbanization and migratory movements from moderate to high altitudes, encouraged by the development of the mining industry. These migratory movements had been accompanied by several modifications in the habits of the native population that negatively influenced altitude adaptation, such as increased smoking rates, impoverished living conditions, sedentary habits, and modifications in the traditional diets, among others. According to Monge Casinelli these migratory movements and changing lifestyles had multiplied the extent of altitude related disorders.[41] Like most of the scientists working in the

Chulec Hospital, Monge Casinelli came from the coastal city of Lima, and had a sort of idealized image of the Andean as a pristine culture that had remained static throughout history. In his assessments of chronic mountain sickness we can observe a criticism of the process of modernization of the Andean population.

Monge Casinelli argued that chronic mountain sickness was due to a loss of acclimatization produced by a combination of hypoxia, modern life in the mines and aging. However, he never went so far as to consider chronic mountain sickness or high altitude pulmonary oedema as industrial diseases or diseases of urbanization. These ailments were mainly "climatic diseases", although aggravated by the exposure to new living and working conditions.

High-altitude regions as "distinct medical spaces"

The notion of a distinct "high-altitude pathology" involved not only "proper high-altitude diseases", but also the modification of the incidence and pathology of other diseases in a hypoxic environment. In 1948 Monge wrote:

> Hygiene has busied itself up to now with a struggle against the aggressions of parasites, infections, poisons, malnutrition, etc.; it does not seem out of place, then, to ask that in America – where millions of people inhabit the high altitudes – doctors and hygienists should direct their attention to the problems resulting from climatic aggression, namely: the incidence and treatment, appropriate to the environment, of tuberculosis, respiratory ailments, illness of the heart resulting from altitude, sudden death, sterility, infertility, etc. All these are factors leading to the deterioration of population and race.[42]

This quote shows how, following corporate interests, the emphasis was placed in environmental conditions rather than social and living conditions. In the 1950s important epidemiological studies were conducted in the high Andes. Although several topics were addressed in these studies, the emphasis was placed in respiratory, haematological and circulatory disorders. The broad disparities found between Coastal and high Andean mortality and morbidity rates were narrowed to geographic explanations, as the consequences of hypoxia. Socio-economic disparities, such as poor housing, deficient nutrition levels, lack of sanitary infrastructure, or higher rates of illiteracy, were only superficially addressed and in some cases completely ignored.

These studies found that certain types of cancer, such as lung cancer and leukaemia, were uncommon among populations living at high-altitudes.[43] On the other hand, heart diseases were found to be much more common. For instance, the incidence of congenital cardio-pathologies, particularly *patent ductus arteriosus*,[44] was found to be ten times higher at 4,000 metres than at sea-level. High rates of congenital cardio-pathologies were attributed to hypertrophy of the right side of the heart produced by oxygen deprivation.[45] Functional nervous disorders were also found to be much more common in highland regions. States of mental liability, weakness and irritability, as well as hysteria, were found to be high in the Andes.[46] While rare in coastal cities, apoplexy was common in Cerro de Pasco, and this was related to the poor nutrition of the brain caused by insufficiently oxygenation of the blood.[47]

Respiratory infections, such as tuberculosis, were particularly common amongst high altitude populations. Tuberculosis was considered to be absent from high altitude regions due to the purity of the air and the effects of low oxygen in the expansion of the lungs. However, studies performed in the 1950s found an elevated rate of tuberculosis amongst the population of Cerro de Pasco, a situation that was attributed to contact with "white" infected people. In an interview with a Rockefeller Foundation officer, a Peruvian doctor explained this phenomenon in the following terms: "Tuberculosis is just getting into high altitudes. For many years the Andeans were more or less isolated, now the white man is going up into high-altitudes with tuberculosis".[48] This vulnerability to TB was attributed to both the fact that Andeans lacked immunity to the disease due to lack of previous exposure; and to the rapid process of urbanization and industrialization and "Western acculturation" produced with the development of the mines, including new dietary and living habits, that were making this population more vulnerable to the disease.

Blood disorders were much more common as well. Sickle cell anaemia became aggravated as elevation increased and while native populations had developed immunity to the disease both Monge senior and Monge junior considered that non-native populations were particularly vulnerable. In 1966, in their influential book *High-Altitude Diseases,* Monge and Monge Cassinelli attributed the absence of a black population in the high Andes to the vulnerability of the "black race" to the effects of hypoxia. For the case of sickle cell anaemia, altitude demography and the black populations they wrote:

> When exposed to high altitude, asymptomatic cases of sickle cell disease
> may develop acute abdominal crises due to splenic infraction. In Peru,
> where there is a fairly large Negro population in sea level areas and where

altitudes of over 4,000 meters (13,200 feet) can be reached in a few hours by automobile or train, conditions for developing this clinical picture are present. This disease may be responsible for the surprisingly small Negro population inhabiting high altitude areas in Peru, and this despite the accessibility of the mountains and a long time presence of Negroes in sea level locations.[49]

In the 1960s two surveys conducted in the Peruvian Andes, one by Pennsylvania State University physical anthropologist Richard Mazess and one by Harvard University physical anthropologist Jean McClung found that while mean birth weight was reduced, neonatal mortality (deaths in the first four weeks of life) notably increased at high-altitudes.[50] These two aspects were related, since low birth weight (less than 2,500 grams) was considered to be a main cause of neonatal mortality. Infants of low birth weight were also more susceptible to fatal respiratory ailments and hypothermia. This higher rate of neonatal mortality was also attributed to the impairment of the mother's ability to properly oxygenate the foetus, the high rate of congenital heart malformations and pathological changes in the pulmonary arteries. The heartbeat of the high-altitude neonatal was also slower than that observed at sea-level.[51] The depressing effect of high-altitude in birth weight (about 400 grams average) was demonstrated to be much higher than that of malaria (300 grams), severe malnutrition (220 grams) or cigarette smoking during pregnancy (220 grams).[52] Together with increased neonatal mortality, post-neonatal mortality (deaths between one and twelve months) also correlated with altitude (being higher at higher altitudes). While in Lima the percentage of post-neonatal mortality was of 13%, McClung found that in the city of Cuzco it was of about 17%. McClung further suggested that post-neonatal mortality was probably higher in rural Andean populations with no access to proper medical assistance. The main causes of post-neonatal mortality were respiratory complications (which accounted for about forty percent of total infant deaths) and cardiovascular malfunction induced by hypoxia.[53]

The fact that hypoxia was found to be the main determinant of infant mortality among the Andean native population, regardless of socio-economic status, maternal care, or obstetrical practices, led to a fatalistic approach to the problem. Medical practitioners considered improbable the possibility of introducing medical technologies capable of counteracting the effects of hypoxia in infant mortality, such as providing supplemental oxygen to pregnant women, moving mothers to lowland hospitals during the last stages of pregnancy and birth, or longer hospitalization for mothers and longer postpartum stays for neonates among the marginalized areas of the high Andes. It was further suggested that as hypoxia was an inherent

element of high-altitude environments, likewise infant mortality was a
natural concomitant of high altitude life.

Scientific research, thus, helped to reinforce in Peruvian social and
political imagination the notion that there were "two countries in one" and
that in addition to cultural differences (such as language, clothing or
lifestyles), there were also hidden physiological and medical disparities
between coastal and Andean populations, reinforcing the notion that while
Andean natives belonged to the highlands, "white-Peruvians" and "Afro-
Peruvians" belonged to coastal locations, bolstering racial stereotypes. A
predominant view throughout the twentieth century was that of the
medical distinctiveness of high altitude environments. This was observed
in the dynamic community of Peruvian physicians and clinical researchers
working at high Andean hospitals during the first half of the twentieth
century, who were convinced that altitude produced unique medical
problems, and hence, demanded a particular body of medical knowledge.
They maintained that medical treatments, surgical procedures, and medical
texts prepared for sea-level populations had to be adapted to the conditions
imposed by a hypoxic environment. There was also the widespread notion
amongst local medical practitioners that this knowledge had to be
produced locally, given that in Europe medical scientists were only
concerned with the effects of hypoxia during short ascents amongst
mountain climbers or pilots, rather than with the effects of hypoxia in
permanent high-altitude settlers.

Conclusion

In this article I have highlighted the relation between the emergence of the
mining compounds of the high Andes and the visibility of high-altitude
diseases. This relation was produced at two levels. The expansion of the
mining compounds of the high Andes encouraged important migratory
movements from moderate to extreme altitudes and the assimilation of
entire peasant communities as a workforce in the mines. In addition to
their transformation from peasants to miners, Andean workers had to
confront extremely poor living and working conditions, inadequate water
supply, high-levels of atmospheric pollution, and rapid changes in their
traditional lifestyles and diet, with significant consequences in their health
conditions. At the same time, the establishment of modern hospitals in
these mining districts provided medical researchers with hundreds of
patients, laboratories and the most sophisticated medical technology in an
otherwise very inhospitable setting for scientific research. The Chulec
Hospital became a leading international centre for the study of high-

altitude related disorders and for medical innovation. In fact, it was in the Chulec Hospital in the late 1920s and 1930s where the first cases of CMS and HAPE were observed. Subsequently, new cases of these diseases were observed in other Andean countries, and in India and Pakistan. The case of the Chulec Hospital is also important because it shows the role private corporations had in shaping scientific research in a "peripheral" country like Peru. Medical doctors working within the Andean mining compounds oriented their research towards the effects of hypoxia in health rather than the effects of pollution or hard working and poor living conditions both because it responded to their scientific interests and because of their aim to maintain positive relations with the North American Cerro de Pasco Corporation.

Notes

[1] Until the early twentieth century there was no such thing as "high-altitude diseases". When nineteenth century medical geographers, climatologists or colonial physicians used the term mountain diseases or diseases from altitude climates, they referred to the increased incidence of disease when occurring in mountainous regions. This was the case, for instance, with goitre and cretinism. The only proper altitude disease was mountain sickness, the name given to describe the temporary ailments, such as headaches, nausea, and fatigue that most people suffer when exposed to rarefied air. High altitude areas, defined as over 2,500 meters, are characterized by numerous atmospheric challenges, such as low temperatures, aridity, high levels of ultraviolet radiation, and, most importantly, low oxygen pressure. A study performed by the World Health Organization in 1960, the closest to my period of study, found that approximately 25 million people lived at high-altitudes. WHO (1968). *Meeting of Investigators on Population Biology of Altitude.* Washington D.C. WHO, 1.

[2] Although medical historians have generally ignored the field of high altitude medicine, a few notable exceptions are West J. (1998). *High Life: A History of High-Altitude Physiology and Medicine.* New York: OUP, which provides a very detailed timeline on the history of the development of ideas in altitude physiology and an account of the most influential high-altitude scientific expeditions. For the introduction of biomedicine in the Himalayan region see McKay, Alex (2005). "The Birth of a Clinic? The IMS Dispensary in Gyantse (Tibet). 1904-1910". *Medical History* 49: 135-54. Marcos Cueto analyses the case of the Institute of Andean Biology, uncovering the institutional, social and political factors that allowed scientists from peripheral countries to produce relevant scientific knowledge. His pioneering work on the social history of medicine in Latin America informs this research. Cueto, M. (1989). "Andean Biology in Peru. Scientific Styles on the Periphery". *Isis* 80: 640-58. Ana Cecilia Rodriguez

analyzes the Mexican case, focusing on the work of Daniel Vergara Lope.
Rodríguez, A. (2002). "Daniel Vergara Lope and Carlos Monge Medrano: Two
Pioneers of High Altitude Medicine". *High Altitude Medicine and Biology* 3: 299-
309.
[3] On the "new history of medicine" in Latin America see Cueto, M. (2006). *Salud,
cultura y sociedad en América Latina*. Lima: IEP; Hochman, G. (2004). *Cuidar,
Controlar, Curar: ensaios historicos sobre saude e doenca na America Latina e
Caribe*. Rio de Janeiro: Fiocruz; Palmer, S. (2003). *From Popular Medicine to
Medical Populism: Doctors, Healers and Public Power in Costa Rica, 1800-1940*.
Durham: Duke; Sowell, D. (2001). *The Tale of Healer Miguel Perdomo Neyra:
Medicine, Power and Ideologies in Nineteenth Century Andes*. Wilmington: SR
Books; Cueto, M. (2001). *The Return of Epidemics*. Aldershot: Ashgate; Birn, A.
and Solorzano, A. (1999). "Public Health Policy Paradoxes: Science and Politics in
the Rockefeller Foundation's Hookworm Campaign in Mexico in the 1920s".
Social Science and Medicine 49 (9): 1197-1213; Peard, J. (1999). *Race, Place and
Medicine: the Idea of the Tropics in Nineteenth Century Brazilian Medicine*
Durham: Duke.
[4] In 1940 the population living at high-altitudes in Peru constituted 66% of the total
national population, estimated in 7 million. By 1961 this percentage had reduced to
55%. Donayre, J. (1966). *Population Growth and Fertility at High-Altitude*.
Washington: PAHO, p. 74.
[5] WHO. *Meeting*, 11.
[6] Carlos Monge (1884-1970) was born in Lima, where he attended the Faculty of
Medicine of the San Marcos University. After graduating he spent a year in the
London School of Tropical Medicine and Hygiene, and in 1924 he obtained the
chair of medical nosography in the San Marcos University. In 1931 Monge
founded the Institute of Andean Biology and Pathology, devoted to the study of the
effects of altitude in the human body. For the life and works of Monge see Cueto.
Andean Biology, pp. 643-644. Monge, C. (1928). "La Enfermedad de los Andes".
Anales de la Facultad de Medicina 11 (1): 1-73.
[7] Cueto. *Andean Biology*, pp. 646-7.
[8] Monge, C. (1940). "Life in the Andes and Chronic Mountain Sickness". *Science*
New Series 95 (2456): 79-84.
[9] My work is partly informed by recent scholarship on the role of science in
shaping national identities. Among others: Koerner, L. (1999). *Linnaeus: Natura
and Nation*. Cambridge: Harvard University; Auerbach, J. A. (1999). *The Great
Exhibition of 1851: A Nation on Display*. New Haven: Yale University Press;
McCook, S. (2002). *States of Nature*. Austin: University of Texas Press.
[10] Abeyta, L., (2005). "Resistance at Cerro de Pasco: Indigenous moral economy
and the structure of social movements in Peru". unpublished PhD Dissertation.
University of Denver, p. 83.
[11] In 1902 the financier James Ben Ali Haggin, the "greatest mine owner on earth",
organized a dinner party in New York with some of the wealthiest magnates in

America, such as J. P. Morgan and Henry Clay Frick, and decided the establishment of the Cerro de Pasco Mining Company (later known as the Cerro de Pasco Copper Corporation). For a history of this corporation see Abeyta. "Resistance at Cerro de Pasco", p. 139. The Pasco Corporation was expropriated by the Peruvian state in 1974 during the military government of Juan Velasco Alvarado.

[12] In 1940, the Cerro de Pasco Corporation accounted for 60% of Peru's silver output and 25% of its gold. A huge figure if we take into account that in 1940 Peru was the fourth largest producer of silver in the world. Another figure that well illuminates the magnitude of the mining operations in Pasco is that of the production of copper. The production of copper increased from 10,000 tons in 1903 to 35,000 tons in 1915, and to 53,000 tons in 1928. In addition, operations also involved the extraction, smelting and refining of zinc and gold. Flores Galindo, A. (1974). *Los mineros de la Cerro de Pasco 1900-1930.* Lima: PUCP, p. 35.

[13] This workforce came mostly from the districts of Jauja, Concepción y Huancayo, located in the central high Andes between 2,500 and 3,000 metres above sea level. Galindo. *Los Mineros*, p. 112.

[14] By 1925 the Corporation had bought around 500,000ha of land previously owned by peasant communities in the central Andes. Ibid.

[15] However, as Abeyta has pointed out, it was a much segmented labour force. North American administrators and engineers had houses separated from the mining workers with all modern facilities, with schools for their children and clubs for themselves. As indicated by Abeyta, while indigenous people were recruited "to go to the bottom of the mines", management was restricted to North American operators. Abeyta. "Resistance at Cerro de Pasco", p. 210.

[16] Galindo. *Los Mineros*, p. 53.

[17] It was estimated that in 1950 around 43% of the population was not originally from Cerro de Pasco. Arregui, A.; León Velarde, F., and Valcárcel, M. (1990). *Salud y Minería. El riesgo del mal de montaña crónico entre mineros de Cerro de Pasco.* Lima: ADEC-ATC, p. 41. Unless stated otherwise, all translations from Spanish are by the author of this paper.

[18] Mayer, D. (1913). *The Conduct of the Cerro de Pasco Mining Company.* Lima, p. 76. Between 1907 and 1920, 527 miners from the Pasco Corporation were reported to have died in work accidents. Abeyta. "Resistance at Cerro Pasco", p. 220.

[19] Mayer. *The Conduct*, p. 12.

[20] Rycroft, S. (1946). *Indians of the High Andes: Report of the Commission Appointed by the Committee on Cooperation in Latin America to Study the Indians of the Andean Highland, with a view to establishing a cooperative Christian enterprise.* New York: Sowers Printing, p. 51.

[21] According to Abeyta changes in worker management and labour relations were implemented according to a standard of what was beneficial to the company rather

than what was beneficial to the worker. Thus, "providing a minimally healthy environment or workplace education for labour were not company concerns until it was shown that such programs could favourable affect profits" as in the need for a more stable population. Abeyta. "Resistance at Cerro de Pasco", p. 220.

[22] Quoted by Kruijt, D. and Vellinga, M. (1983). *Estado, clase obrera y empresa trasnacional. El caso de la minería peruana, 1900-1980*. Mexico: Siglo XXI Editores, p. 55.

[23] The Chulec Hospital was actually built not in the city of Cerro de Pasco but in the smaller town of La Oroya, which was also part of the mining compounds of the Cerro de Pasco Corporation.

[24] Misses Smith, Morris, Baduacio and Gavino. *Los Andes*, Cerro de Pasco, 901 (1921) 3.

[25] The dental clinic and the ward for infants were built in the 1950s. *La Antorcha* Cerro de Pasco (1954), 2.

[26] "Ayuda Mutua". *El Serrano*. La Oroya 5 (1950), 31.

[27] "Harold Crane, chief surgeon". *El Serrano*. La Oroya 4 (1949), 1-2.

[28] The term HAPE was first utilized by North American physician Charles Houston in 1960. West. *High Life*, p. 161.

[29] Crane, H. (1927). "Soroche. Mountain sickness anoxemia". *Anales de la Facultad de Medicina*, 2: 306-12.

[30] Hurtado, A. (1937). "Aspectos Fisiológicos y Patológicos de la Vida en Altura". *Revista Médica Peruana* 9: 3-52. According to John West an earlier description of high-altitude pulmonary oedema appeared in Thomas Ravenhill, T. (1913). "Some experiences of mountain sickness in the Andes". *Journal of Tropical Medicine and Hygiene* 16: 313-20. However this article went unnoticed. Thomas Ravenhill worked between 1909 and 1911 as medical officer to the Poderosa Mining Company in the Chilean Andes. West. *High Life*, pp. 148-50. This condition was found to be particularly common during the conflict between India and China that took place in the Himalayas in the 1960s. West. *High Life*, pp. 148-9.

[31] Hurtado. *Aspectos Fisiológicos*; West. *High Life*, p. 148.

[32] Hurtado, Alberto to Allan Gregg. Oroya, March 8, 1936. RAC, RF, RG 1. 1., Series 331. Folder 5, Box 1.

[33] Cueto. *Andean Biology*, p. 646-7.

[34] Stein, S. (1980). *Populism in Peru: The Emergence of the Masses and the Politics of Social Control*. Wisconsin: University of Madison-Wisconsin Press.

[35] Walker, W.F., Assistant Manager of Operation, to Alberto Hurtado. 28 August 1946. Alberto Hurtado's papers, Catedra Alberto Hurtado, UPCH. He also received economic support of the Corporation for the construction of a research station in Morococha some years later. Alberto Hurtado to Julian Smith, Cerro de Pasco Corporation. June 30, 1953. Alberto Hurtado's papers, Catedra Alberto Hurtado, UPCH.

[36] Guerra Garcia, R. (2001). *Alberto Hurtado*. Lima: UPCH, p. 73.

[37] Hurtado to Ross, May 2, 1949. Alberto Hurtado's papers, Catedra Alberto Hurtado, UPCH.

[38] Lizarraga, L. (1954). *Soroche: Edema Agudo del Pulmón*. Lima: Universidad de San Marcos. A copy of Lizarraga's dissertation is located in the High-Altitude Medicine and Physiology Archives of the University of California, San Diego.

[39] Ibid., "Preface".

[40] Carlos Monge Cassinelli was born in Lima in 1921. He attended the Medical Faculty of the San Marcos University, and followed postgraduate studies at Johns Hopkins University. Once he returned to Lima he became Professor at the San Marcos University, researcher of the Institute of Andean Biology and later Director of the Cayetano Heredia Medical University Institute for High-Altitude Research. He was also named Chancellor of the Cayetano Heredia Medical University.

[41] Monge Cassinelli, C. (1966). "Natural Acclimatization to High-Altitudes: Clinical Conditions". In Donayre. *Population Growth*, pp. 46-9.

[42] Monge, C. (1948). *Acclimatization in the Andes. Historical Confirmations of Climatic Aggression in the development of the Andean Man*. Baltimore: The Johns Hopkins Press, p. 12.

[43] Hurtado. *Aspectos Fisiológicos*; Kruger, H. and Arias Stella, J. (1964). "Malignant tumours in high altitude people". *Cancer* 17: 1340-47; Marticorena, E. (1959). "Influencia de las grandes Alturas en la determinacion de la persistencia del canal arterial". *Revista de la Asociación Medica de Yauli* 4 (1-2): 37.

[44] PDA is a congenital heart problem whereby there is an abnormal circulation of blood between two of the major arteries of the heart, the aorta and the pulmonary artery.

[45] Hellriegel, K. (1956). "Ductus arterioso persistente: factores de la altura influencian la incidencia de ciertos defectos congénitos del sistema cardiovascular y en especial del conducto de Botal". *Revista de la Asociación Médica de la Provincia de Yauli* 2: 16-21; Recavarren, S. and Arias Stella, J. (1962). "Right Ventricular Hypertrophy in Native Children living at High Altitude". *American Journal of Pathology* 41: 55-64.

[46] Monge, C. and Monge Cassinelli, C. (1966). *High Altitude Diseases: Mechanism and Management*. Illinois: Charles Thomas Publisher.

[47] Ibid., pp. 67-74.

[48] Visit to the Lima laboratory of the National Institute of Andean Biology. GWG interview with Alberto Hurtado, August 18, 1953. RF, RG 1.1, Series 331, Box 2, Folder 16.

[49] Monge and Monge. *High Altitude Diseases*, p. 59.

[50] In 1965 Richard Mazess found that neonatal mortality in the highland districts of Peru was almost double than in the lowland regions, around 50-60 deaths per 1000. Mazess, R. (1965). "Neonatal Mortality and Altitude in Peru". *American Journal of Physical Anthropology* 23: 209-14; McClung, J. (1969). *Effects of High-Altitude on Human Birth*. Cambridge: Harvard University Press, p. 9.

[51] McClung. *Effects of High-Altitude*, p. 47.

[52] Ibid., p. 75.
[53] Ibid., p. 76.

PART V

SCIENCE IN THE PERIPHERY

SCIENCE IN THE PERIPHERY

XIMO GUILLEM-LLOBAT

The categories of centre and periphery have been invoked for decades by scholars from many disciplines. Historians, anthropologists, sociologists, economists and political scientists have used these terms to analyse the relations between different geographical, political and economic areas, as well as between institutions and social groups.[1] In the history of science, the standard version of the centre-periphery model considers that there is a narrow set of countries – namely France, Germany and Britain, and also the USA in the 20th century – which are the centre of innovative research in science. The role of the other countries – the periphery – is limited to a passive reception of the science produced in the centre, and the quality of their science is measured by the faithfulness of the reproduction.

Although this model of analysis is still – explicitly and too often implicitly – commonly used, it has also been thoroughly criticized. The papers included in this part are meant to contribute to this debate by providing particular case studies where these categories are challenged or problematized. This introduction intends to frame them by providing a synthetic and analytical genealogy of this debate. Accordingly, I will first outline how the centre-periphery model has evolved during the last decades and discuss its validity as a model to explain scientific practices and their development. This framework aims at answering a genuinely historical question: is the centre-periphery model a historical object, based on the perceptions of historical actors, or is it instead an analytical construct based on intellectual, cultural and political trends in current scholarship?

It is a difficult question – and out of the pretentions of this chapter – to determine the origins of the centre-periphery model. However, it seems certain that its use in the 1960s and 1970s in disciplines such as economic geography and international policy contributed to the promotion of its use in history of science.[2] In this period, outstanding sociologists supporting functional structuralism, such as Edward Shils, applied these categories to the study of society, and in particular to international relations driven by colonialism.[3] From the 1970s, *Minerva* – a journal devoted to international science policy, founded and closely controlled by Shils as its editor[4] – became a locus for studies using the centre-periphery model as analytical frame. Many of them had a strong historical tone and habitually focused on France, Germany, and Britain.[5] The sociologist Joseph Ben-David wrote a paper in Minerva – later enlarged into a book – in which the centre-periphery model was used to compare nineteenth-century academic structures in France, Britain and Germany.[6] Ben-David's work was widely read by historians of science, and had a lasting influence in the field.[7] Between the 1970s and the 1990s, this model became a common frame adopted by historians of science with a wide range of ideologies.[8]

The realm of the centre-periphery model is not strictly limited to that of scholars using this terminology in their works. As already mentioned, its use is very often implicit rather than explicit. In fact, the active creation/passive reception process was already promoted through a highly influential paper, "The Spread of Western Science", by historian of technology George Basalla, published in Science in 1967.[9] Like the centre-periphery model, Basalla's three-stage model of scientific development subdued scientific development to the reproduction of the cultural values of the centre in the periphery, applied in this case to former colonies of a restricted set of European countries responsible of defining science as a Western phenomenon.

Although Ben-David's and especially Basalla's models had a particularly important focus on colonial science, they also defined – often more implicitly than explicitly – a similar divide within European countries and strongly influenced the development of history of science studies on both sides of the divide. Accordingly, historians were either interested in the individual or the group from which ideas spread (in that case, they talked of "transmission" in terms of "influence") or in the "receptors" and the process of "reception".[10] In both cases scientific ideas and practices were considered as immutable entities produced in the centre and transmitted to a periphery deprived of any active role. Centre and periphery were presented as invariable in time and space, and their identity was ambiguously defined, not clearly distinguishing between geographical, political, economic, technological, social, cultural or intellectual features. Moreover, transmission of knowledge to the periphery supposed that any change introduced was seen in a negative manner and instead, faithfulness was the measure of success.[11]

From the 1990s, however, the centre-periphery model and transmission/reception studies were explicitly criticized.[12] Roy MacLeod urged historians of colonial science and medicine to abandon old assumptions about centre and periphery,[13] and instead to reconceptualize them in the frame of the "moving metropolis", the dynamic and scattered assembling and dismantling of imperial projects.[14] In this sense, Paolo Palladino and Michael Worboys have questioned the unidirectional flow of knowledge, pointing out that imperialism shaped the development of metropolitan science itself.[15] Changes brought by imperialism were not only due to the access to the natural resources of colonies and the appearance of new political interests, but also to the contribution of the colonized cultures. Palladino and Worboys argue, for example, that Indian researchers produced scientific knowledge that was highly valued in the metropolis, and which was influenced by indigenous Indian traditions of natural knowledge.

As underlined by Kapil Raj, nowadays historians of science and empire have come to conceive knowledge as a co-construction involving both the metropolis and the dominated colonies. Since the late 1960s, some historians have also used the anthropological notions of "acculturation" and "cultural encounters". Today, this perspective considers encounters between cultures as interactions in which it is necessary to take into account all agents – in one and other side of the divide – in order to understand cultural change. Originally applied to the Western interaction with primitive people, then to imperial metropolises and their colonies, this approach could now find its way in the analysis of the making of knowledge within Europe itself.[16]

In considering the "European periphery", the influence exerted by the periphery over the centre seems even more evident. Roberts has recently pointed out that "the omnipresence of travel and exchange in the history of science shows 'centres' to exist in a state of productive co-dependence with their 'peripheries'".[17] And it is worth to acknowledge how scientists from the "European periphery" who travelled to the "European centre" influenced those scientists and institutions in which they stayed. Even though those "peripheral" scientists perceived their travels as a means of acquiring the tools of modernization, they were also carriers of a knowledge that may have been influential in their respective destinations.[18]

The active role of the European "periphery" in the production and appropriation of knowledge is also one of the main foci of international research programmes such as Science and Technology in the European Periphery (STEP). Scholars from countries traditionally considered to be peripheral in the development of science, such as Denmark, Sweden, Russia, Turkey, Belgium, Greece, Spain and Portugal, came together in this project to challenge the current predominance of hegemonic historiographies.[19] According to the STEP foundational document, "there are many centres and many peripheries, and they change in time, spaces, and disciplines. Depending on the subject one is discussing – a place may at one and the same time be both centre and periphery. A centre may, over time, change into a periphery, and vice versa. And a single country may contain both centres and peripheries, thereby making purely national distinctions of dubious use".[20] As stated in the introduction to part IV of this volume, national boundaries are too often taken for granted by scholars, producing accounts where the "national" is an uncritical category implicitly accepted, and the "international", a category too often implicitly dismissed, without proper historiographical justification. In parallel, it is fundamental to understand how scientific knowledge circulated within and across national borders, who the major agents were in this communication process,[21] and how this in fact conditioned the making of scientific knowledge.

As argued in the general introduction to this book, James Secord's proposal of making the communication of science central to the conceptual frame of history of science can contribute to renew and to constitute a more sophisticated and accurate frame for the historiography of science. In this context, "transmission" and "reception" can be replaced with "communication" and "appropriation". The former involves an interactive process without a clearly privileged direction. The latter focuses on the ways in which knowledge is used, considering that these are active processes creating new meanings and products, independently of the original context of production. Both concepts are fundamental tools to revise and challenge the centre-periphery model. As stated by Gavroglu, by focusing in the process of appropriation historians can abandon the emphasis on how fully the centre is being reproduced in the periphery.[22] Instead, the "receptor" of those ideas and techniques (the periphery) recovers the active role denied by the traditional model. Accordingly, invention and appropriation are given an equal epistemological status, and they are connected through communication.

In this perspective, the history of medicine and technology in Spain offers excellent ground on which to confront the various aspects of the centre/periphery debate and test solutions contributing to the development of the historiography of science in an international context. Traditionally, the study of history of science in this country is frequently driven by the perception of its being peripheral to the practice of science. A longstanding controversy over the backwardness of Spanish science, which can be traced back to the eighteenth century, stills exerts a considerable influence among historians and in general on the public discourse about science. This controversy, which is also deeply interwoven with ideological confrontations,[23] is not restricted to the study of Spanish science, as it is a notorious symbolic source for Spanish scientists, who are – and have been for many decades – using it to stress their situation as members of a weak peripheral community. In recent decades, different scholars have tried to overcome the intellectual constraints caused by this controversy. On the one hand, some historians focused on recovering Spanish scientific heritage through rigorous empirical research and considering actors and sources in relation to both their local and their international significance.[24] More recently, a new generation of historians of science has tackled the question by relying on the methods of the new social and cultural history of science.[25]

The Iberian studies considered in this chapter can shed new light on the centre/periphery controversy following the terms of the discussion previously exposed. I will briefly highlight the main topic of each of the following papers, and will then discuss some of their principal contributions to the debate. Even though the three case studies composing this chapter are local in that they focus on particular contexts within the Spanish state, their relevance goes beyond their locality. According to David Chambers:

> If we think of science not as a knowledge factory successfully established only in certain strongholds but rather as a network of individuals, institutions and practices, reaching around the world and subject to many and diverse influences, then science and its cultural and socioeconomic relations, constitutive and contextual, if to be understood at all, must be understood in every locality in which it appears.[26]

Certainly, local studies have sometimes implied a loss of global perspective and have complicated the comprehension of the way knowledge travels.[27] However, by working in a comparative frame, these local studies offer the possibility of overcoming the constraints of locality.[28] This is a major aim of this part.

In his paper, Ximo Guillem-Llobat deals with the process of industrialization of food production and with the problems in food quality which appeared linked to this new system of production. His paper focuses on the Valencian Country but also deals with the traditionally considered "European centre". In its turn, Néstor Herran analyses the development of a radioactivity programme in twentieth-century Spain, with important applications to agricultural development. His study carefully places the work of chemist José Muñoz del Castillo – a central figure in this movement – in the European context of radioactivity research. Xavier Mañes, focuses instead on the development of crystallography as a research subject in Spain during the first half of the twentieth century, in the context of important social and political changes in Spanish society, including the Civil War and the establishment of an authoritarian regime.[29]

By focusing on their respective case studies all three papers contribute to problematize the centre/periphery divide, breaking with the transmission/reception model usually linked to the traditional conception of this divide. The "peripheral" scientists who appear in all three cases did not restrict their activity to passive perception tasks. They were active producers of original knowledge that strongly influenced science and technology in their communities. Therefore, although they were certainly dependent on knowledge developed in foreign communities – mainly located in the traditional centre – their activity is better defined through appropriation than through reception.

Appropriation took place not only in relation to new ideas but also to technological innovation. In this sense, Herran and Mañes show how active these "peripheral" scientists were in designing and reinventing the instruments they needed for their research. Moreover, the papers included in this chapter argue that the flow of information between centre and periphery was not unidirectional. For instance, Guillem-Llobat's paper reveals the contribution of Valencian scientists to the international science community through their participation in international conferences. As it is stated in Herran's and Mañes' papers, Iberian scientists often worked in isolation from the international community of scientists. However, these "peripheral" scientists were sometimes capable of contributing to first line debates, as argued by Guillem-Llobat.

The essentialist conception of the centre/periphery divide is also questioned in all three papers. Some of the approaches on which this divide is based have been highlighted in all three cases.[30] This divide was not innate and linked to the essence of each community but instead it had an origin in social, economic and political dynamics. In this sense, Herran considers some of the factors that prompted the origin of the peripheral condition of the development of radioactivity in Spain and the role played by its promoters. Thus he concludes that the special dynamics of Spanish institutions were determinant in reducing Spanish scientists to a peripheral condition. In Mañes' paper, the Spanish political framework is considered to be the major factor defining the role played by Spanish scientists in the international community. The change from the republican regime to the fascist dictatorship becomes an important factor in order to explain the isolation in which crystallography was developed in Spain. Moreover, the new economic organization linked to the fascist regime, the autarky, is also highlighted for its influence on the development of this discipline.

Finally, all three papers question the national basis of the centre/periphery divide. Mañes and Guillem-Llobat show that one place could be both centre and periphery at the same time depending on the region under consideration. Even as one local community was peripheral to another local community, it could also act as a centre for a third local community. Hence, the case studies by Mañes and Guillem-Llobat contribute to a relativization of the centre/periphery divide by showing the existence of double peripheries. This leads, in conclusion, to prompt a non-essentialist conception of the centre/periphery divide, where communication, appropriation and multi-polarity become some of the central guidelines.

Notes

[1] Fox, R. (2003). "Introduction". In Fox, coord. "Centre and Periphery Revisited. The Structures of European Science". *Revue de la Maison Francaise d'Oxford* 1 (2). [www.mfo.ac.uk/Publications/Revue%20Fox/Sommairefox.htm] (accessed 26 July 2007).

[2] For an overview of the topic see Gavroglu, K., et al. (2008). "Science and Technology in the European Periphery. Some historiographical reflections". *History of Science* 46: 1-23.

[3] Shils, E. (1961). "Centre and Periphery". In *The Logic of Personal Knowledge: Essays Presented to Michael Polanyi on his Seventieth Birthday 11th March 1961*. London: Routledge & Kegan Paul, pp. 117-130, and (1975). *Essays in Macrosociology*. Chicago: University of Chicago Press.

[4] Bulmer, M. (1996). "Edward Shils as a Sociologist". *Minerva* 34: 7-21.

[5] Relevant examples of these are: Ben-David, J. (1970). "The Rise and Decline of France as a Scientific Centre". *Minerva* 8 (1-4): 160-79; Gizycki, R. v. (1973). "Centre and Periphery in the International Scientific Community: Germany, France and Great Britain in the 19th Century". *Minerva* 11 (4): 474-93; Nye, M. J. (1975). "The Scientific Periphery in France: The Faculty of Science at Toulouse (1880-1930)". *Minerva* 13 (3): 374-403; Singer, B. (1982). "The Ascendancy of the Sorbonne: The Relations between Centre and Periphery in the Academic Order of the Third Republic". *Minerva* 20 (3-4): 269-300; Chayut, M. (1994). "The Hybridisation of Scientific Roles and Ideas in the Context of Centres and Peripheries". *Minerva* 32 (3): 297-308; Ashby, L. (1996). "Centre and Periphery in Academe: Some Personal Reflections". *Minerva* 34 (1): 95-101; Presas i Puig, A. (2005). "Science on the Periphery. The Spanish Reception of Nuclear Energy: An Attempt of Modernity?" *Minerva* 43 (2): 197-218.

[6] Ben-David, J. (1971). *The Scientist's Role in Society: A Comparative Study*. Englewood Cliffs, N.J.: Prentice-Hall.

[7] See introduction to part II and Faidra Papanelopolou's paper in this volume.

[8] Homburg, E. (2004). "Shifting Centres and Emerging Peripheries: Global Patterns in Twentieth-Century Chemistry". *Ambix* 52 (1): 3-7. See for example: Pallo, G. (1987). "Some Conceptual Problems of the Center-Periphery Relationship in the History of Science". *Philosophy and Social Action* 13: 27-32, and Polanco, X. (1990). *Naissance et développement de la science-monde. Production et reproduction des communautés scientifiques en Europe et en Amérique Latine.* Paris: Editions La Découverte.

[9] Temkin, O. (1977). "Comparative Study in the History of Medicine". In Temkin, ed. *The Double Face of Janus, and other Essays in the History of Medicine.* Baltimore: Johns Hopkins University Press, pp. 126-36; Ziman, J. M. (1969). "The Rutherford Memorial Lecture, 1968: Some Problems of the Growth and Spread of Science into Developing Countries". *Proceedings of the Royal Society of London. Series B, Biological Sciences* 174 (1034): 69-89.

[10] Dolby, R.G.A. (1977). "The Transmission of Science". *History of Science* 15: 1-43; Fox. "Introduction".

[11] Gavroglu et al. "Science and Technology in the European periphery".

[12] Chambers, D. W. (1993). "Locality and Science: Myths of Centre and Periphery". In Elena, A.; Lafuente, A., and Ortega, M. L., eds., *Mundialización de la ciencia y cultura nacional.* Madrid: Ediciones Doce Calles, pp. 605-17, and MacLeod, R. (1993). "The Worldwide Diffusion of Science". In Ibid., pp. 735-7.

[13] MacLeod, R. (1987). "On Visiting the 'Moving Metropolis': Reflections on the Architecture of Imperial Science". In Reingold, N. and Rothenberg, M., eds. *Scientific Colonialism: A Cross-Cultural Comparison.* Washington, D.C: Smithsonian Institution Press, pp. 217-49.

[14] Anderson, W. (2003). "How's the Empire? An Essay Review". *Journal of the History of Medicine and Allied Sciences* 58 (4): 459-65.

[15] Palladino P. and Worboys, M. (1993). "Science and Imperialism". *Isis* 84 (1): 91-102

[16] See for example Glick, T. F. and Pi-Sunyer, O. (1969). "Acculturation as an Explanatory Concept in Spanish History". *Comparative Studies in Society and History* 11 (2): 136-54; Raj, K. (2007). *Relocating Modern Science: Circulation and the Construction of Scientific Knowledge in South Asia and Europe, 1650-1900.* Basingstoke: Palgrave Macmillan, and Buklijas, T. and Lafferton, E. (2007). "Science, medicine and nationalism in the Habsburg Empire from the 1840s to 1918". *Studies in History and Philosophy of Biological and Biomedical Sciences* 38: 679-86.

[17] Roberts, L. (2005). "Circulation Promises and Challenges". In *The Circulation of Knowledge and Practices: The Low Countries as an Historical Laboratory.* Workshop, Woudschoten, 27-28 May. [http://www.gewina.nl/werkgroep/]

[18] Simoes, A.; Carneiro, A., and Diogo, M.P., eds. (2003). *Travels of Learning: A Geography of Science in Europe.* Dordrecht: Kluwer.

[19] See [http://www.cc.uoa.gr/step/]

[20] STEP Founding Document, cited in Nieto-Galan, A. (2003). "Networks and Peripheries: Jean-Francois Persoz (1805-1868) and the 'Republic of Chemist-Dyers'". In Fox. "Centre and Periphery Revisited".
[21] Simoes; Carneiro, and Diogo (2003). *Travels of Learning.*
[22] Gavroglu, K. (2003). "The Centre from the Periphery: Appropriating Scientific Ideas During the Eighteenth Century". In Fox. "Centre and Periphery Revisited".
[23] Garcia Camarero, E. (1970). *La polémica de la ciencia española.* Madrid: Alianza Editorial; Barona, J. L. (1994). *Ciencia e Historia. Debate y tendencias en la historiografía de la ciencia.* València: Seminari d'estudis sobre la ciencia, and Nieto-Galan, A. (1999). "The Images of Science in Modern Spain. Rethinking The 'Polemica'". In Gavroglu, K., ed., *The Sciences in the European Periphery during the Enlightenment.* Dordrecht: Kluwer, pp. 73-94.
[24] For more information see, for example, López Piñero, J.M. (1979). *Ciencia y técnica en la sociedad española de los siglos XVI y XVII.* Barcelona: Labor.
[25] Nieto-Galan, A. (2006). "The History of Science in Spain: Imperial Past, Peripheries and the Making of the Modern State". *Neusis* 15: 50-74 (in Greek).
[26] Chambers. "Locality and Science".
[27] Secord, J. A. (2004). "Knowledge in Transit". *Isis* 95 (4): 654-672. See the general introduction to this volume.
[28] Jacob, M.C. (1999). "Science Studies after Social Construction: The Turn toward the Comparative and the Global". In Bonnell, V. E. and Hunt, L., eds. *Beyond the Cultural Turn: New Directions in the Study of Society and Culture.* Berkeley and Los Angeles: University of California Press, pp. 95-120.
[29] During the twentieth century the Restoration was followed by a first dictatorship, a republican regime, a civil war and finally a second dictatorship.
[30] In all three cases there was a certain delay in the initial steps of research. Scientific and technological dependence was more frequent in those "peripheral" communities, and some of the main scientists discussed in these papers are presented as professionals who perceived themselves as peripheral scientists.

FOOD QUALITY CONTROLS IN THE EUROPEAN PERIPHERY:
VALENCIAN SCIENTISTS AND LABORATORIES IN THE LATE NINETEENTH CENTURY

XIMO GUILLEM-LLOBAT

The second half of the nineteenth century was a period of change in European food production. These changes, which occurred alongside the processes of industrialization, posed new problems in food quality which galvanized scientists, politicians, and judges. The debate on food quality prompted new scientific studies, and drove innovation in this particular field of science.

The international dimension of this topic, and particularly the fact that food quality control affected a large number of very different countries, makes it central to the centre/periphery debate. Food analysis was developed in each European country in a different context. However, in the end, both the methods used and the organization of local services for food fraud detection were quite similar. This was probably due to the existence of several international scenarios where food fraud was discussed and negotiated. In this paper I examine the development of food fraud regulation in three regions traditionally considered part of the scientific centre (Great Britain, France and Germany) and one "peripheral" region (the Valencian Country). To do so, I consider how the problem was first identified, how new laws were passed and how new professionals and centres became responsible for food quality control. Later on, I consider how these changes occurred in each of the aforementioned regions, and I deal with the international scenarios where food fraud regulations were discussed. The comparative study and the analysis of the principal features of these international scenarios finally allow for a criticism of the validity of the centre/periphery divide for this particular case study.

Food quality control: an international perspective

During the nineteenth century, the food chain, including every step from food production to consumption, was deeply modified by industrialization. This effect was produced at different speeds all over Europe. With the development of new foodstuffs, faster transport systems and better methods of food conservation, the food chain spread through larger geographical areas and incorporated new steps. One of the consequences of these changes was the impoverishment of food quality due to the increase in food adulteration and alteration.[1]

The impoverishment of food quality was due to a number of factors. The increase in the number of wholesalers intervening in the food supply process from production to consumption favoured the practice of adulteration. Moreover, further knowledge about chemical substances opened new and more sophisticated avenues for adulteration. The alteration of food was also aggravated in this new context. Foodstuffs were exposed to the environment for longer periods, and the likelihood of bacteriological contamination was therefore higher.

As food quality became a problem for public health and fair trade, several authors in various countries started to report on this situation. The debates they contributed to had various outcomes. It seems that even though the first denounces of food impoverishment were not followed by measures of control, by the second half of the century some changes had taken place. I will now briefly consider some of these changes.

One of the authors frequently highlighted for his early contribution to the debate on food adulteration was Frederick Accum. In 1820, his famous *Treatise on Adulterations of Food and Culinary Poisons* was published in England.[2] The book had great success, its first edition being sold out within one month.[3] Notwithstanding this, the social impact of this treatise was not followed by any change in English food legislation. In other European regions, like France or Germany, other books on food adulteration were published in the following years.[4] However, apparently none of them had relevant consequences for state food regulations.

Only by the mid-nineteenth century did some works start to awaken the interest of certain local authorities. This was the case, for example, with the reports published in *The Lancet* between 1851 and 1854 by the Analytical Sanitary Commission.[5] These reports, which dealt with the adulterations practised in Great Britain, were not substantially different from other papers published in previous years. However, just one year after the last report was published, a parliamentary select committee was created in order to investigate food purity, and in 1860, the first food act was passed. This act was later followed by several others, the most important being the 1875 Sale of Food and Drugs Act.

In France, some major changes also took place, from the beginning of the second half of the nineteenth century onwards. Jean-Baptiste-Alphonse Chevallier was the central figure in the fight against food adulteration in this case. For several years this influential pharmacist waged an active campaign to enact more severe food and drug laws.[6] Finally, the National Assembly responded to Chevallier and his supporters by passing the 1851 special law. This law probably meant a step forward in the French regulation of food quality but there was still a long way to go. Only with the passing of the general food law of 1905, did France have a legal instrument comparable to the British 1875 Sale of Food and Drugs Act at its disposal.

The implementation of food regulations also occurred in the German Empire during the second half of the nineteenth century. In 1879 a general food law was published and for the first time uniform control of food and its adulterations was guaranteed all over the Reich.[7]

The rhythm at which food laws were passed in different regions gives us a hint of the differences in response to the problem of food adulteration and alteration in different economic, political and cultural contexts. If we compare the texts of these laws, we can also find important differences in the approach and the role played in food control by the authorities of each state, city or town. Some laws were stricter than others, some included more details in the definition of adulteration or in the establishment of mechanisms of control, and others gave a greater regulatory role to the market. Nevertheless, despite the differences, there was a general change in food legislation in "the big three"[8] which should be highlighted.

Those involved in food quality control also changed over the period. As I have already stated, food quality had been controlled for centuries. These controls had been developed by different historical actors. Inspectors dealing with food quality control had frequently had low academic credentials and they had depended on their knowledge of the organoleptic features of foodstuffs in order to analyse their quality. Even when physicians or pharmacists were involved in the analysis of foodstuffs, they mainly supported their study by organoleptic analysis. Their limited chemical knowledge rarely qualified them to attempt different approaches to food analysis.

During the second half of the century, the development of chemistry and bacteriology offered new possibilities in the study of the quality of foodstuffs. These new possibilities were very useful when dealing with the highly refined adulterations that appeared with the description of many new chemical compounds. However, the superiority of the chemical analysis over the organoleptic one was not always obvious.[9] In fact, in the transit from one type of analysis to the other some other factors intervened, such as the professionalization of chemistry with the emergence of societies, academies, publications and university studies; the adoption of a higher social status by professional science; the consolidation of an "expert" culture for the control of food quality; and the rise of the culture of precision measurement with instruments in science.

With the development of chemistry, new historical actors took a central role in food quality control. In France, for example, two new societies became important for food quality: the *Société Scientifique d'Hygiène Alimentaire*, involved in the publication of a journal and in the organization of several international meetings, and the *Société des experts-chimistes de France*, which was also published a professional journal. Meanwhile in Great Britain, the *Society of Public Analysts*, set up in 1874, was central to the introduction of a system of control where public analysts were the main agents in charge of food quality.[10]

The appearance of new experts in food quality control was common to "the big three". However, the kind of professionals involved in the fight against food adulteration and alteration was not always the same. For example, while veterinarians were very important in food quality control in Continental Europe, in Great Britain they had very few responsibilities in this field.[11] Nevertheless, when we take into account the main historical actors in charge of food quality, a general tendency can be established, despite the existence of several differences.[12]

Finally, before moving on to the Valencian case, I will describe one of the main changes in the food controls of the late nineteenth century: the creation of municipal laboratories. In the 1870s the first of these municipal laboratories were created in some European cities. These laboratories were established in order to fight food fraud and other kinds of commercial fraud; on some occasions they were also in charge of other duties related to public health.

The municipal laboratory of Brussels was one of the first created in Europe. This laboratory inspired the creation of many French laboratories in the late 1870s and early 1880s. In Germany, after the passing of the law of 1879, municipal authorities also prompted the establishment of a municipal service of this nature.[13] Meanwhile, in Great Britain the analysis of samples was organized around public analysts, especially after the passing of the 1875 Sale of Food and Drug Act, which included the necessary appointment of these professionals by municipal authorities.[14]

The creation of municipal laboratories mainly depended on the initiative of municipal authorities even though the state also promoted, in some way, the establishment of these centres. Once they were established, the municipal authorities were again the ones responsible for their operation. This fact led to great heterogeneity in the control of food quality not only between countries but also within countries themselves. The laboratories were responsible for the analysis of the samples sent either by professional inspectors or by consumers and merchants. Normally, their staff included chemists and pharmacists who could undertake the chemical analysis of the sample with the help of sophisticated instruments. Sometimes, physicians, engineers or veterinarians were also part of the staff. However, the staff was usually composed of only a couple of professionals, with a chemist being responsible for the detection of food fraud.[15]

We have now seen the response given to the rise of food fraud in countries traditionally considered part of the "centre". But did the "periphery" have a similar response? In the following narrative I shall attempt to answer this question by considering the case of the Valencian Country.

The Valencian food quality control

In the second half of the century, the Valencian Country, on the east coast of the Iberian Peninsula, still had an economy mainly based on agriculture. In rural areas, families mainly fed on the foodstuffs produced by themselves or their neighbours. Even in València city itself, the main occupations of the population were related to agriculture. Therefore, in comparison with more industrialized societies, Valencians had greater knowledge of the origin of much of the food they ate. In addition, the Valencian Country did not yet have a food industry like the one developed in British, French or German regions, and the consumption of industrially processed foodstuffs was arguably much lower. The Valencian context did not favour food adulteration and alteration as described for the aforementioned European countries. In fact, Valencian authors from the second half of the century were glad to assert the Valencian lower level of food fraud in comparison with France, Great Britain or Germany.[16]

However, the Valencian Country also saw an increase in food adulteration and alteration during the second half of the century. This impoverishment of food quality alarmed several Valencian professionals in the public health field. Thus, in the Valencian case, as I have done for the aforementioned European countries, I am going to expose the importance of the complaints raised by local actors in relation to the seriousness of food fraud.

The major voice in this context was Joan Baptista Peset Vidal (1821-1885). Peset Vidal was a Valencian physician and a member of one of the most famous families involved in the Valencian public health services in the nineteenth and twentieth centuries. He was a member of several societies and academies, and a distinguished member of the Institut Mèdic Valencià (the Valencian academy of medicine). One of his most relevant books was *Topografía Médica de Valencia y su zona.*[17]

This book covered many different aspects of Valencian society, including its geographical context, its diseases, its industry and its traditions. There was also a section devoted to the city's food supply. This section not only described the usual diet of Valencians but also dealt with the dramatic impoverishment of food quality in the city. The author highlighted the increase in fraud, paying special attention to the adulteration of flour, chocolate, milk, wine, vinegar, liqueurs, and food condiments. He pointed to several possible frauds suffered by these foodstuffs and considered which of these frauds were usual in Valencian markets. Peset Vidal argued that the situation was not yet as dramatic as it was in other European countries. Many of the adulterations he included in his text were not practised, in his opinion, in València.

Topografía Médica de Valencia y su zona was arguably an influential book in the Valencian professional context of medicine. However, we do not know exactly how influential the section dedicated to food adulteration was. Despite this fact, we can imagine that either this section or Peset Vidal's interest in food quality was probably a determinant factor in stimulating his son, Vicent Peset Cervera, and that this influence was not superfluous. As I will show, Peset Cervera was to become the most important actor in the consolidation of the new Valencian food quality control service.

We can assume that Peset Vidal's work had some influence on food adulteration over the Valencian food quality control service. However, this influence did not prompt a general food law similar to those we considered in the aforementioned countries. In València and in several towns of the region, some municipal regulations were passed in order to improve the food control system. However, it was only by the end of 1908, that a general food law was published. This was the first major law of this kind at a state level, and it had a lasting influence. The influence of Peset's ideas, over a law that was in fact passed decades after the first edition of his book, was therefore probably non-existent.

Moreover, Peset Vidal's book lacked relevant information needed to identify food adulterations. It gave an interesting picture of the situation of food fraud in the city of València but it did not deal in detail with existing mechanisms of identifying and prosecuting food fraud. In the Valencian Country, there was not a book of this kind until the publication of José Soler i Sanchez's *Analisis y ensayos de los alimentos y las bebidas y los condimentos.*[18] Peset Vidal's and Soler Sanchez's books were published locally and distributed nationally, a fact displaying the Valencian interest in food quality control and the creative contribution of Valencian authors to the Spanish context. Nevertheless, Valencian authorities and analysts were also informed by other books written in Spanish in previous years.

The number of books dealing with food analysis published in Spanish during the second half of the nineteenth century was quite high. The analysis of several medical bibliographical repertories, shows a list of almost a hundred titles directly related to this topic, published in this period. This fact can give us a clue of the importance of the topic in Spanish society. However, one point we must consider for our future discussion on the centre/periphery divide is the origin of these books. Even though many of the authors of the mentioned list were Spanish, the first major books published in Spanish were mainly translations of French works. In particular, I am referring to the treatise by Garnier and Harel, translated and published in 1846, and Chevallier's treatise, translated and published in 1854.[19] Nonetheless, translation and authorship are closely related practices and, in this case for instance, the translation of Garnier and Harel's treatise into Spanish constituted a new work, as Magín Bonet, doctor in pharmacy, and professor of industrial chemistry in Madrid, besides doing the translation, made additions to the work that, according to contemporary commentators, contributed to transform it into a complete treatise.[20]

Another question already addressed when dealing with "the big three" was the changes in the profile of the actors involved in food quality control. In the Valencian Country we can also identify the general tendency previously described for the aforementioned European countries. During the second half of the nineteenth century and mostly in the last two decades, new experts were usually appointed by municipal authorities as municipal pharmacists, chemists or physicians. They were in charge of the chemical analysis of food samples and, where a municipal laboratory existed, they were usually part of its staff. Veterinarians also gained increasing recognition, mainly in relation to fraud detection in meat and other foodstuffs obtained from animals. As time passed, they became more and more involved in food inspection and, with the creation of abattoirs in many towns of the Valencian Country, they found their place not only in inspection but also in sample analysis.

Therefore, contrary to what happened in former times, academic credentials became a prerequisite in order to hold major positions in municipal food control. Furthermore, where previously when a food sample was of doubtful quality just any physician or pharmacist was called, this was to change in the last decades of the century. Only special professionals who had passed a public examination, which included food analysis in its list of topics, were usually allowed to analyse food.

Of the new municipal posts in food quality control, the most important ones were those in the municipal laboratory. In the Valencian Country, apparently only València and Alacant had such laboratories. However, their contribution to food quality control was not limited to the cities where they were set up. Sometimes, these laboratories also analysed samples from smaller towns of the province. As I will show, these laboratories were essential for Valencian food quality control.

The Municipal Chemical Laboratory of València was created in 1881. Its creation was prompted by a complaint by several traders of the city market who reported that other traders were obtaining more competitive prices by adulterating the food they were selling. As a result, the city council hired a chemist who was given the task of detecting adulterations. Subsequently, this led to the creation of the laboratory.[21]

In its first years, the Laboratory was run by its director, the pharmacist Domingo Greus, and by the chemist and physician Vicent Peset Cervera.[22] They both worked from 9:00 a.m. until noon, analysing different kinds of samples. The number of food samples tested during those first years was not very high. With such a short working day and with two professionals only, sometimes assisted by the porter, Leandro Sear, no miracles could be worked. As would be expected, the number of tests was substantially smaller than the amount performed in Paris during the same period. If Peset in 1882 mentioned 40 food analyses in the Valencian laboratory from August to November,[23] in Paris more than ten thousand tests were undertaken in the same year.[24]

However, if we compare the volume of food tests done in València with the number of tests carried out in other towns in the Valencian Country, we can appreciate the importance of the task developed by the Valencian laboratory. In fact, as already noted, the second municipal laboratory was only then opened in Alacant.

In order to favour the creation of further municipal laboratories, the state and provincial authorities put pressure on other towns with a large number of inhabitants. The first Spanish general food law, passed in 1908, obliged the municipal authorities of cities with more than 10,000 inhabitants to set up a municipal laboratory. In the Valencian Country, cities like Xàtiva were already in this situation, so a municipal laboratory should have been immediately created. Moreover, the important influence of Xàtiva over the towns of the region could have justified this need still further. However, not only did the authorities of Xàtiva fail to fulfil their new duties, but they were also insensitive to the specific demand made, some years later, by the Spanish health authorities to the city mayor.

In Alacant, the municipal authorities did finally agree to create a municipal laboratory, although in this city, again, the establishment of the laboratory was not an easy job. The creation of the municipal laboratory of Alacant was basically a personal project impelled by the pharmacist José Soler i Sanchez. He had already been in charge of food analysis in Alacant and, when the laboratory was set up in 1887, he became its director. The laboratory was not as active as the Valencian one but, at least at the beginning, it remained quite busy. During the first eight years, almost one thousand food tests were performed. This level of analysis was again far from the number of food tests conducted in big laboratories like the Parisian one or even smaller ones like the laboratory of Brussels.[25] However, probably no higher figures could be expected with a part-time staff composed only of Soler i Sanchez and his assistant.

When Soler i Sanchez was replaced by his son as director, the activity of the laboratory decreased still further. At the turn of the century, the budget allocated to the laboratory became smaller and smaller, and, therefore, food quality control in Alacant worsened until new municipal projects were initiated in the second decade of the century.[26]

The measures impelled in order to control food quality were different in each town and city of the Valencian Country, as we have seen in the various examples of this paper. However, some general features can be drawn from these examples. The budgets allocated to these control services were very small. No municipal laboratories were created in towns where they should have been established in order to comply with the food laws in force. Moreover, when the municipal authorities were finally sensitive to their creation, the new centres were rather unstable. For instance, the staff complained about the budget and the poor infrastructure on which they were reliant. In addition, instability was also evident from frequent changes in the location of both municipal laboratories.

The laboratories of Alacant and València were both forced to move from one building to another on several occasions during the first years of their existence. In some of the premises the professionals did not seem to complain much about the facilities. However, if we consider, for example, the first years of the Valencian laboratory, we can identify, for each of its first three locations, a number of complaints raised by Vicent Peset Cervera. Peset Cervera criticized the small size of the rooms, the poor signage, which made them invisible to passers-by, the need for natural lighting, and so on.

Short budgets, the limited size of the staff, and the part-time nature of the employment, along with unstable locations, meant that the Valencian food quality control service was rather limited. Even so, its activity surely improved food quality in some way.

Valencian and Spanish scientists had a limited contribution to international forums on food quality. Nonetheless, they participated in them and international discussions and models had an effect in their design of laboratories and experimental agendas, and in the preparation of state and municipal policies.

International scenarios for food quality discussions

I will now consider some international scenarios which were probably determinant factors in defining the aforementioned food control services. As I have already noted, even though many differences can be found in the services established in each country and even in each city, important concurrences in the general design of these services can be noted. Such coincidences did not always occur by chance. During the last decades of the nineteenth century there were many international initiatives to coordinate a global response to food adulteration and alteration. In the following, I will introduce some of these international projects, which influenced the evolution of control services in each locality.

The European Federation of Permanent Analytical Controls was created at the end of the nineteenth century. This network of professionals was established with representatives from Great Britain, Austria, France, Italy and Holland and, in the words of the organisation, it assumed the following objectives:

1.- To create a system of mutual support.
2.- To popularise the Permanent Analytical Control System[27] in civilized countries.
3.- To bring about the unification of analytical methods.
4.- To supply authentic information and advice to firms whose products are under permanent control.

5.- To collect and collate facts relating to adulteration and sophistication of all kinds, and to supply information tending to the prevention or suppression of these practices.[28]

It is not clear for how long this network functioned. Its existence is mainly visible through the advertisements included in several periodical publications specializing in food safety at the turn of the century. A few years later, these advertisements were replaced by adverts of the British, the French or the Dutch services. International coordination had probably come to an end.

Another international initiative aimed at fighting food adulteration was the journal *Revue Internationale populaire et scientifique des falsifications des denrées alimentaires*. This journal was directed by P. F. van Hamel Roos, an important figure in the international food quality arena.[29] The journal included articles by authors from very different countries, and it was closely linked to the international meetings on hygiene and demography. In fact, it soon became the voice of a special international commission created through conferences to coordinate the fight against food adulteration.[30]

International conferences were in fact the major vehicle of communication between international professionals in the food quality field. During the late nineteenth and early twentieth centuries, a certain number of internationally successful scientific and medical conferences were held, of which those on hygiene and demography were especially important for the coordination of international food quality control.

At the International Conferences on Hygiene and Demography, scientists had important debates on different issues of either of the two main topics. Furthermore, politicians participated as delegates sent by their respective governments, evidencing the importance of the political dimension when dealing with public health. Finally, another of the main objectives stated in the proceedings of these meetings focused on a third target: public opinion. The conferences had to be useful for the promotion of the social reforms introduced following scientific development.

The first meeting of the international conferences on hygiene and demography took place in Brussels in 1852. The next of these meetings was held in the same city in 1876 and from then on there was a new meeting in a different European city every two or three years, including not only the "big three" but many other countries.[31] These meetings were well attended: contemporary press articles give numbers of between one thousand and three thousand participants. Dozens of countries sent delegations and, depending on the venue, the relative proportions of different nationalities varied.

The international conferences included debates on many different aspects of public health, including those related to food hygiene. At almost every meeting there were several sessions on food matters, and their significance grew, from the third meeting held in Geneva onwards. At this meeting, Paul Brouardel [32] forced the passing of a statement which encouraged the representatives of each state to prepare a report on the food quality control services of their respective countries. These reports were important in order to configure international measures against food adulteration.

At the next meeting, held in The Hague, the delegates declared not having the information required to discuss the topic yet and, therefore, new efforts had to be made. In order to prepare the debate for the following meeting, a special international commission was created. The commission was directed by Brouardel, and included one delegate from France, England, Spain, Belgium, Austria, the United States, Switzerland and the Netherlands, respectively.

Finally, in the meeting held in 1887 in Vienna, some reports were presented, and intense debate on food safety developed. The discussion was supposed to be continued in the following meetings. However, for some reason, although food hygiene was still in vogue, there was never again a debate of the magnitude attained in Vienna. The scientists involved in the Commission continued contributing to the *Revue Internationale populaire et scientifique des falsifications des denrées alimentaires* and they probably continued working on food safety, but they dropped their initial project until the following century.

Nonetheless, conferences on hygiene and demography were not the only meetings in which food adulteration and alteration were discussed. There were other international conferences where special contributions were made in the redefinition of food adulteration and the regulation of food quality. In some of these meetings, the topic was presented as a continuation of the debates which were taking place in the conferences on hygiene and demography, their existence being well noted in the proceedings of the former. This is the case with the *Congrès pharmaceutique de Bruxelles,* held in 1885, or the international conferences on applied chemistry, especially the second meeting of this kind, held in Berlin. On the other hand, subsequent specific meetings on food adulteration were organized by the White Cross of Geneva; but these were held in the twentieth century.

The control of food adulteration was therefore not only a local problem; it also had an international context of discussion which led to a certain homogeneity in the measures introduced across Europe. The participation of representatives from many different European countries permitted the establishment of a fairly common model for facing food adulteration. However, this model was not necessarily the outcome of a horizontal debate between scientists from each European country. In fact, there were important differences in the degree of involvement of different European regions in these international scenarios. In the following section I will focus mainly on the participation of Valencian scientists. Subsequently, I will analyse the dependence relations between different countries in the European context, in relation to food quality control.

Valencian contributions in the international scenarios

The participation of Valencian scientists on the international scene can be measured in different ways. Here I will mainly consider the participation of these scientists in the international meetings and networks, which I have previously described, and their international mobility and short stays in foreign laboratories where they could obtain particular knowledge.

The participation of Valencian scientists in international meetings and networks where food adulteration and alteration was discussed was quite insignificant. The work of Valencian practitioners went unnoticed in the European Federation of Permanent Analytical Controls and, at the international conferences on hygiene and demography there was Valencian participation in only one meeting – that held in Madrid in 1898.

In this meeting, Peset Cervera presented a paper with the title: "L'adultération des médicaments, dénoncée par les rayons X".[33] The paper discussed original research in the application of the recently discovered X rays to the detection of adulteration, mainly in drugs but also in food. In this same meeting, another Valencian scholar, José Pérez Fuster, a member of the staff of the Bacteriological Laboratory of València[34] also presented a paper on food quality. His paper, entitled "Les eaux de plus grande consommation à Valencia sous le point de vue de l'hygiène publique" focused on drinkable water.

Not much can be said about the participation of Valencian scientists in these international meetings. However, we can take a broader perspective and consider the participation of Spanish scientists in the international scene. It must be noted that, even though the services established for food quality control were mainly municipal, the state also contributed in some way to a slight coordination of municipal initiatives. Some laws were passed at a state level, and the municipal laboratory of Madrid sometimes played a role as a central laboratory. Therefore, the participation of Spanish scholars in these international meetings did influence the Valencian context to a certain extent.

In fact, overall, Spanish scientists also had a rather low to average representation in those meetings;[35] they always had some representation but normally it was rather poor. Perhaps only two exceptions should be noted: many Spanish scientists participated in the meeting held in Madrid; and one particular Spanish scientist, Angel Fernández Caro, was highly involved in food adulteration debates. Fernández Caro, who was an important member of the Spanish Society of Hygiene, became part of the special commission established in the international congresses on hygiene and demography in order to coordinate a response to food fraud. He was, therefore, an important bridge for the transmission of knowledge on food quality control between Europe and Spain.

Yet, Fernández Caro's case was probably just an exception to the general law; an exception to the usual minimal Spanish influence in international contexts. In fact, this minimal influence is not only noticeable in the present analysis of nineteenth-century international meetings. Spanish scientists of the period were also prone to criticize their low level of participation in international meetings; for instance, Doctor José Ubeda Correal in 1894 was very negative as to the international mobility of Spanish scientists:

> I can assure you that I do not remember any chemists from our country being designated by any official centre to participate in any of the international meetings on food or hygiene held in recent years; this participation, however, could have contributed valuable data to the country's hygiene.[36]

Maybe he was too negative in his statement, but his conclusion is highly relevant in relation to scientists' own contemporary views of the role Spanish science played on the international scene. Furthermore, this declaration was a clear vindication of Spanish chemists' desire for an increase in their professional and social status, and most probably recrimination directed at the Spanish state for their lack of support for science or the upkeep of efficient policies in this field.[37]

Nonetheless, the Spanish government did in fact design an ambitious programme promoting short stays for Spanish students in foreign laboratories and academic centres. In 1907, the *Junta de Ampliación de Estudios* (JAE), an institution especially created to run this programme, was founded.[38]

The links with foreign centres promoted by the JAE were not fostered in the same way with every country. France and Germany were the major poles of attraction for Spanish students. Important colonies of foreign scientists and students existed in countries such as those. This was not the case for the Valencian Country or Spain.

During the first third of the twentieth century, thousands of Spanish students funded by the JAE were sent as "pensionados" to foreign centres in order to study different subjects including food quality control. One of them was Luisa Cruces Matesanz, who stayed first in Wiesbaden and then in Paris in 1912, in order to work in different laboratories where food adulteration was analysed and detected. Another Spaniard who visited Paris to study the new service for the detection of food adulteration, after the passing of the law of 1905, was Angel del Campo Cerdán. The report he wrote once he was back in Spain shows how useful he found his visit and how determined he was to try to adapt the French service to the Spanish reality.

However, the number of Spanish, and particularly Valencian, researchers in food quality control benefiting from this programme was low. The only Valencian scientist linked to the municipal laboratories who seems to have benefited from these grants, and studying matters related to food quality, was José Pérez Fuster. In 1911, as the director of the Bacteriological Laboratory, he visited France and Italy to study bacteriology in general, but with a special interest in the study of the treatments applied to drinkable water.

With this last example, we can identify a new country of destination: Italy. "The big three" were possibly the main focus of attraction, but they were not the only destinations of interest for scientists. As I discuss in the following paragraphs, even if we accept the existence of a centre/periphery divide, this divide must be understood without the rigidity of the traditional model.

Conclusion: The centre/periphery divide in food quality control

Throughout this paper I have described some important characteristics of the food quality control systems established in several European countries. These descriptions will now be useful in order to discuss the traditional model of the centre/periphery divide and to study the responsibility of each region in knowledge production and circulation.

I have first described the emergence of several authors who published influential works on food adulteration. Their complaints were firstly important for society to accept the aggravation of food adulteration and alteration, and finally became the basis in prompting the passing of new food laws. In Great Britain, France and Germany, some of these works on food adulteration were already important in the first half of the nineteenth century, while in the Valencian Country, the first important works on food adulteration were only published by the end of the 1870s. Moreover, if we consider the case of Spain, and not only that of the Valencian Country, although some general books on food adulteration can be found shortly before the mid-nineteenth century, these first published books were translations from French works. Therefore, the precocity of "the big three" in complaining about, and searching for solutions to, food adulteration and alteration is well corroborated.

However, this precocity must not only be linked to the existence of a stronger scientific community. As I pointed out, food adulteration and alteration became an important problem in "the big three" before it was an issue in the Valencian Country. The higher level of food frauds was probably a determining factor in forcing the mobilization of British, French and German scientists. This fact would also be important for the passing of several general food laws which in Spain were only passed some years, or even decades, after.

The appearance of the first works on food fraud and the passing of the first general laws make the difference between the two realities in our case study. However, these two realities are not only determined by the level of development of their scientific communities, as traditionally argued. In the definition of these two realities, some economic factors played an important role too. The differences in the models of food production were of overwhelming importance in establishing different speeds in food control innovation.

However, some differences between scientific communities may have also influenced the divergence between both realities. For instance, we should consider the existence of important differences between the levels of organization of professionals working on food quality control in different countries.

When I focused on the change of historical actors working on food control, I highlighted the existence of well-organized scientific societies in several countries. These societies were mainly French and British and, certainly, no society of this kind was found either in the Valencian Country or in Spain. The existence of these societies was also probably very important in order to adopt a faster rhythm of innovation in food fraud detection in the countries where they existed.

Scientists from "the big three" were not only better organized, but they also had more relevant participation in international meetings where food fraud was analysed. In these scenarios, they were much more influential than Valencian and even Spanish scientists. The delegations of "the big three" were larger and their scientists were more active in the meetings not only as lecturers but also in the debates and political negotiations that followed each lecture. However, again, this divergence between both realities cannot be exclusively put down to differences in the level of education or initiative of the scientists. Some other factors may have been significant, for example, the language used in these meetings, mainly French, made it easier for French scientists to participate. German and English were also used, even though that was not so common. However, Spanish was only used to a slight extent in the meeting held in Madrid and Catalan was not used at all. Valencian scientists would therefore have had an added difficulty to overcome in order to participate in the conferences.

When I dealt with municipal laboratories, I also briefly described the situation in all the aforementioned countries. As I said, the situation was different between countries but also among them. Food fraud detection services had been established mainly as municipal services and were therefore different, to some extent, in each town. This fact certainly shows the complexity of the topic we are dealing with. In food quality control, it is difficult to use countries as units of a comparative history approach due to the variability of such units. Even more difficult is to stake the boundaries, in a rigid way, of a European centre and periphery. However, if we consider an average description of several regions, some divides can be drawn.

Municipal laboratories were created in many cities and towns during the 1870s and the 1880s. They had similar objectives and most of them were composed of a similar staff. However, if we focus on the Valencian laboratories, we can identify some of the features usually assigned to peripheral science. Valencian laboratories had limited budgets, they had unstable locations, and in almost every case serious deficiencies in infrastructure were suffered. The instability of these projects is in contrast with well consolidated services like the one in Paris.

The foreign origin of most instruments used in the municipal laboratory of València [39] was not common in regions traditionally considered part of the centre. While Valencians had to contact Parisian makers in order to obtain instruments, things were much easier in Paris. They could establish a closer relationship with producers from the same city and therefore adapt instruments to their particular needs. [40]

There is also evidence of the important influence that some European centres exerted in establishing the municipal laboratory of València. This was quite obvious in the laboratory project prepared by Peset Cervera. He had a declared interest in the food quality methods tested in France and Germany, but also, to some extent, in other countries such as Russia, Italy and Portugal. The first two countries, however, caused special admiration. Peset referred to France as "our Giant neighbour" and to Germany as the "Real centre of modern science, a shiny star around which the rest of satellite nations rotate, drawing more or less narrow orbits".

From the influence of foreign experiences in the preparation of the Valencian laboratory, we can first draw one major conclusion. The centre/periphery divide is corroborated in two ways. It is obvious that the innovations developed by German and French scientists were very important in order to progress in food adulteration detection in València. They were both a model to follow when preparing the project of the Valencian laboratory. In addition, the centre/periphery divide is reinforced in the words of Peset Cervera when he refers to Germany and France with so much admiration. His words show that this divide, regardless of its real existence, was obviously perceived by peripheral scientists like Peset.

I have highlighted several features of the innovation in food fraud detection which confirm the existence of a centre/periphery divide between the countries considered in this paper. I have questioned that, as was traditionally accepted, this divide was only inspired by differences in the level of development of the corresponding scientific communities. Economic and political factors, leading to the predominance of certain countries, their cultural patterns and their language, could also be important in order to draw the centre/periphery divide.

As I have already mentioned other countries, not included in "the big three", were also models in preparing the project of the Valencian laboratory. Countries such as Russia, Italy and Portugal were not particularly mentioned in Valencian works on food adulteration if compared to the countries in the "the big three" group. However, they did appear in the laboratory project and on other occasions.[41] This shows that the centre/periphery divide cannot be considered a clear separation of two groups: one creating knowledge and another one following the leaders. In nineteenth-century Europe, some countries were more influential than others in food quality control, but this capacity for influencing others had many different degrees. The bipolar conception of the divide is thus in question.

Moreover, it should be noted that the laboratory of Madrid also appeared as a model to be followed in the projects of València and

Alacant. The staff of the laboratory of Madrid had been inspired by the laboratory of Paris in its first years of existence,[42] but at the same time it had been a source of inspiration for València and Alacant. Each region, city or country could be centre and periphery at the same time, depending on which other region, city or country it is compared with.

If we compare Madrid and Valencia in the international context, we can also underline the existence of a double periphery. Spain participated in the international scene with a more passive role than the one played by "the big three". But this Spanish representation was not an even representation of each region of the country; Madrid was usually much better represented. Madrid had both the principal laboratory and the best prepared scientists in Spain, and their budget was much higher than the one of the Valencian service. Thus, the Valencian Country could be seen as the periphery of the periphery, or a part in the double periphery.

Finally, I will briefly question one last feature of the traditional model of the centre/periphery divide. In this paper, we have seen that peripheral scientists like Peset Cervera or Perez Fuster were also capable of developing original research and presenting their results in an international context. Valencian scientists were therefore more active than expected from the traditional model. In fact, examples of this kind have impelled historians to shift from studies of the reception of scientific theories to the study of the appropriation of new theories. In this way, the active role of the periphery is better understood.

Appropriation is also present in the early Spanish literature on food adulteration. As we have seen, Spanish scientists like Magín Bonet created new knowledge by appropriating a French text, and modifying it in relation to local needs. This need for appropriation was also very clear in the report drawn up by one of the Spanish "pensionados" whom I already mentioned. Campo Cerdán travelled to Paris where he was overwhelmed by the effectiveness of the Parisian service of food quality control. However, when he considered the possibility of exporting the service to Spain, he was concerned with the need to transform the service in order to resolve the particular food frauds practised in Spain.[43] He knew the issue was not simply how to present the ideas which inspired the French service, but rather to know how to work with those ideas in order to create the service needed in Spain.

Thus, although there were differences in food quality control in different countries, and although economic and political factors are important to understand these, they are not the only aspects to be taken into account. The traditional centre-periphery model is narrow in that it reflects its origin in the fields of economics and politics studies. However, historians need to take into account broader cultural phenomena. In spite of this, it is clear that regions like the Valencian Country were strongly dependent on other European regions such as France and Germany, revealed through the Valencian acquisition of scientific instruments, texts, experiences and ideas produced abroad. Nonetheless, firstly, this dependence was not only with regard to the "big three"; secondly, the observation of foreign developments was a common practice performed as well by leading countries;[44] and thirdly, this communicative exchange is not at all a passive "reception", but is based in appropriation, that is, the creative use of foreign knowledge in relation to the needs of local contexts.

Notes

[1] While food adulteration was caused in a conscious manner by producers or merchants of foodstuffs who wanted to obtain a bigger profit by cheating the customer, food alteration was not usually a food fraud consciously plotted. It was usually due to improper handling: poor storage of foodstuffs allowing exposure to bacteriological contamination, use of saucepans and pots containing dangerous metals, etc.

[2] Accum, F. (1820). *A treatise on adulterations of food, and culinary poisons. Exhibiting the fraudulent sophistications of bread, beer, wine, spirituous liquors, tea, coffee, cream, confectionery, vinegar, mustard, pepper, cheese, olive oil, pickles and other articles employed in domestic economy; and methods of detecting them.* London: Printed by J. Mallett.

[3] Burnett, J. (1979). *Plenty and Want: a Social History of Diet in England from 1815 to the Present Day.* London: Scholar Press.

[4] Ferrières, M. (2002). *Histoire des peurs alimentaires: du Moyen Âge à l'aube du XXe siècle.* Paris: Editions du Seuil.

[5] The Comission was formed by several authors but recent historiography has highlighted the role played by Arthur Hill Hassall. This fact has recently been critized by Berris Charnley. Charnley, B. (2005). *Dr. Arthur Hill Hassall, the Analytical Sanitary Commision and the Origins of Food Analysis: A Re-examination of the "Food Adulteration Crisis" in the 1850's.* unpublished MA thesis. Leeds: University of Leeds.

[6] Berman, A. (1978). "J. B. A. Chevallier, Pharmacist-Chemist: A Major Figure in Nineteenth-Century French Public Health". *Bulletin of the History of Medicine.* 52 (2): 200-13.

[7] Teuteberg, H. (1994). "Food adulteration and the beginnings of uniform food legislation in late nineteenth-century Germany". In Burnett, J. and Oddy, D., eds. *The Origins and Development of Food Policies in Europe.* London: Leicester University Press, pp.146-70.

[8] A term commonly used to designate Great Britain, France and Germany.
[9] See Hamlin, C. (1990). *A Science of Impurity. Water Analysis in Nineteenth-Century Britain.* Berkeley: University of California Press, and Stanziani, A. (2005). *Histoire de la qualité alimentaire: XIXe-XXe siècle,* Paris : Editions du Seuil.
[10] See Oddy, D. J. (2006). "Food Quality in London and the Rise of the Public Analyst, 1870-1939". Paper presented at the *Ninth Symposium of the International Commission for Research into European Food History.*
[11] Hardy, A. (2003). "Professional Advantage and Public Health: British Veterinarians and State Veterinary Services, 1865-1939". *Twentieth Century British History.* 14 (1): 1-23.
[12] It is also worth to pinpoint a general tendency in the passing of food laws.
[13] Teuteberg. "Food adulteration".
[14] Oddy. "Food quality".
[15] Note, however, that there are also important exceptions to this general feature. For example, Paris had a municipal laboratory that in the 1880s maintained a staff of over sixty professionals.
[16] See Soler Sanchez, J. (1897). *Análisis y ensayos de los alimentos, las bebidas y los condimentos.* Alicante: Establecimiento Tipográfico de Vicente Botella, and Peset Vidal, J. B. (1878). *Topografía Médica de Valencia y su Zona.* València: Imprenta de Ferrer de Orga.
[17] Peset Vidal. *Topografía Médica.*
[18] Soler Sanchez. *Análisis y ensayos.*
[19] Garnier, J., and Harel, Ch. (1846). *Falsificación de las sustancias alimenticias y medio de conocerlas* [translation, synopsis and added texts by Magín Bonet]; and Chevallier, M. A. (1854-1855). *Diccionario de las alteraciones y falsificaciones de las sustancias alimenticias* [translated by Ramón Ruiz Gómez].
[20] *Archivo Biográfico de España, Portugal e Iberoamérica,* I, 260-3.
[21] Barona Vilar, C. (2006). *Las políticas de la salud: la sanidad valenciana entre 1855 y 1936.* València: Universitat de València.
[22] Peset Cervera, who was son of the aforementioned Peset Vidal, was not only a key figure in food analysis once the laboratory was established but he was also responsible of the elaboration of the project for the creation of the laboratory.
[23] Peset Cervera, V. (1882). "El Laboratorio Químico Municipal de Valencia". *Gaceta de los hospitales, revista quincenal de Medicina y Cirugía prácticas.* 1: 409-14.
[24] Vera Lopez, V. (1885) *Laboratorios municipales de salubridad.* Madrid: Imprenta de F. Maroto e Hijos.
[25] Where, for example in 1883, nine-hundred analysis were done in only one year.
[26] Guillem-Llobat, X. and Perdiguero, E. (2006). "Fighting adulteration in early European food industrialisation. The case of Alicante". Paper presented at the *Conference on the History of the Food Chain-From Agriculture to Consumption and Waste* in Godollo (Hungary).

[27] This system consisted of a private initiative that favoured pure food industries in their respective national markets. It offered to manufacturers the opportunity to voluntarily submit their products to scientific examination. Then, by advertising the list of pure food producers and by allowing any consumer to submit samples of the mentioned producers to be analysed, the system contributed to making consumers confident of these products. Thus, consumers and producers of pure food benefited from the service.

[28] Included in the *British Food Journal* of October 1900.

[29] He seems to appear in almost every international initiative during the last decades of the 19th century. He ran an international journal specializing in food adulteration and participated in the European Federation of Permanent Analytical Controls. Moreover, he participated at several international meetings where food adulteration was discussed; meetings like those of the International Conferences of Hygiene and Demography.

[30] In the following section I will refer in more detail to this special commission.

[31] The meetings continued in the twentieth century but in the period considered in this paper they took place in the following cities (in chronological order): Paris, Turin, Geneva, The Hague, Vienna, Paris, London, Budapest, Madrid and Paris.

[32] Paul Brouardel was Professor and Dean of the Faculty of Medicine in Paris, member of the Académie de médicine, and President of the Comité consultatif pour l'hygiène du Ministère du commerce en France.

[33] Although the Spanish participants in the Madrid conference arguably presented their papers in Spanish, it is significant that the titles of their papers often appear in French in the subsequent conference proceedings.

[34] Created in 1894 next to the Chemical Laboratory, it always had a close relationship with the latter, and although it was not so active in the analysis of food it had some importance, mainly in the analysis of drinkable water but also of meat.

[35] In the meeting held in Brussels in 1876, the Spanish delegate apologised for the low attendance of Spanish scientists, explaining that Spain had recently suffered important internal political conflicts preventing the normal participation of Spanish scientists. In the next meeting, held in Geneva, the Spanish delegation continued to be rather small but this certainly changed in the meeting in The Hague in 1884. Nevertheless, it was never one of the largest delegations; the Madrid meeting was really an exception.

[36] For more information, see: Ubeda Correal, J. (1894). *Falsificaciones de los alimentos. Discursos leídos en la Sociedad Española de Higiene*. Madrid: Enrique Teodoro. All translations from Spanish are by the author of this paper.

[37] The vindication of more state support to the practice of science, and the use of the international context as a way to prompt this participation is a characteristic phenomenon in the nineteenth century. See Faidra Papanelopoulou's and Josep Simon's papers in this volume for a description of the French case.

[38] Sanchez Ron, J. M. (1988). *1907-1987. La Junta de Ampliación de Estudios e Investigaciones Científicas 80 años después*. Madrid: CSIC

[39] In one report of the activity of the laboratory prepared by Peset Cervera there is information on expenses on French instruments. See Peset. *Laboratorio Químico*.

[40] As in fact they did, for example, with microscopes. For more information on this issue, see Vera. *Laboratorios*.

[41] For example, when saccharin regulation was discussed in Spain, Italy was the model followed to pass a new law forbidding its use.

[42] For more information see Vera. *Laboratorios*.

[43] Campo Cerdan, A. (1910). "Métodos de Análisis de Alimentos". *Anales de la Junta para ampliación de estudios e investigaciones científicas*, 2 (6): 312-31.

[44] On this topic see Josep Simon's paper in this volume.

WATERS, SEEDS AND RADIATION:
RADIOACTIVITY RESEARCH IN EARLY TWENTIETH-CENTURY SPAIN

NÉSTOR HERRAN[*]

1904 was an eventful year for the new science of radioactivity. The awarding of Nobel Prize in chemistry to Henri Becquerel, Pierre Curie and Marie Curie launched wide public interest in their research, helping Pierre Curie to get a professorship at the Sorbonne and paving the way for the emergence of the radium industry. At the same time, Rutherford's theory of the atomic disintegration began to be widely accepted, not least by the wide diffusion of his book *Radioactivity*. In the Austro-Hungarian Empire, uranium ores were secured by the government, in order to control this suddenly valuable resource.

In the wake of these events, chemist José Muñoz del Castillo established the first laboratory devoted to radioactivity in Spain. Increasingly gathering resources from the Spanish government, the centre's researches received wide attention by the local press and members of the Spanish scientific community. However, Muñoz's radioactivity was largely ignored in the main scientific centres and departed from what, in scientific and historical terms, have been considered the mainstream theories and practices of radioactivity. Inspired by "classical" physico-chemical models and built on heterogeneous notions taken from "popular radioactivity", his theoretical insights and experiments on the low effects of low radiation doses on humans and plants were "out of touch" with the researches carried out by the Curies, Rutherford and other radioactivists, which have been retrospectively considered as mainstream in the discipline.

Recent revisionist approaches to the history of radioactivity, led by the interest of challenging the linear and teleological character of standard historiography, have uncovered the diversity of techno-scientific practices and cultural meanings of early radioactivity research.[1] As part of this renewed historiography, this paper aims to understand Madrid's laboratory activities in its cultural context, tracing the emergence of the centre/periphery dichotomy from a non-essentialist point of view. I will argue that this asymmetry arose from a series of factors, such as the socio-economic and industrial structure of the country, the organization of scientific excellence in the local context, the constitution of expert knowledge communication system, and the ways in which scientific instruments were built, appropriated and used.

This article is organized in five sections. The first one deals with the early appropriation of radioactivity in Spain, and with Muñoz del Castillo's success in becoming Spain's main radioactivist.[2] The second section recounts the first years of the Laboratory of Radioactivity and its early engagement in the analysis of the radioactivity of waters. In the third, I analyse the laboratory incursion in the field of radioactive agriculture and its effects after 1911. The fourth section is devoted to the centre's material culture and its transformation in a metrological institution in the mid-1910s. Finally, I consider the causes of the Laboratory's decline, and present the main conclusions of my case study for our discussion of centres and peripheries.

José Muñoz del Castillo and the appropriation of radioactivity in Spain

Radioactivity research originated in late nineteenth century France from Henri Becquerel researches on the phosphorescence of uranium. His observations on the emission of invisible rays by uranium were enlarged and generalised by Pierre and Marie Curie, who assigned the name "radioactivity" to these phenomena and employed it in the task of identifying new chemical elements. [3] In the early 1990s, several radioactivity laboratories were created in different European and American cities. The establishment of radioactivity as a discipline was also coupled with increasing public interest (Becquerel and the Curies' Nobel Prize playing an important role in this respect), and the emergence of industrial facilities to produce radium – by the 1910s it was the most expensive chemical element – and other radioactive elements.

News about radioactivity research reached Spain by means of both the popular and the scientific press. However, unlike experiments with X-rays, which were soon replicated in academies and university and private laboratories, public interest on radioactivity did not encourage empirical research. Forefront radioactivity studies depended on the availability of radioactive materials, and Spain lacked industrial facilities to undergo the costly process of separation. The attitude of Spanish physicists and chemists towards radioactivity was epitomized by José Echegaray, president of the Spanish Royal Academy of Sciences, Madrid (RACEFN) and spokesman for academic science in Spain. Echegaray, a prolific popularizer, limited himself to theoretical ruminations and expressed his resistance to accept what he considered to be more "radical" interpretations portraying radioactivity as an atomic disintegration. Political controversies overlapped the scientific debate in Spain, and radioactivity, loosely defined as "revolutionary science", confronted more conservative and "moderate" approaches.[4]

The first research programme on radioactivity in Spain was established by Professor José Muñoz del Castillo in 1903. Born in Sevilla in 1850, he had started his academic career as secondary school teacher in Logroño, a small province capital in northern Spain. His leading role in the creation of a network of meteorological observatories and participation in campaigns against the phllyoxeric pest promoted him to a professorship at the University of Zaragoza in 1880.[5] In 1892, after a short period as professor in the General Preparatory School for Engineers and Architects in Madrid, he gained a position at the University of Madrid. There, he combined teaching of chemical mechanics with studies on water purification. In 1901 Muñoz was admitted in the RACEFN. In this position he felt for the first time free to develop a research programme, which would be focused on the research and production of chemical elements, an area he ambitiously christened as "estequiology".[6]

Muñoz's interest on the production of chemical elements brought him naturally to radioactivity, a discipline whose techniques were in the international context of the time, the last frontier in the discovery of new elements. In 1903, he aimed his first efforts to procure pitchblende for the production of radioactive elements, as the Curies had recently done. However, the attempt was frustrated by the Austro-Hungarian embargo. Deprived of raw materials and lacking an adequate laboratory, Muñoz's early studies were mostly devoted to theoretical questions. Submissively following Echagaray's views of the atom, he favoured an interpretation of radioactivity as a chemical phenomenon. His hypothesis, in harmony with other "classical" accounts supported by older chemists such as Becquerel and Mendeleyev, circulated in Spanish scientific journals and was even presented in the International Congress of Radiology held in Liège in 1905. There, he met the scepticism of his foreign colleagues, who favoured Rutherford's hypothesis considering radioactivity as deriving from the disintegration of the atomic nuclei.[7]

Despite his isolation from radioactivity's leading theories and actors, Muñoz was recognised as main expert in this field by the Spanish government, a position which helped him to recruit abundant human and material resources. Well connected to political and academic power,

Muñoz transformed the laboratory associated to his chair in the University of Madrid into the Laboratory of Radioactivity.[8] A measure of Muñoz success in institutionalizing the discipline is that, by 1908, this facility counted three assistants, and a budget similar to that of other university departments in Spain, gathering about 8% of the state's investments in scientific material.[9] This money was used to acquire radioactive material (about 10 mg of high activity radium bromide) and off-the-shelf instrumentation for his experiments. This choice made laboratory research line dependent on the provision of foreign apparatus. Hence, innovation could only proceed by ingenious manipulation of black-boxed tools.

Radioactive waters

After his unsuccessful incursion into theorizing, Muñoz found his way into the study of environmental or "atmospheric" radioactivity.[10] These researches combined natural-history-style scientific work with analytical determinations of the activity of mineral waters. The core belief sustaining these practices was that radioactivity had stimulating properties, and could account for the supposedly beneficial but obscure effects of mineral waters and spas on the human body. Far from being a marginal topic, the measurement of radioactivity of waters involved prestigious scientists and institutions. Researchers such as Pierre Curie, André Laborde and William Ramsay measured the radioactivity of springs, and the topic featured extensively in early journals devoted to radioactivity.[11]

Muñoz decision to study the radioactivity of waters was motivated by the endemic shortage of radium in his Laboratory. Moreover, lack of advanced experimental skills played against his involvement on disintegration experiments such as those carried out by Rutherford in the Cavendish Laboratory. Muñoz previous involvement in water analysis provided continuity with his earlier techno-scientific activities and relations with the still influential community of Spanish medical hydrologists. Long considered as medical speciality, by 1900, medical hydrology counted about 200 state-appointed doctors, who managed a network of spas attended by a large and wealthy clientele.[12]

Muñoz's interaction with the community of medical hydrologists was carried out either by contributing to their society's journal or, more importantly, by performing radioactive analysis at their request. A lively correspondence and exchange of water and mineral samples was established between directors of spas and Muñoz's laboratory, which helped him to draw the first "map of Spanish radioactivity". According to Muñoz, the presence of high activity in waters indicated the existence of radioactive minerals in the subsoil, and even the existence of new chemical elements, such as the "galaico", a new element – of the actinides – allegedly found in the analysis of Lerez's spring waters, in the Spanish north-western region of Galicia. However, the existence of this element was never accepted by the international scientific community, and Muñoz's researches were generally ignored or discredited as inaccurate.

As a result of the activities of Muñoz's laboratory, radioactivity was frequently included among other parameters in the customary physico-chemical descriptions of mineral waters in Spain. Indeed, the presence of high levels of radioactivity was also currently employed in advertising springs and spas, as it was claimed that it provided additional medical properties to these waters. This kind of analysis became so relevant that, between 1905 and 1915, Muñoz's Laboratory of Radioactivity routinely produced certificates assessing the radioactivity of waters for commercial spas and springs. These kinds of analysis were not unusual in other European laboratories, such as the Radium Institute in Vienna or the Curie laboratory in Paris.[13]

Madrid's laboratory research on radioactive waters produced two dissertations, written by Muñoz's pupils Eugenio Morales Chofré (1904) and Faustino Díaz de Rada (1906).[14] It was also the subject of 68 articles in the journal *Anales de Física y Química*.[15] However, Muñoz's articles were largely invisible to his foreign peers. Indeed, as the *Anales* raised the standards of its contents, he was gradually ostracized. Consequently, the laboratory established in 1909 its own publication, the *Boletín del Laboratorio de Radioactividad*, aimed at presenting the members' researches to an international audience (in order to be readable by this audience, articles by Spanish scholars were written in French), and presenting foreign advances in radioactivity in Spain. This was carried out by translating key articles of the perceived leading researchers in this field, such as Pierre Curie, André Debierne, Ernest Rutherford and more experts on the radioactivity of waters. This format, however, did not last long. Muñoz illness and lack of resources motivated a change in the editorial line. Since 1910, all articles were published in Spanish and the journal became by and large self-referential.

An agricultural wonder

In 1911, the Laboratory of Radioactivity was upgraded to become a research Institute. This transformation, which was associated to a significant increase in the centre's resources, motivated the emergence of new research lines focused on radioactive agriculture. Muñoz's interest in this topic can be related to at least four factors. First, previous researches on the analysis of mineral waters had established the centre's expertise around the measurement of low levels of radioactivity, which required simpler instrumental and technical adaptation. Second, Muñoz counted with experience on agronomic research, dating from his previous works on the phllyoxeric pest. Third, radioactive agriculture did not require expensive amounts of raw materials. Availability of samples seemed assured by the emergence in France and Portugal of an industry of radioactive fertilizers in the early 1910s, which was set up by radium producing companies in order to make the most of the low activity residues of their factories. Last, but not least, the Spanish economy was in this period largely relying on agriculture. About 60% of the population worked in this sector and there was strong political concern about the modernization of farming. Muñoz, who in 1910 had been appointed senator, knew perfectly well from the political debate that the topic could decisively attract the interest of government officials.

Radioactive agriculture was not an isolated Spanish development. As well as pioneering the use of radium for medical applications, Armet de Lisle was also one of the first industrialists to be involved in this market. His company, the Banque du Radium, established facilities for the production of radioactive fertilisers in the early 1910s.[16] The company star product, "Excitor Agral", seems to have had some impact among French farmers, who possibly trusted reports of the Society of Civil Engineers of France declaring in 1913 that low activity radioactive fertilizers could be beneficial for agriculture.[17] Radio-agricultural research was also common in other European and American countries. Among early practitioners were Julius Stoklasa, who studied the effect of pitchblende-based fertilizers at the University of Prague, and agronomists of the Harper Adams Agricultural College at Newport (United Kingdom), who stated the profitability of radioactive fertilizers at the request of the British Ministry of Agriculture. In the United States, radioactive fertilizers with names such as "Nirama" and "Liquid Sunshine" were also marketed in the 1910s with some success.[18]

Madrid's Institute of Radioactivity involvement in radioactive agriculture dates back to early 1912, when the centre enlarged its outbuildings with the addition of an "observatory of radioactivity", a "radioactive fertilizers test field" and, in January 1913, a station to measure the radioactivity of farmland.[19] Agricultural research soon became the main activity of the Institute, and can be considered as an extension of his earlier researches on the radioactivity of waters. Both were based on Muñoz's belief that radioactivity had a beneficial and measurable effect on plants, and involved the study of the relationship between radioactivity and meteorological parameters and experiments with radioactive fertilizers. The Institute researchers systematically recorded the activity of air and earth, which were tabulated with atmospheric pressure, temperature, and wind velocity. [20] They also compared the growth of vegetal species exposed to different radioactive elements, [21] or studied the effect of different elements on the growth of nitrifying bacteria. [22] However, the most common experiments tried to demonstrate increases in plants' growth due to the action of radioactive fertilizers, which in general terms reported that low doses of radioactivity produced increases of 20% in plant size. [23]

According to Muñoz, the Institute's most important result was the development of "thorionization", a technique based on the addition of thorium to farmland. Seeds exposed to this radioactive element were collected and planted again. After several cycles, a "variety completely different from the original one" was obtained. [24] According to the Laboratory experiments, the "thorionized plants were more homogeneous" in all stages of development, "their leaves were wider, and greener (…) [and] fructified somewhat earlier". [25] Thorionization can also be considered as an ingenious adaptation to the Institute's lack of radioactive materials (thorium being a relatively abundant substance), which resonated with a nationalistic and utilitarian agenda to develop a Spanish industry of radioactive elements. This objective is most evident in Muñoz's frequent speeches advertising his research in different national forums. [26] An early example was during the 4th congress of the Spanish Association for the Advancement of Science, [27] held in Madrid in June 15-20th, 1913. Muñoz, who was a prominent member of the association, used this space to present the Institute to the Spanish scientific community, arguing about its usefulness for the country's economic progress. [28] A similar claim was made in every edition of the Institute's "radio-agricultural weeks", an initiative aimed at propagating the use of radioactive fertilizers among Spanish farmers by means of annual courses. [29]

Favourite topics of such speeches were Muñoz's proposal on the nationalization of Spanish radioactive minerals (called by him "the Spanish radioactive wealth") [30] and the extension of radio-agriculture in Spain under the control of the Institute of Radioactivity. Appeals to state intervention in this area were common, and were justified with the argument that the nationalization of radioactive ores would reduce the price of radioactive fertilizers. Muñoz also argued for the relevance of the Institute's service for the radioactive analysis of farmland [31] as a tool to fight fraud, [32] establishing "a barrier against foreseeable greed of unscrupulous middlemen, in case they forged or reduced the virtual efficacy of such fertilisers".[33] The service actually advised farmers on the radioactive fertilizers best suited to the physical features of their terrains, gathering data on a wide diversity soils.[34] Data collected by the service did not produce a new map of radioactivity in Spain, but they were probably considered by Muñoz as a useful resource for radioactive minerals prospecting. These activities, together with Muñoz's frequent appeals to "mineralogists, chemists, miners and anybody that studies nature or searches for exploitable minerals" requesting samples, and the analysis of the collections of the Museum of Natural Sciences in Madrid performed by his collaborators, show that Muñoz's early interest in establishing an industry of radio-elements in Spain was still alive.[35]

Instruments and the Spanish radium standard

By 1917, the Institute of radioactivity enjoyed a privileged position in the Spanish system of science and technology. Its facilities had been extended to occupy a whole building, and Muñoz had five research assistants and two technicians. Routinely engaged in analysis of water, minerals and agricultural soils, the laboratory had stabilized its research programme around the study of biological effects of low levels of radiation. However, its results had been almost invisible abroad. Muñoz's old-fashioned ideas on the origin of radioactivity, his lack of connection with leading radioactivists; and the Institute's peculiar research line meant that his activities were ostracized. Moreover, World War I ravaged Europe from 1914, leading to a sudden decrease of civil scientific activity, as well as a collapse of international cooperation. As the country remained neutral, Spanish science benefited from the war-induced national economic growth, but indirectly suffered the effects of this disjointed climate by the suppression of research stays abroad and decreased international communication.

The most remarkable effect of war in the Institute was the incorporation of Bela Szilard, a Hungarian radioactivist who fled from Paris during conflict.[36] After completing a degree in chemistry in Hungary, Szilard had worked in Marie Curie's laboratory and had subsequently been involved in the development of radioactivity-related instruments, such as electrometers, surveying devices and radioactive lighting conductors. After Szilard moved to Spain, Muñoz helped him to obtain a temporary position in the Institute of Radioactivity until the end of the war. His work at the Institute involved giving advanced lectures on radioactivity and, most importantly, the development of instrumentation. As the Institute did not count with a workshop, Szilard assembled the instruments in the Torres Quevedo Automatic Institute.[37] After Szilard's return to Paris, the instruments remained in the Institute of Radioactivity.[38]

Szilard's contribution to improve the instrumental resources of the centre was significant because the laboratory had relied for a long time on instruments acquired in the previous decade. The Institute's involvement in the analysis of water and soils had shaped its material culture to include measuring devices such as Engler and Sieveking's fontaktoscope or Elster and Geitel's electroscopes, and using as unit of radioactivity the volt-hour. These devices and standards belonged to the diverse panoply of instruments and units typical of early radioactivity, suffering problems of calibration and comparison between different settings and laboratories. This situation had motivated several unification and regulation attempts, which coalesced in the 1910 negotiations among British, French, Austrian, German and American radioactivists to establish a radioactivity standard. The solution reached by an international committee was to establish the curie (defined as the emanation in equilibrium with a gram of radium) as the standard unit of radioactivity.[39] This unit was huge compared to the availability of radioactive substances of contemporary laboratories, but reflected Marie Curie hopes that the radium market would expand and made this element plentiful. However, the needs of researchers studying atmospheric radioactivity of radioactive mineral prospecting were disregarded, and did not fit into this new scheme well. In 1911, two standard radium tubes were built in Paris and Vienna, the French remaining the primary standard after calibration, and the Austrian as the secondary one.

In Madrid, the establishment of the radium standard was considered not much as a nuisance, but as an opportunity to be in the same wavelength as foreign laboratories of radioactivity. Muñoz's first attempts to acquire a radium standard dated back to late 1917, when he contacted Curie's laboratory to procure a tube. By then, eight countries (France, Austria, United Kingdom, Germany, United States, Sweden, Japan and Portugal) had primary or secondary standards and had established facilities or boards to organize the metrology of radium. In following these examples, Muñoz's aim was to transform the Institute of Radioactivity into the equivalent Spanish metrology institution, adding the tasks of calibrating instruments and medical radium to the Institute's services for the measurement of mineral waters, minerals, soils and radioactive fertilizers.

The correspondence between Muñoz and Marie Curie reveals the difficulty for acquiring a secondary radium standard in Spain, mostly related to its cost, which was higher than the Institute's annual budget. After intensive lobbying of the government – which involved the donation of 3 milligrams of radium by the Portuguese radioactive fertilizer-producing company Henry Bournay and Co. – the Institute acquired two radium-filled tubes late in 1919. The acquisition of these standards, and their certification by Curie's laboratory, allowed the establishment of an official service for measuring radioactivity in the Institute.[40]

Rise and fall of Madrid's Institute of Radioactivity

The transformation of the Institute of Radioactivity into a centre devoted to metrology was coupled with the zenith of its radioactive agriculture research programme. The inspection of contemporary journals and newspapers show that the Institute popularizing campaigns had convinced only a few Spanish agronomists.[41] For example, H. Gorria, professor of the Barcelona Agriculture High School, argued for the use of radioactive fertilizers on the following terms:

> Considering the positive effects of radioactivity when applied to agriculture, it would be most convenient that these experiments were repeated in our country, an also that radioactive ores and strings were investigated, as they could become real sources of wealth if applied to agricultural exploitation, either by stimulating vegetation or increasing our fields' productivity.[42]

There were also attempts to reproduce the Institute's results, such as the experiments "about the influence of radioactivity on vines" performed by Eloy Martínez Philippest in Logroño, or the studies on commercial radioactive fertilizers carried out in Tortosa. [43] However, the most important test of radioactive fertilizers outside the Institute of Radioactivity was carried after King Alfonso XIII received Muñoz in February 1920. In the course of this interview, which was announced in the *Boletín* as "a patriotic initiative of His Majesty the King Alfonso XIII, as Highest Farmer of Spain", Muñoz explained to the monarch that he aimed at setting his researches on a larger scale. [44] According to the *Boletín*, the king offered Muñoz the facilities of Alfonso XII's Agronomic Institute, and passed an edict urging Bernardo Sagasta, director of this centre, to provide land and resources for these experiments.

The Moncloa experiments were conceived as a large-scale test, which would be performed in plots of land with and without addition of radioactive substances. Muñoz considered them as a great opportunity to promote radioactive agriculture, and reacted accordingly.[45] The press was called to attend a conference, in which Muñoz showed the Institute's most spectacular results, such as the development of a variety of giant corn "four times thorionized, which [brings about] higher, fruitier and leafier plants".[46] By stressing the king's support as warranty of his programme viability, Muñoz urged the press to spread thorionization and to encourage capitalists in investing in thorium mining. If not, such wealth could end in hands of "foreign capitalists". At the same time, he encouraged journalists to inform farmers that his Institute provided free advice on radioactive fertilizers, and invited them to follow the development of the Moncloa experiments by means of guided visits.

Muñoz's conference was reported in several newspapers, triggering off a myriad of letters from farmers in request of thorium. A common topic of this correspondence, which established contacts between Muñoz and at least 200 people, was the scarcity of thorium compounds in Spain. Muñoz tried to resolve this problem by supplying this product from the Institute. Accordingly, he published in the *Boletín* that anyone interested in replicating his experiments would receive from the institute a dial with thorium chloride, with a leaflet with instructions. However, these samples were not enough to sustain large experiments, so he increased his appeals to Spanish industrialists to invest in radioactive mining. His articles in the *Boletín* urged them even to acquire foreign ores of thorium minerals, such as those of Brazilian monacite. These proposals were never taken into account.

In May 1920, Muñoz gave his last lecture as professor of inorganic chemistry at the University of Madrid. His retirement also implied his resignation as director of the Institute of Radioactivity, which was allocated to his disciple Faustino Díaz de Rada.[47] This replacement unchained reactions which would affect the following development of the Institute. As we have seen before, Muñoz position was much related to his good connections with politicians, a situation which contrasted to his lack of prestige among his peers. His old-fashioned views on radioactivity and the Institute's activities were not much respected in the Faculty of Science, where a young generation of scientists, led by Blas Cabrera, was on the rise. Cabrera had confronted Muñoz's theories inside the Spanish Society of Physics and Chemistry, and battled for tuning Spanish physics and chemistry into the science produced in leading European centres.

Presumably, these attitudes contributed to the Faculty's change in its policy regarding the Institute, leading to the requisition of the centre's instruments to provide material for other laboratories. The Institute never recovered to its previous levels of activity. Indeed, Díaz de Rada never got a professorship and the *Boletín* was no longer published with the same frequency. After three years interruption, the journal reappeared in 1923, but in a weakened style. [48] The Institute also interrupted the research programmes on radioactive agriculture, and no further mention of these experiments appeared in its journal issues. As records of the Institute of Radioactivity and of Alfonso XIII Agronomic Institute are lost, we do not know if these changes were related to a dramatic failure of the Moncloa experiments. The only conclusion that can be drawn is that, after the Institute's demise, radioactive agriculture had no longer an impact on Spain. No trace of radioactive fertilizers appeared in agronomy journals during the 1920s and no mention to supporters or practitioners of radioactive agriculture was kept in reference books or official provisions. Radioactive agriculture vanished faster than it appeared, and even its memory has been long forgotten. [49]

Radioactivity in Spain: An invisible science?

The history of Madrid's Institute of Radioactivity reveals interesting insights on the nature and construction of scientific peripheries, revealing the existence of underlying factors which could better explain the emergence of the asymmetry centre/periphery. The first factor, which is related to the lack of resources and staff, is the socio-economic structure of the country. In the case of Muñoz's researches, it emerged as the inability to develop an industry of radioactive elements in Spain. Recall here that, since the beginning of his researches on radioactivity, Muñoz had tried to procure radioactive elements either by purchasing abroad or by requesting investment in prospecting of what he called the country's "radioactive wealth". However, his proposals were systematically ignored. Lack of a solid industrial basis, together with the inability of the Spanish state to generate a community of radioactivists, meant a lack of professional opportunities for Muñoz's students, precariousness of material and instrumental exchanges, and absence of alternative research groups, which could reproduce or refute Muñoz's experiments. At the same time, radium scarcity fostered the development of researches about low doses of radioactivity, which were tied to important measurement problems and which were in the end quite marginal in the discipline.

These circumstances, however, are not enough to explain the whole story. The Cavendish laboratory and other centres could contribute to mainstream radioactivity with scarce material resources. Indeed, radium scarcity could have been overcome by means of smooth relations with foreign groups. This was not possible because of Muñoz's scientific prestige. The marginality of Muñoz in the international scene contrasts with his solid position in Spanish academia, and brings up two additional underlying factors in the construction of the periphery: the organization of scientific excellence of the local context and the configuration of expert knowledge communication devices.

In relation to the first, we should note here that in his local context, Muñoz's history was temporarily one of success. He had a notable ability to gather resources from the government, succeeding in the institutionalization of the discipline and its promotion before different audiences.[50] These efforts helped him to appear in front of the Spanish state as the main expert in radioactivity, gathering resources and official support for his researches. This privileged situation, reinforced by an academic establishment giving a bonus to seniority and respect to hierarchy, made him independent of the international excellence system and informal networks between European radioactivity laboratories. Muñoz dependence on political power, not on prestige among peers, was also reflected in the structure of his research school. Following organizational models typical of the Spanish state administration, it was marked by his director self-centredness and his students' submission to his paradigm, allowing few spaces for personal development and alternative approaches. These features, together with structural elements such as the lack of research positions in Spain, worked together against the emergence of a strong community of radioactivists. The paradoxical situation that Muñoz's institutional position grew stronger as his international prestige decreased is representative of the isolation of the Spanish research system, incapable of maintaining endogenous research lines for a long time.

The self-centredness of Spanish academic structures relates to the structure of scientific communication. As we have seen, Muñoz forcefully tried to put scientific journals under his control. After a first try with the *Anales*, he established the *Boletín del Laboratorio de Radioactividad* as a device to communicate his researches to the scientific community and, later, as a tool for spreading radioactive agriculture in Spain. Muñoz strategy to not publish abroad was not unusual in the context of nineteenth- and early twentieth-century science. What is relevant here is his failure to connect with the core of the radioactivists' community, and to be taken into account in contemporary discussions in the discipline. And, when these contacts were established, like in the case of the Spanish radium standard, they were not relationships between equals, but of subsidiarity.

Finally, I will point out one of the foremost important elements in this history, which is related to instrumentation. Since the establishment of Madrid's laboratory of radioactivity, its researches were carried out with foreign instruments, and these devices were barely altered or improved. By the mid 1910s, Muñoz's group was using the same devices that were bought ten years before. By contrast, during this period radioactive instrumentation evolved significantly, both by improving traditional electrometers and electroscopes and by introducing new techniques such as scintillation and "electrical method" based devices, which would later be known as Geiger counters. Here, we should note the episode of Szilard's involvement with the Institute, where he assumed the role of expert improving the Institute material resources instead of scholar trying to learn from local developments. Finally, and more importantly still, Muñoz was completely disconnected from negotiations which led to the establishment of the Curie, the standard of radioactivity, in 1911. The establishment of this standard involved again the marginalization of Madrid's laboratory to the rank of customer of built-on scientific commodities, and required the readjustment of all measuring procedures to fit the new standard.

In 1919, after the exchange of correspondence between Muñoz and Marie Curie in relation to the secondary standard and her visit to Madrid to inaugurate the First National Congress of Medicine,[51] Curie was appointed honorary director of Madrid's Institute of Radioactivity. In informing Curie of his action, Muñoz requested her to send a portrait of her that would be exhibited in a small gallery in the Institute together with pictures of other "political figures distinguished because of their support to the Institute".[52] This attitude reveals that Muñoz and his laboratory peripheral condition was not only a category constructed in the process of legitimizing and creating scientific knowledge, but also a constituent element of his self-perception as scientist.

It is possible that contexts of justification and recognition could be locally established, creating heterogeneous perceptions about what is central or peripheral, and that different members of a scientific community have divergent views on scientific relevance, placing differently the frontier or main contributors in a particular area of research. However, the existence of centres and peripheries is not only a question of point of view, but a persistent feature of modern science. In this sense, my proposal to check for underlying factors (economy, academic organization, communication and instruments) to account for the emergence of centres and peripheries should be also understood as a contribution for resolving the most persistent tension in history of science: that between the local creation of science and the universal validity of its claims.

Notes

* This paper was a result of research project HUM2005-02437/HIST, funded by the Spanish Ministry of Education and Science. I would like to thank Simone Turchetti for advice and criticism.

[1] Traditional approaches, which historian Jeff Hughes categorises under the label "bomb historiography", used to situate radioactivity in a linear chain of events leading from Rutherford's theory to nuclear physics and the development of the atomic bomb and nuclear industry. This narrative has been recently challenged by works such as Hughes, J. (1990). *The Radioactivists: Community, Controversy and the Rise of Nuclear Physics*. University of Cambridge. unpublished PhD thesis, and Boudia, S. (2001). *Marie Curie et son laboratoire. sciences et industrie de la radioactivité en France*. Paris: Editions des archives contemporaines.

[2] I borrow the term from Hughes. *Radioactivists*.

[3] The term "radioactivity" appeared originally in a paper published by Marie Curie in 1899: "Les rayons de Becquerel et le polonium". *Revue Générale des sciences* 10: 41.

[4] Echegaray, for instance, talked of radioactive elements as "subversive elements". On the political readings of radioactivity in Spain, see Herran, N. (2006). *Radioactividad en España. Ascenso y declive del Instituto de Radiactividad.* unpublished PhD thesis. Barcelona: Universidad Autónoma de Barcelona..

[5] Muñoz popularized and implemented methods to fight the phllyoxeric pest, a parasite of the vine which ruined European vineyards in the 1890s.

[6] Muñoz del Castillo, J. (1901) *Discursos leídos ante la RACEFN en la recepción pública del Ilmo. Sr. D. José Muñoz del Castillo el día 3 de febrero de 1901.* Madrid: Imp. de L. Aguado.

[7] There is no complete account of the reception of Rutherford's theory. For a local survey, centred on American chemists, see Badash, L. (1979). *Radioactivity in America. Growth and Decay of a Science.* Baltimore: The John Hopkins University Press.

[8] His patrons included José María Rodríguez Carballo and Manuel Allendesalazar, for example. Rodríguez, professor of estereotomy in the University of Madrid, had worked with Muñoz in the Preparatory School of Engineers and was transferred with him to the Sciences Faculty, where he became Dean. In this position, he supported the establishment of Muñoz's Laboratory of Radioactivity. Allendesalazar was an agronomist and conservative politician. He held many public offices in the 1900s, such as Mayor of Madrid (1900), Minister of Agriculture (1903), Minister of Government (1905) and State Secretary (1908), as well as the direction of different public companies, such as the Bank of Spain and Tabacalera (the Spanish tobacco monopoly).

[9] The laboratory budget for acquiring scientific material was 17,500 pesetas in 1908. This same year, the global state expenditure for this was 200,000 pesetas. Muñoz del Castillo, J. (1908). *Cuarta memoria acerca del Laboratorio de*

radiactividad. Madrid: Hijos de M. G. Hernandez. Data on the state expediture
comes from *Gaceta de Madrid* June 17, 1908 (p. 1294)
[10] There is scarce historical literature on the relation between radioactivity and
hydrology. The most useful material has been published in Boudia, S. (1998). "Les
'fontaines de jouvence': thermalisme, eaux minérales et radioactivité. In Bordry,
M. and Boudia, S., eds. *Les rayons de la vie. Une histoire des applications
médicales des rayons X et de la radioactivité en France, 1895-1930.* Paris: Institut
Curie, pp. 126-30, and Boudia. *Marie Curie.*
[11] See, for example, Curie, P. and Laborde, A. (1906). "Sur la radioactivité des gaz
qui proviennent de l'eau des sources thermales". *Le Radium* 3: 195. A survey of
the scientific literature shows that these researches came out of fashion in the
1910s, after the establishment of the industry of radioactive elements.
[12] By 1900, spas were visited by 85,000 people a year, about 1% of the Spanish
population at the time. The public visibility of these practices was even more
important, as patients belonged to the wealthiest families. On the history of
medical hydrology in Spain, see Rodríguez Sánchez, J. (1998). "Los usos
regeneracionistas de la simbología del agua: entre la decadencia balnearia y el
moralismo Kneippista". *Dynamis* 18: 107-26.
[13] On the Vienna Institute of Radium, see Rentetzi, M. (2004). "Gender, Politics,
and Radioactivity Research in Interwar Vienna. The Case of the Institute for
Radium Research". *Isis* 95: 359-93. On the activities of Marie Curie's laboratory,
see Boudia. *Marie Curie.*
[14] Morales Chofré got a PhD in 1904 with a dissertation on radioactive surveying
and the use of the electroscope. Díaz de Rada doctored in 1906 with a dissertation
on the Contribution of radioactivity to chemical analysis. After staging in Paris
with a Spanish government grant, Díaz devoted his career to the analysis of
mineral waters.
[15] *Anales* was the journal of the recently created Spanish Society of Physics and
Chemistry (SEFQ). Established in 1901, the Society struggled for involving
Spanish scientists in research by providing its journal as regular basis for
communicating of their results. Despite Muñoz was the most prolific author in the
first five years of the journal, we should also note that such accomplishment
reveals more of his determination to control the society–in which he held the
presidency and a vice-presidency in the period 1907-8–than of his scientific
prestige among his peers. On the history of the SEFQ and its journal, see Valera-
Candel, M. and López-Fernández, C. (2001). *La física en España a través de los
Anales de la Sociedad Española de Física y Química.* Murcia: Servicio de
publicaciones de la Universidad de Murcia.
[16] The extension of the Banque du Radium facilities, located in Mouy dans l'Oise
and taking up a surface of 13,000 square meters, suggest a large-scale R+D
endeavour. Unfortunately, historical research on radio-agriculture is still
fragmentary and incomplete. For a short survey, see Boudia. *Marie Curie.*

[17] Gorria. H. (1914). "Radiocultura". *Ibérica* 32: 94-6. Paris Academy of Sciences also supported these researches, allocating many grants, of 2,000 francs each. See Petit, G. (1920). "La fertilité du sol et la radio-activité". *La Nature* 15 Oct, pp. 246-51.

[18] On the French and British cases, see Fournier, P. and Fournier, J. (1999). "Hasard ou mémoire dans la découverte de la radioactivité?". *Revue d'histoire des sciences* 52 (1): 51-79. 's works are collected in Stoklasa, J. and Penkava, J. (1932). *Biologie des Radiums und Der Radioaktiven Elemente*. Berlin: Von Paul Parey. On the American case, see Badash. *Radioactivity*, p. 147.

[19] La Redacción (1913). "El Instituto de Radiactividad en el primer semestre de 1913". *Boletín de Radiactividad* 5: 5-11. Muñoz del Castillo, J. (1913). "Primeros ensayos para el establecimiento de una técnica destinada a la determinación práctica de la radiactividad de las tierras de cultivo". *Boletín de Radiactividad* 5: 33-6.

[20] Muñoz del Castillo, J. (1913). "Comparación de la curva barométrica del mes de febrero de 1913 con la de las variaciones de la radiactividad del subsuelo del Observatorio del Instituto en el propio mes a la profundidad de 0,50 metros y a las cuatro de la tarde". *Boletín de Radiactividad* 5: 14-7.

[21] On the basis of his own theories, Muñoz distinguished between radioactivity of radium and radioactivity from other elements. The first was translated into Spanish as "radioactividad", while the generic term of the phenomena was called "radiactividad". This is the reason why his laboratory, institute and bulletin were called respectively of "radiactividad". Muñoz status as main expert in this field made that radiactividad was the form accepted by the Spanish Academia, diverging from the form of the French original term and subsequent translation into other languages.

[22] Some comparative experiments were published in Muñoz del Castillo, J. and Díaz de Rada, F. (1915). "Algunos datos de laboratorio acerca del influjo de las disoluciones rádicas y de las emanaciones radiactivas sobre varias substancias bioquímicas". *Boletín de Radiactividad* 7: 53-4, and Castrillo Martínez, A. (1915). "Diferencia de efecto de los nitratos de Tono y de Uranio, en el cultivo de trigo tremesino de la Estación Agronómica de Madrid (experiencia de 1915)". *Boletín de Radiactividad* 7: 179-84. Experiments on bacterial cultures were described in Gómez Fantova, M. (1915). "Primera nota acerca de mis experimentos relativos a la influencia de las radiactividades radica, torica, actinica y mixtas sobre las bacterias nitrificantes del suelo". *Boletín de Radiactividad* 7: 219-20.

[23] Muñoz del Castillo, J. (1913). "La radioactividad como factor de la vegetación". In *Actas del cuarto congreso de la AEPC*. Tomo 1. Madrid: Imprenta de Eduardo Arias, pp. 211-26.

[24] Muñoz del Castillo, J. (1920). "Los compuestos del metal radiactivo torio como maravilla agrícola y la prensa española corno factor social para un gran bien a la agricultura". *Boletín de Radiactividad* 12: 49-52.

[25] Thorionization techniques were increasingly refined, introducing technical terms such as "thorionization degree", measuring the number of cycles a plant lineage had been exposed to "thorionized" soils. Castrillo Martínez, A. (1915). "Experimentos de torianización efectuados en 1914 en el Instituto de Radiactividad". *Boletín de Radiactividad* 7: 57-62.

[26] Thorionization, like other Muñoz results, was unknow out of Spain. I have only found one reference in the international literature, in the Italian *Bulletin of Agricultural Intelligence.*

[27] It is remarkable that associations of this type, aiming at supporting science and vindicating more state and social support to its practitioners, were created all over Europe between the nineteenth and twentieth centuries. Their differences and particularities are still in need of a comparative study and I do not imply that their similar names implied equivalent functions. Nonetheless, we can cite as illustrative examples: the Gesellschaft Deutscher Naturforscher und Ärtze (1822), the British Association for the Advancement of Science (1831), the American Association for the Advancement of Science (1848), the Association Française pour l'Avancement des Sciences (1872) (later fusing with the Association scientifique de France, created in 1864). The Spanish Asociación para el Progreso de las Ciencias was established later, in 1908. On the British, German and American associations, see Bensaude-Vincent, B. (2003). "Chemistry". In Cahan, D. ed. *From Natural Philosophy to the Sciences.* Chicago: The University of Chicago Press, pp. 196-220, on p. 213.

[28] Muñoz. "La radioactividad como factor de la vegetación". On the AEPC, see Ausejo, E. (1993). *Por la Ciencia y por la Patria: La Institucionalización Científica en España en el Primer Tercio del Siglo XX. La Asociación Española para el Progresos de las Ciencias.* Madrid: Siglo XXI.

[29] The institute held five radio-agricultural weeks between 1915 and 1919 in five different Spanish provinces (Madrid, Logroño, Sevilla, Badajoz and Valladolid). Including an introduction to radioactivity and its measuring techniques, they were mostly devoted to promote radioactive fertilizers, thorionization and other techniques developed in the institute. The courses were attended by more than 1000 people, an audience mostly made up of professionals on the fields of agriculture (farmers, agronomists, etc.) and education (students, teachers, secondary school and university professors).

[30] Muñoz del Castillo, J. (1916). "La riqueza nacional radiactiva y los abonos radiactivos". *Boletín de Radiactividad* 8: 5-6.

[31] The service was officially recognised in 1914. Muñoz del Castillo, J. (1914). "Técnica oficial del Instituto de Radiactividad para la determinación práctica de la radiactividad de las tierras de cultivo". *Boletín de Radiactividad* 6: 36-9.

[32] On the control of fraud in this period, in Spain and other countries, in relation to foodstuffs, see also Ximo Guillem-Llobat's paper in this volume.

[33] La Redacción (1920). "Nota sobre las muestras-tipo de Fertilizador radiactivo, facilitadas por la Sociedad Henry Burnay y Compañía". *Boletín de Radiactividad* 12: 45.

[34] In my inspection of the *Boletín* between 1914 and 1920, I found references to analysis of soils from Toledo, Jaén, Madrid, Huesca, Oviedo, Badajoz, Las Palmas, Tenerife, Pamplona, Salamanca, Murcia, Palencia, Huelva, Sevilla, Valencia, Cádiz, Zaragoza, Coruña, Granada y Ciudad Real. The Institute performed on a regular basis analysis of minerals' radioactivity, such as Gil Bermejo's systematic study of uranium minerals of the Iberian Peninsula.

[35] See, for example, La Dirección (1916). "Circular que el Instituto de Radiactividad dirige a los señores mineralogistas, farmacéuticos, mineros y a cuantos, con uno u otro fin, dedican su esfuerzo al estudio de la naturaleza, o a la busca de minerales explotables". *Boletín de Radiactividad* 8: 168-70.

[36] On Szilard, see Palló, G. (1981). "Szilárd Béla tudományos életrajza". *Századok*, 115: 770-798. I thank the author for personally translating parts of this article into English.

[37] On the Torres Quevedo Institute and scientific instrumentation in Spain, see Xavier Mañes's paper in this volume. See also Romero, A. (1998). "Dos políticas de instrumental científico: el Instituto del Material Científico y el Torres Quevedo". *Arbor* 160 (631-2): 359-386.

[38] After being restored by historians of the CSIC, the instruments are currently kept at Madrid's National Museum of Natural Sciences. On the instruments' restoration project, see Moreno, R.; Romero, A., and Redrajo, F. (1996). "La recuperación de la instrumentación científico-histórica del Consejo Superior de Investigaciones Científicas". *Arbor* 153 (603): 9-54.

[39] In the 1930, the curie was redefined as the quantity of any radioactive substance in which 3.7×10^{10} disintegrations occur each second.

[40] The official announcement is found in *Gaceta de Madrid* 37 (February 6, 1920), p. 444.

[41] Some examples are Benjamín Navarro articles on radioactivity in the popular science journal *Ibérica*, or Leonardo Peris' book *La radiactividad y la electricidad en la agricultura*, (Radioactivity and Electricity in Agriculture), which argued for the economic viability of radioactive fertilisers and for the industrialisation of agriculture. Navarro, B. (1917). "De radioactividad". *Ibérica* 178: 345-50, 187: 75-8, and 180: 109-11. Peris, L. (1918). *La radiactividad y la electricidad en la agricultura*. Valencia: Tip. JM Destruels.

[42] Gorria, H. (1914). "Radiocultura". *Ibérica* 32: 94-6. All translations from Spanish are by the author of this paper.

[43] Muñoz Burjau, V. (1916). "Experiencias con abonos radiactivos efectuados en Tortosa en el año de 1915". *Boletín de Radiactividad* 8: 58-60.

[44] Muñoz del Castillo, J. (1920). "Una iniciativa patriótica de S. M. el Rey Don Alfonso XIII, como primer agricultor de España". *Boletín de Radiactividad* 12: 43-44.

[45] The experiments were mainly performed on corn and potato crops, and included thorionization tests with different doses of this element. Muñoz del Castillo, J. (1920). "Preparación de experiencias en la Moncloa (Instituto Agrícola de Alfonso XII) en virtud de la Real orden de 11 de marzo de 1920". *Boletín de Radiactividad* 12: 56-64.

[46] The press conference was held in the headquarters of the Press Association in March 30, 1920. Muñoz del Castillo, J. (1920). "Los compuestos del metal radiactivo torio como maravilla agrícola y la Prensa española corno factor social para un gran bien a la agricultura". *Boletín de Radiactividad* 12: 49.

[47] Díaz was appointed in July 1920 (Real Orden July 7, 1920). A year before (March, 1919), Muñoz had requested Diaz's appointment as vice-director, by putting forward that he needed assistance because of sight problems caused, in his opinion, by his continued stay in "activated" premises.

[48] Díaz de Rada made his living from at least three different jobs: he worked as lecturer in Madrid's industrial school, as assistant in the Faculty of Sciences at the University of Madrid, and as assistant chemist in the Institute of Radioactivity. *Díaz de Rada personal files*, AGA (5) 31/15633.

[49] See, in particular, Maylin, A. (1921). *Los abonos industriales*. Barcelona: Soler; Soroa, J. M. de (1921). *Los abonos baratos*. Barcelona: Calpe; Navarro de Palencia, J. (1921). *Los abonos del trigo*. Barcelona: Calpe, and the *Hojas divulgadoras de la dirección general de agricultura, minas y montes*.

[50] Muñoz efforts to promote science, and his emphasis on utilitarianism and insertion in a nationalistic discourse as consolidation strategy are analogous to those of scientists from other countries, such as Marie Curie.

[51] The congress was held from May 20th to 25th, 1919

[52] Imaz Eraso, F. (1919). "Mme. Curie, Directora honoraria del Instituto de Radiactividad". *Boletín de Radiactividad* 11: 148-54

MAKING SCIENCE IN THE PERIPHERY:
DETERMINATION OF CRYSTALLINE STRUCTURES IN SPAIN, 1940-1955

XAVIER MAÑES

This paper uses a case study, the development of crystallography in Francoist Spain, to reflect on the historical use of the centre-periphery model. I will argue that, although the dictatorial regime, international isolation and a lack of international leadership in any branch of science did affect the country in this period, by no means did they force a passive and uncreative attitude amongst Spanish researchers. The latter were not mere recipients of the science and technology produced in leading foreign countries, as traditionally portrayed by centre-periphery models. In spite of their lack of resources and their dependence on research produced abroad, Spanish scientists actively appropriated foreign research, and made original contributions in their fields of research, adequate to their needs and practices, although they seldom found recognition outside their local national context.

Moreover, Spanish crystallographers' professional practices contributed to the configuration of Determination of Crystalline Structures (DCS) as a discipline in an original way, different to its development in other countries.[1] In particular, I will argue that the traditional use of the centre-periphery conceptual frame does not take into account of the complexities of scientific practice. The categories of centre and periphery are subject to temporal variation, inexact if applied to a compact national unit of analysis, and dependent on the perspectives of both historical actors and historians.

The first section of this paper introduces the subject of DCS, and its practice in Spain. After presenting the work performed before the Spanish Civil War (1936-1939), I will briefly describe subsequent developments, strongly marked by the economic policy of the first years of dictator Franco's regime. I will present the different active research groups in this period, and analyse their strategies to acquire technology and the deficient role of local industry in that process. Finally, I will frame Spanish research on DCS in the international context, and will study the appropriation of foreign developments in this field by Spanish scientists, as well as their original work.

An historical introduction to the Determination of Crystalline Structures: centres of production and the shaping of a new discipline

In 1912, the German physicist Max von Laue discovered that crystals could diffract X-rays, causing a revolution in crystallography. Until then, this field had been limited to the study of the external morphology of crystals, in a close relationship with mineralogy. X-ray diffraction made it possible to analyse crystalline structures, in order to reveal their atomic organization. [2] During the following years, the structures of many inorganic crystals were elucidated, and many research groups began to study those of organic crystals, which were more difficult to determine. [3]

During the 1920s and the 1930s, along with the study of more and more complex compounds, more powerful X-ray diffraction equipments were developed. Some landmarks were the Weissenberg machine, built in Germany in 1924,[4] and the precession machines, developed in the United States during the 1940s. Those machines, like many others built, were required for the data collection process previous to the determination of the structures. The configuration of this new field of study thus favoured a high degree of technological dependence.

Together with the precision and power of the machines involved, calculation methods were the other critical point in the determination of structures. In order to use all the collected data, different methods were developed to make the elucidation process of a structure easier. One of the landmarks was the use of Fourier series, which began to be systematically applied in the United Kingdom in the late 1920s. Thanks to this and other improvements in calculation methods, it was possible during the 1940s to conclude the determination of two-dimensional structures and to start to elucidate three-dimensional ones.

During these years, Germany, the United Kingdom and the United States were the countries that led this new field. This was due to their tradition of research, the extent of the resources devoted to it, and the existence of a well-established industry. In that sense, it is worth noting that the material importance of metals such as copper, aluminium and, above all, iron and its alloys, contributed to favour a close connection between X-ray diffraction and industry. This was especially remarkable after the Second World War, and it resulted in an important increase in the links between private companies and academic institutions. [5]

The new technique quickly took a central role in many industrial laboratories, for example, in areas such as organic chemistry, metallurgy and the textile industry.[6] The increasing use of X-ray diffraction offered important advantages in DCS research, and new computer programs were rapidly adapted to crystallographic research once the first computers were developed in the mid 1950s.

Among scientists, physicists were the most involved in the development of these new studies, followed by geologists and chemists.[7] In fact, the increasing interest in this research subsequently led physicists to try to integrate the new body of knowledge into the field of solid state physics.[8] In parallel, other scientists tried to include DCS in the field of geology. This led to the foundation, in 1947, of an international organization devoted to the new trends of crystallography, the International Union of Crystallography (IUC). The creation of the IUC is a good example of the tensions that took place between physicists and geologists to obtain the monopoly on DCS research,[9] illustrating that the process of professionalization and of the configuration of the discipline was not a simple one. As I will show, the interdisciplinary development of DCS was also a feature of the Spanish case. However, the professional development of DCS in Spain also offered highly distinctive features.

The introduction of DCS in Spain

The introduction of X-ray diffraction techniques in Spain took place in a way that illustrates the scientific backwardness of the country at that time, and the perception Spanish scientists had of it. At the beginning of the twentieth century, a Spanish physicist referred to the practice of his discipline in these terms:

> We can be sure that physics is the branch of science in which Spanish spirit has done the least; it is extremely rare to find a Spanish name connected to original studies.[10]

This feeling of backwardness among Spanish intellectuals – the central issue of the so-called Spanish science *polémica*[11] – a generalized feeling of irremediable decadence and endemic incompetence in science, in fact led the country to a movement of regeneration. As a consequence, Spanish research experienced a certain improvement during the 1920s and the 1930s, a period that has been referred to as the "silver period" of Spanish science.[12]

The first Spanish scientist to present Laue's discovery to the public was Francisco Pardillo, geologist and professor in the Faculty of Sciences at Barcelona.[13] Pardillo's response occurred only a few months after Laue's work. Although he had been perfectly aware of the importance of X-ray diffraction for the future of crystallography since the early 1910s, he had to wait until 1925 to have even rudimentary X-ray diffraction equipment at his disposal, mainly built by himself "with only a few pieces acquired in the market, with no money from the University".[14]

We have almost no details about the construction of that equipment, for no work related to its use was published, but Pardillo's limited resources were only too evident. Only in the 1930s did Pardillo have access to a more suitable X-ray diffraction machine. Even so, the lack of assistance and the start of the Spanish Civil War meant that only two papers using the new machine were published by the Catalan team during the pre-war years.

Pardillo was not the only Spanish scientist who tried to work on DCS. Another team was created in 1922 around the physicist Julio Palacios in Madrid, at the Sección de Rayos X (X-Rays Section) of the Laboratorio de Investigaciones Físicas (LIF), the institution created by the Spanish government to promote physics and chemistry in the country.[15] The X-ray diffraction facilities at the LIF were built at the same time as those in Barcelona, but Palacios had access to much greater resources than Pardillo. First of all, he received funds from the Rockefeller Institution, holding a funding program for science in Spain. Moreover, the team benefited from an important donation from an Argentinian institution, the Spanish Cultural Institution of Buenos Aires.[16] This organization gave the money to promote Spanish research in a leading branch of science. The five-year funding program lasted from 1928 to 1933 and took the name of "Cátedra Cajal" (Cajal Professorship). Funded by the Republic of Argentina, the Professorship was intended to be "run by a renowned foreign scientist, helped by Spanish professors", and called for the assistance of the Spanish government in providing the required laboratory space and instrumentation.[17]

The foreign scientist chosen to hold the Professorship was the then Director of The Polytechnic School in Zurich, Paul Scherrer, and the subject chosen was DCS. Other foreign scientists such as W. L. Bragg made visits to Madrid, helping Scherrer and Palacios with the research activities of the Professorship. The strong promotion of DCS at the LIF, with reasonably well-equipped laboratories, enabled such studies to achieve, in a few years, a clear central role in Spanish science. Taking the *Anales de la Sociedad Española de Física y Química* as a sample,[18] DCS studies represented 68% of the articles published during the years 1932-1936, almost all of them from the Madrid group. This team produced twenty-four papers in which X-ray diffraction machines were used, which provides a striking comparison with the team from Barcelona. This was also due to the large number of collaborators in Palacios' team, the availability of external funding allowing him not only to hire renowned scientists, but also to give several grants to young researchers.[19] Although the period of time in which Palacios' team functioned was quite short,[20] both the quality of the work done and the contacts with renowned foreign scientists allowed them to influence, to some extent, the international DCS debate.[21]

The Civil War and its consequences: Spanish isolation and Franco's autarky

Not surprisingly, the Civil War (1936-1939) had very important consequences for the whole of Spanish society, science included. Apart from the scientific break that the war itself imposed, post-war shortages made it difficult to resume scientific work. Moreover, the new regime implemented an economic policy that was not precisely suitable for the practice of science. The economic context of the autarkic period (1939-1959) is essential in understanding the evolution of science and technology during the beginning of Franco's dictatorship. This economic policy was based on the idea of self-sufficiency and the launching of an economic development plan independent of the international context. This choice of direction was reinforced, once the Second World War was over, by Spanish political isolation in the international sphere, characterized by a blockade that made it extremely difficult to import goods, including equipment needed for industrial and scientific research.[22]

The blockade also affected the movement of Spanish scientists, who had serious difficulties when it came to travelling, therefore preventing their attendance at most international meetings. As a result, direct contact with foreign scientists, which had proven to be very important during the Cajal Professorship years, was practically cut off. That isolation remained until the 1950s, when the political strategies of the Cold War turned Franco's anti-communism into a close ally of the United States, leaving the way to a policy of a certain amount of openness.

The first years of Franco's regime were characterized by a discourse in favour of the industrialization of the country, with a close bond between national defence and industrialization. As a result, the technological development model promoted by the regime was limited to certain fields like those of fuel, minerals and transports, leaving other areas aside. In addition, the autarchic policy was not able to achieve the pre-war levels of production until the mid 1950s, causing chronic strangulation of Spanish economy, the effects of which lasted until the 1980s.[23]

Thus, it is clear that both Franco's economic policy and the shortages caused by the Civil War were very detrimental to scientists' interested in acquiring the technology needed for their research. This was the case for DCS research, deprived of international funds like those that the Cajal Professorship had secured. By the 1950s this situation started to change slightly, due to an increase in the budget assigned by the government to the Consejo Superior de Investigaciones Científicas (CSIC), the institution founded by the Franco regime, from the ashes of the republican Junta de Ampliación de Estudios, to promote science in Spain.[24] Nevertheless, this was not the case for the universities, which still had their budgets limited to teaching. As a result, the degree in which each scientific team was linked to the CSIC was extremely relevant for the amount of resources available to them.

The resumption of DCS studies: Spanish teams in 1940-1955

Franco's repression of Spanish academia has been quite well studied in the last thirty years.[25] In the case of physics, some of the most outstanding researchers during the pre-war years either went into exile or were marginalized by the new authorities. This was true of Blas Cabrera (electromagnetism), Arturo Duperier (meteorology) and Miguel Catalán (spectroscopy). However, Palacios and Pardillo did not had few problems in this sense;[26] the former declared support for Franco, while the latter kept a strategic equidistance with the two fronts during the war. Both scientists went on with their DCS research once the war had finished. DCS studies in Spain during the period 1940-1955 were organized around three groups, in Madrid, Barcelona and Seville.

In Madrid, the "X Rays Section" resumed its work in the Institute Alonso de Santa Cruz under the direct sponsorship of the CSIC. By 1940, six researchers were working on DCS under Palacios' direction.[27] Until 1947 the size of the team remained stable; subsequently, it started to grow considerably, configuring a stable group of about fifteen scientists by 1955. This group was primarily composed of chemists, plus several physicists and some engineers. In Barcelona, meanwhile, Pardillo resumed his research at the University of Barcelona with the help of only three researchers. Like Palacios' group, Pardillo's team increased in size by the mid 1940s, reaching a stable size of about ten scientists by 1955. Apart from some chemists, nearly all researchers in this team were geologists, since the group was situated in a Department of Geology.[28] A third team joined these two groups in the mid 1940s. Located in the Chemistry Section at the University of Seville, the team was formed around the physicist Luis Brú just after his arrival. Due to the characteristics of the department, all the scientists that helped Brú were chemists. For the same reason, they did not start DCS investigations from the beginning. As a result of the fact that the limited facilities of the department were only considered suitable for chemistry, the team had to wait until 1946 to have the minimum equipment to do research in DCS. Finally, like the other two teams, the Seville group had to wait until the mid 1950s to have a stable team with a critical mass of researchers.

The supremacy of chemists in Spanish DCS contrasts with the prominence of physicists in the majority of foreign countries where these studies were conducted. Apart from the singular case of the Seville team, two reasons contribute to explain this fact. Firstly, research in DCS did not have a strong connection with solid state physics from the beginnings of its institutionalization in Spain.[29] Secondly, the CSIC employed, in the post-war years, a much greater number of chemists than any other kind of scientist.[30]

Also relevant is the quantity of resources available for each group. As stated before, the degree of connection to the CSIC was crucial. The Madrid team was in a privileged situation with respect to the other two teams, which were both located in a university and only linked to the CSIC as Local Sections.[31] The geographical distribution of the CSIC in Spain was such, that for the period 1940-1955, Madrid agglutinated 68% of research centres [32]– this was the prevailing centralism of the moment. To some extent, we can refer the situation of the Barcelona and the Seville teams as one of a double periphery, as in addition to forming part of a peripheral and isolated country, they had fewer resources than the Madrid centre, which obviously had an impact on their possibilities to conduct research. However, it would be a mistake to ascribe the strong centralism and CSIC privileges only to Franco's regime. This Spanish political geography has a longer history. As I have already discussed, all the funds from international organizations in the pre-war years were not assigned to Pardillo's team, but to that of Palacios in Madrid.

Another striking fact is the growth of all the teams during the 1950s. This was mainly due to the increase of CSIC sponsorship in those years. Once the economic shortages of the post-war period had partially decreased, Franco's government made a slight effort to promote science in Spain, through the increase of grants for young researchers.

By 1955, Spanish teams in DCS had reached some stability in its number of members, thus moving forward to a certain degree of consolidation. However, the production of the three groups clearly shows their backwardness in relation to research produced in the countries that kept the leadership in this subject.[33] For instance, no three-dimensional structures were elucidated in Spain during the entire period, something which was quite common in the leading countries. This backwardness was also perceived by Spanish researchers, as shown in their contemporary perspectives:

> In our country, although DCS research has been promoted in these last years, the use of X-ray diffraction techniques is still quite limited ... X-ray facilities, set up in several places in Spain, are used for scientific and technical purposes, but the small number of laboratories hampers greater development of these activities.[34]

Having stated that Franco's repression did not seriously affect pre-war DCS scientists, we have to look for the causes of this setback in the economic and technological policies of the new regime, which were influenced by the new political situation. However, it is worth noting that, despite the different composition of the teams, the direction of the research conducted in the three Spanish groups was rather similar, all of them being interested not only in the determination of structures, but also in the development of technology and associated calculation methods. The only difference was the existence, in the geologists' team, of simultaneous DCS work along with research in crystallography and mineralogy, something that is easily explained if we take into account their own research tradition.

The issue of technology and the role of industry in regards to it

The Institute Leonardo Torres Quevedo of Scientific Material (ITQ) was the branch of the CSIC created to promote the autonomous technological development applied to its research centres. The link between the philosophy of the ITQ and the regime's autarky policy was clear:

> This Institute is in charge of all equipment. Anything broken is to be repaired in our workshop. With regard to new equipment, the centres have absolute freedom to purchase them where they prefer, and only in case of not finding them in the market [the Spanish market, well understood], this Institute will design and produce them. This is the way to proceed, in order to not damage our national industry of scientific equipment and to achieve,

at the same time, the creation of new equipment in our Nation. As a result, we increase our scientific, industrial and economic potential and, besides, it allows the restriction of scientific equipment import.[35]

Despite the fact that Spanish development of technology was basically focused on matters related to national defence, the self-sufficiency discourse did not have a large impact in the area of scientific and technological advancement at the ITQ.[36] In addition, and by contrast to what happened in the leading countries, industry hardly supported technological development in DCS in Spain. So, with regard to DCS technology, the initiative of individual researchers organized in teams were much more important than the efforts of State and industry. For example, in order to mitigate the lack of equipment in the Barcelona team, the geologist Josep Lluís Amorós proposed to the ITQ the construction of several X-ray diffraction machines from his own modified designs. Amorós' success led him to take part in a program for the construction of crystallographic equipments at the ITQ.[37] In addition, he collaborated with the "Sección de Electricidad y Radiaciones" (Section of Electricity and Radiations) of the CSIC, where two physicists helped him with theoretical work for the construction of another X-ray diffraction machine.

Although the Madrid team kept some of the equipment from the Cajal Professorship, they also urgently needed new machines to keep pace with international research, as DCS technology developed very quickly during the 1930s and the 1940s. The physicist Fernando Huerta was the most active member of this team in the development of technology, designing three different X-ray diffraction machines. Although he did not commercialize them, he published the designs, believing that his contribution would be appreciated. In his most successful machine, Huerta included original work, replacing the translation movement of the standard Weissenberg's machine film, with a helicoidal movement. However, this idea was assessed years later in the following terms:

> *An unfortunate anachronism occurred in the development of the Weissenberg method.* Annoyed by the non-symmetrical appearance of symmetry in the Weissenberg photograph, F. Huerta showed that, by giving the cylindrical Weissenberg camera a screw motion, rather than only a translation motion, the intersecting lines of symmetry of reciprocal space could be preserved as true symmetry in the Weissenberg photograph. But this was only achieved by complicating an already complicated mechanism. Meanwhile *precession cameras* with their more faithful symmetry reproduction *were already replacing Weissenberg cameras, so the helicoidal Weissenberg motion never became popular. It probably would have achieved popularity if presented twenty years earlier.*[38]

The precession machines, developed in the United States during the mid 1940s, were not introduced in Spain until the 1960s. Huerta's was a very good idea, but it could not overcome the constraints of Spanish backwardness and communication difficulties. This case represents a good example of how peripheral countries could produce new knowledge by the appropriation of ideas originating in other countries. Huerta's work does not fit the traditional centre-periphery model; on the contrary, it reinforces the idea of the periphery as a dynamic – although isolated and resourceless – place.

With regard to equipment for facilitating DCS calculations, in 1953 the Madrid team was the first of the three Spanish groups to develop a mechanical adding machine for the calculation of Fourier series terms. This is another good example of the magnitude of the backwardness in Spain, where until three years earlier no one had used technological methods for the reduction of the calculation involved in the elucidation of crystalline structures, as remarked by one contemporary:

> Until 1950 nobody in Spain had used mechanical methods to solve DCS calculation operations, and the few studies done by the means of Fourier series, had been made directly, with only the help of a simple calculation machine.[39]

The Barcelona team also did some work on this subject, producing a simple mechanism that determined the exact location of the atoms in the last structures' calculations in its two-dimensional projection,[40] and they proposed some modifications in tabulator machines to adapt them to the calculation of crystalline structures.[41]

Another important contribution in this field was made by the Seville group. They devised and successfully applied several modifications in the adding machine, which had previously been developed by the French scientist Gerald von Eller. Apparently, these were not minor improvements, according to the inventors' own presentation of their design to the Sociedad Española de Física y Química:

> Eller created an optical device to calculate the structure's factors for molecules, where atoms have an atomic factor represented by the same function, except for one constant. This limitation turns out to be of little help to the procedure. We now present a modification, which makes it possible to calculate the structure's factors for all the forms of the atomic factor function of its atoms. We can now calculate quickly and simply the structure factors' signs.[42]

Furthermore, according to the Seville team, Eller admitted the importance of their improvement:

> In the 1956 International Symposium of Crystallography, Eller told us that we had benefited more from his machine than even himself. That's how we used to exhaust the poor resources that we had.[43]

However, the policy of openness that came with the 1950s made technological imports possible again. From then on, the definitive openness of Spain to the world discouraged attempts like these, and a long process of technological dependence started.

As already remarked, the practical importance of X-ray diffraction made it so that, especially from the 1940s on, its use was extended to several areas of industry in many countries with a dynamic interaction between industrial and academic institutions. However, this was not the case in Spain, where the role of industry was very limited, as remarked by one of the main Spanish DCS researchers:

> Research work is now in progress in Spain, with financial support from the CSIC, on many theoretical and practical problems, although there is not, at present, much technological application of the methods of X-ray crystallography in industry.[44]

Aside from the lack of input from the state, the private sector also did little to encourage the work of Spanish DCS research groups.[45] The first industrial donations to these began, in small proportions, in the mid 1950s. This was due to the Spanish post-war economic situation, apart from its industrial tradition and the regime's strong protectionism, which did not precisely encourage industry's involvement in scientific research. Apart from some minor collaboration between the Madrid team and the ceramics sector, Barcelona was the only group that had some remarkable connections with industry, which would later represent the beginning of an important collaboration with the Catalan pharmaceutical industry. Regardless, we must insist on the limited relevance of those collaborations during the period 1940-1955, which consisted in the awarding of a few modest grants and some funding to buy laboratory equipment.

Conclusion

DCS research in Spain during the period 1940-1955 can be regarded as a continuous effort of scientific updating in a field of a rapid growth. Obviously, the Civil War and the shortages that it caused, with the

isolation of Spain from the international scientific community, a deficient
economic policy and very serious difficulties in importing foreign
technology did not help Spanish scientists in their efforts to keep pace with
international research. As a result, the state of DCS research in Spain was
characterized during this period by backwardness, despite the good work
done in the pre-war years. Thus, Spain certainly did not lead DCS
research, remaining dependent on scientific production and advancements
taking place in other countries.

However, it would be a mistake to consider the leading countries in
this field as the only centres of production. As the strategies of Spanish
scientists to develop DCS-oriented technology suggest, they tried to
overcome their material shortages by themselves, developing X-ray
diffraction machines and calculation methods technology with their
limited resources. In fact, Spanish DCS investigators, to some extent,
appropriated foreign DCS technology; for example Huerta's designs,
which could have resulted in useful technologies, although they finally
failed to do so. But this also means that Spanish DCS scientists produced
some original knowledge, remaining peripheral with regard to the
scientific mainstream but not merely devoted to the reception of new
advancements, thus prompting us to ask ourselves about the validity of the
traditional centre-periphery model. Paradoxically, the lack of a well
established national industry, and of its connection with scientific and
technological research, was a relevant factor for the initiative taken by
Spanish scientists.

Adverse economic, technological and communicational conditions
obstructed scientific research but, at the same time they encouraged
creative work, despite its difficulty. As remarked in the case of Pardillo,
this scientific creativity also took place in the pre-war years, but
paradoxically it increased when the isolation of the local community was
greater. The Spanish context also offers originality in relation to the
particular professional configuration of DCS as a discipline, since research
in this field mainly conducted by chemists, by contrast with the supremacy
of physicists in other countries – a fact worth noting for prospective
comparative studies of its international development.

Furthermore, in spite of the contemporary perspectives of Spanish
intellectuals, who in general assigned an intrinsic character to Spain's
backwardness in science, the peripheral status of the country in this field is
characterized by a greater complexity and variations over time. Although
Spain had a peripheral position in DCS from the 1910s to the 1950s, its
relation to the leading countries in this field changed with time. During the
short period of the Cajal Professorship, the availability of more resources

and the close communication with renowned foreign scientists prompted DCS in Spain to approach the scientific level of most developed countries.

Finally, the definition of Spain as a peripheral state shows that it is problematic to use the concept of nation as a unit of analysis, as every country encapsulates regional differences. The deeply-rooted Spanish political and economic centralism, exemplified by the very different budget assigned to the several DCS teams, warns us about the risk of making bold generalizations for an entire country. The Barcelona and Seville teams suffered from a double periphery in terms of economic and technological resources, not having the same means to conduct research as teams established in the Spanish capital.

Notes

[1] I will use the concept of 'discipline' as framed by Rudolf Stichweh, considering that their formation is both related to the configuration of well-defined sets of problems and to the establishment of clearly defined professional profiles. See Stichweh, R. (1994). "La structuration des disciplines dans les universités allemandes au XIXe siècle". *Histoire de l'éducation* 62 (mai): 55-73.

[2] For a critical and complete study on Laue's discovery and its use in the determination of crystalline structures, see Forman, P. (1969). "The Discovery of the Diffraction of X-rays by Crystals. A Critique of the Myths". *Archive for History of Exact Sciences* 6: 38-71.

[3] The first inorganic structure was determined in 1913; the first organic one elucidated in 1923.

[4] Although, until 1934, the technique involved was not simplified enough to commercialize the machine.

[5] The importance of the links between research, industry and the military in the American case, led to the post-war situation, where the US clearly took the leadership in the new discipline.

[6] Examples of the research produced under those interests include Rooksby, H. P. (1941). "Application of the X-ray powder method in industrial laboratory problems". *Journal of Scientific Instruments* 18 (5): 84-90; Hattiangli, G. S. (1949). "Characterization of some commercial soaps by X-ray diffraction". *J. Research National Bureau of Standards* 42: 331-41; Susich, G. (1950). "Identification of organic dyes by X-ray powder diffraction". *Analytical Chemistry* 22: 425-30.

[7] On the international evolution of X-ray diffraction and DCS investigations, and for the technical details involved, see Ewald, P. ed. (1962). *Fifty years of X-ray diffraction.* Utrecht: N. V. A. Oosthoek's Uitgeversmaatschappij.

[8] In the 1930s in the USA, solid state physics began to acquire a well defined identity inside the vast field of physics, and this process was completed towards

the end of the 1950s. The Division of Solid-State Physics (1947) of the American Physical Society was the first institution exclusively devoted to solid state physics. For a complete exposition about the evolution of solid state physics, see Weart, S. (1992). "The Solid Community". In Hoddeson, L. ed. *Out of the Cristal Maze: Chapters from the History of Solid-State Physics.* Oxford: Oxford University Press., pp. 617-62.

[9] These tensions prompted the International Union of Pure and Applied Physics (IUPAP) to oppose the foundation of the new institution. Kamminga, H. (1989). "The International Union of Crystallography: Its formation and early development". *Acta Crystallographica* A45: 6.

[10] González Martí, I. (1917). "Estado de la enseñanza de la física en las Universidades de España". *Actas del Sexto Congreso de la Asociación Española para el Progreso de las Ciencias* I: 35-57. All translations from Spanish are by the author of this paper.

[11] On the Spanish *polémica*, which goes back to the 18th century, see López Piñero, J. M. (1969). *La introducción de la ciencia moderna en España.* Barcelona: Ariel; García Camarero, E. (1970). *La polémica de la ciencia española.* Madrid: Alianza. For a revision of the question, see Nieto, A. (1999). "The Images of Science in Modern Spain. Rethinking the *Polémica".* In Gavroglu, K. ed.. *The Sciences in the European Periphery during the Enlightenment.* Dordrecht: Kluwer, pp. 73-94.

[12] On the Spanish scientific "silver period" see the several chapters included in Sánchez Ron, J. M. (1999). *Cincel, Martillo y Piedra: Historia de la Ciencia en España (siglos XIX y XX).* Madrid: Taurus. For the specific case of physics, see Sánchez Ron, J. M. and Roca, A. (1993). "Spain's first school of physics: Blas Cabrera's Laboratorio de Investigaciones Físicas". *Osiris* 8: 127-55.

[13] Pardillo, F. (1913). "Descubrimientos recientes sobre la estructura de los cristales". *Boletín de la Real Sociedad Española de Historia Natural* 13: 336-39.

[14] Candel, R. (1955). "El Doctor Francisco Pardillo Vaquer (1884-1955)". *Archivo Rafael Candel Vila (1903-1976).* Barcelona: Museu Geològic del Seminari de Barcelona. Rafael Candel was one of the few scientists who worked with Pardillo.

[15] About LIF, see Sánchez Ron and Roca. "Spain's first school". The LIF formed part from the Junta para Ampliación de Estudios e Investigaciones Científicas (JAE), the institution founded by the government to promote Spanish development in all areas of knowledge.

[16] It seems highly relevant that the main impulse to such investigations in Spain was due to foreign capital.

[17] "Cátedra Santiago Ramón y Cajal para Investigaciones Científicas: Homenaje de la colectividad española de la República Argentina", document dated on 1922, December the 1st. *Archivo de la Secretaría de la JAE (1907-1939).* Ref. 154-31. Madrid: Archivo de la Residencia de Estudiantes.

[18] *Anales* was the journal of the Sociedad Española de Física y Química (SEFQ), founded in 1903, being until the 1960s the main vehicle for the spreading of work in physics and chemistry produced in Spain.

[19] While Pardillo had only a couple of assistants, Palacios and Scherrer worked with an average of twelve collaborators during the years that the Professorship lasted.

[20] The Professorship finished, as had been agreed, in 1933. Nevertheless, the equipment and the momentum of the team encouraged a high level of production of papers until the beginning of the Spanish Civil War.

[21] Although not without controversies, it has been stated that a method of detection attributed to J. M. Robertson, the multiple film method, has parallels in some of the work of Madrid's team, whose results were published in the *Chemical Abstracts*. For more details about the work done by Palacios' team and the rest of the groups mentioned in this section, see Mañes, X. (2005). *Determinación de estructuras cristalinas en España: Inicios, desarrollo y consolidación (1912-1955)*. Barcelona: Universitat Autònoma de Barcelona. unpublished MA dissertation.

[22] On autarky see for example Edgerton, D. (2006). *The Shock of the Old: Technology and Global History science 1900*. London: Profile Books, pp. 115-9.

[23] Sanz, L. and López, S. (1997). *Política tecnológica versus política científica durante el franquismo, Documento de Trabajo 97-01*, Madrid: CSIC, p. 3. For a more complete report on the Spanish economic context, see Clavera, J. ed. (1973). *Capitalismo español: De la autarquía a la estabilización (1939-1959)*. Madrid: Edicusa.

[24] See n. 15.

[25] On scientists' repression in Spain, see Claret, J. (2004). *La repressió franquista a la universitat espanyola*. Barcelona: Universitat Pompeu Fabra. unpublished PhD dissertation; Otero, L. (2001). "La destrucción de la ciencia en España. Las consecuencias del triunfo militar de la España franquista". *Historia y Comunicación Social* 6: 149-86. On the question of the Spanish exile, see Barona, J. L. and Lloret, J. B. (2000). "La historiografía sobre el exilio científico tras la II República". *Cronos* 3 (2): 393-408.

[26] Although a Barcelona team member, Rafael Candel, had to go into exile, and could not come back to Spain until 1950. Another scientist who had problems with the new regime was Luis Brú. A Palacios' team member, Brú had won a professorship at the University of La Laguna in 1935. After the war, in 1942, Brú was permitted to move to the University of Seville.

[27] Only one of these scientists had worked with Palacios at the LIF, with the average number of members at the LIF's team being about twelve.

[28] At the University of Barcelona, the Natural Sciences Section of the Faculty of Sciences gave place, in 1953, to the Faculties of Geology and Biology.

[29] A "Sección del Estado Sólido" (Solid State Section) was created in the same Institute Alonso de Santa Cruz in 1953. Since they were interested mainly in theoretical research, this team had almost no connection with the "X Rays Section" at the same Institute.

[30] Concretely, 42% of the CSIC's researchers were chemists, but only 11.5% were biologists and 1.9 % were physicists. To explain this, it has been suggested that

several of the leading members of the CSIC were chemists. González, P. and Jiménez, J. (1979). "La investigación en el Consejo Superior de Investigaciones Científicas. Estudio de un grupo significativo durante el período 1940-1955". In González, P.; Jiménez, J., and López Piñero, J. M., eds. (1979). *Historia y sociología de la ciencia en España*. Madrid: Alianza, p. 155.

[31] For example, the budget (in pesetas.) assigned by the CSIC to the three teams was, in 1950, as follows: 25,570 for Seville, 46,749 for Barcelona and 256,165 for Madrid (to share with a modest Optics Section). *Memorias CSIC (1950)*, 413-18.

[32] González, Jiménez, and López Piñero. *Historia y sociología*.

[33] This feeling of backwardness, that goes back to the Spanish *polémica* (see n. 11) was common in Spanish science during the 20th century. See Santesmases, M. J. (2001). *Entre Cajal y Ochoa: Ciencias Biomédicas en la España de Franco, 1939-1975*. Madrid: CSIC.

[34] Asensio, I. (1954). "Instalación de un laboratorio de rayos X". *Boletín de la Sociedad Española de Historia Natural*, 52: 135. Asensio was a Madrid team member.

[35] *Memorias CSIC* (1940-1941), pp. 249-50.

[36] On Franco's technological policy, see Sanz and López. *Política tecnológica*.

[37] Amorós designed four different X-ray diffraction machines during the period 1948-1955.

[38] Lima-De-Faria, J. ed. (1990). *Historical Atlas of Crystallography*. Dordrecht: Kluwer, p. 113. Italics are mine.

[39] Agudo, M. C. (1950). "Sobre el cálculo de estructuras cristalinas por medio de las series de Fourier". *Anales* 46 (A): 205.

[40] Font, M. (1951). "Nueva regla de cálculo para la determinación de factores de estructura". *Anales* 47 (A): 133-38.

[41] Pach, A. (1954). "Simplificación de las operaciones de cálculo de estructuras cristalinas mediante la máquina alfanumérica modelo 405-IBM". *Publicaciones del Departamento de Cristalografía y Mineralogía* 1 (2): 93-104.

[42] Act from the SEFQ session (Seville's Section) of 1952, December the 10th. *Anales*, 49 (A): II, 10.

[43] Márquez, R. (1997). "Semblanza general de D. Luis". *Memorias de la Real Academia Sevillana de Ciencias* 5: 331. Rafael Márquez was a member of the Seville team.

[44] Amorós, J. L. and Lonsdale, K. (1950). "Crystallography in Spain". *Nature* 4218 (166): 392.

[45] During those years, there was practically no research and technological development activity in Spanish companies, with the exception of some pharmaceutical Catalan groups and machine-tools companies in the Basque Country. Sanz and López. *Política tecnológica*, p. 5.

CONTRIBUTORS

Mónica Blanco, PhD, is assistant lecturer in mathematics and statistics at the Universitat Politècnica de Catalunya (Barcelona). Her doctoral thesis, focused on an international comparative analysis of mathematics educational books, providing new insights on the history of calculus in eighteenth-century Europe. She is currently working on the algebraization of calculus and its communication in eighteenth-century Europe, with special regard to educational books on this subject.

Patrick J. Boner, PhD, is a postdoctoral fellow at the Kommission für die Herausgabe der Werke von Johannes Kepler, Bayerische Akademie der Wissenschaften. His research on the conception and critical reception of Kepler's cosmology has resulted in several published articles, most recently "Kepler v. the Epicureans: causality, coincidence and the origins of the new star of 1604". His current study of Kepler's correspondence is anticipated for publication as a full-length monograph.

Daniele Cozzoli, PhD, is research fellow at the Universitat Pompeu, Barcelona. His research has dealt with seventeenth-century optics and philosophy of science, on which he has published a number of papers. He is currently preparing a book on Descartes' method. His research interests have recently moved to the study of twentieth-century biomedical sciences. He is interested in the relations of science and industry in the making of contemporary pharmacology, and conducts research on the scientific career of Daniel Bovet, the discovery of antihistamines drugs and the history of the Istituto Superiore di Sanità (Italian Institute of Health).

Matiana González-Silva is PhD student at the Centro de Estudios de Historia de la Ciencia, Universidad Autónoma de Barcelona (UAB). Her research focuses on science in the media and the public images of science, technology and medicine. Her dissertation deals with journalistic coverage of human genetics in the Spanish newspaper *El País* during the last three decades of the twentieth century. She is also interested in related topics such as the role of media in the definition of scientific policies, the shaping of particular models of scientific popularization and the use of history in debates on contemporary technoscience. She writes a column on scientific issues for the Mexican newspaper *Público-Milenio*.

Ximo Guillem-Llobat, PhD, has pursued postgraduate studies in Scientific Communication and History of Science at the universities of Barcelona and València. His research focuses on the history of food in the nineteenth and twentieth centuries, and especially, on the study of the implementation of industrialisation processes and systems of food quality control. He also works on agricultural innovation and on nutrition popular movements, in the early twentieth-century. He has participated in several international conferences and published several articles in the mentioned topics.

Néstor Herran, PhD, is associate lecturer in the history of science at the Universitat Autònoma de Barcelona (UAB). His research on the history of physical sciences and technology in the twentieth century has resulted in several articles, focusing on the early development of radioactivity in Spain and the development of isotope-based technoscience after World War 2. His current research involves the study of the development of physics and computing in twentieth-century Spain, and the interaction of the physical sciences and political discourses in the public sphere.

Tayra Lanuza-Navarro, PhD, is a postdoctoral fellow at the European University Institute, Florence. Her research has dealt with the history of early modern astrology in seventeenth century Spain, as part of the practices of astronomers, physicians and cosmographers, in the social, political and religious context of this period. Her current research projects have the purpose of comparing the role of astrology in early modern Spain, England and Italy, studying networks of communication and the dissemination of information as well as ideas of popular medicine in astrological texts.

Jorge Lossio, PhD, is associate lecturer at the Pontifical Catholic University of Peru and the San Ignacio de Loyola University in Lima. His research has dealt with the history of climate and health in the nineteenth and twentieth centuries, publishing several articles about medical perceptions on the role of the environment in health, high-altitude physiology, and medical debates about life at high-altitudes. He is also Executive Secretary of the Peruvian Network for the History of Science, Technology and Health.

Tiina Männistö-Funk is a PhD student in Finnish history at the University of Turku, Finland. She has also studied as a postgraduate student in Forschungsstelle für Sozial- und Wirtschaftsgeschichte at the

University of Zurich. Her master's thesis on the production of female body in guidebooks for young women has been published as a book in Finland. Her current research deals with the history of everyday technologies in Finland, especially the history of bicycles, gramophones and photography.

Xavier Mañes is a PhD student at the Universitat Autònoma de Barcelona (UAB). His research deals with the history of science and technology in twentieth-century Spain. His doctoral thesis is devoted to the study of the introduction of X-rays in Spain and the development of its associated technology and research. He is member of a Spanish government-funded research project on the history of physics in twentieth-century Spain.

Faidra Papanelopoulou, DPhil, is a Marie Curie postdoctoral fellow at the Maison des Sciences de l'Homme, Paris. She received her DPhil on "The emergence of thermodynamics in mid-19th-century France" in 2004 from the University of Oxford, and held a post-doctoral position at the Centre Alexandre Koyré, Paris. Her main interests lie on the history of the physical sciences in the nineteenth and early-twentieth centuries, and the popularization of science. She is currently working on the history of low-temperature research and the emergence of physical chemistry, as well as on the public image of science in early-twentieth-century Greece.

Pedro Ruiz-Castell, DPhil, is a postoctoral fellow at the Universitat Autònoma de Barcelona. He pursued postgraduate studies in the history of science at the universities of València and Oxford with a thesis entitled "Astronomy and astrophysics in Spain, 1850-1914", published by Cambridge Scholars Publishing. His research mainly focuses on the history of physics and allied sciences during the nineteenth and twentieth century, science in the public sphere, and scientific instruments.

Josep Simon is MPhil in history of science and medicine (València), MSc in history of science (Oxford), and PhD student (Leeds), with the thesis "Communicating Science in 19th Century Europe: The production, distribution and use of Ganot's physics textbooks". His research brings together the history of nineteenth-century physics, the history of education, book history, and the study of scientific instruments, in an international scenario of knowledge communication. He has recently contributed to three collective books, on the Baillières and Franco-British communication (published by BIUM and by Vrin, Paris), and on science popularization (Ashgate, Aldershot).

SELECT BIBLIOGRAPHY

Anderson, B. (1991). *Imagined Communities: Reflections on the Origin and Spread of Nationalism*. London: Verso.

Arnove, R. F. (1999). "Introduction: Reframing Comparative Education. The Dialectic of the Global and the Local". In Arnove, R. F., and Torres, C. A., eds. *Comparative Education: The Dialectic of the Global and the Local*. Lanham: Rowman & Littlefield Publishers, Inc., pp. 1-23.

Basalla, G. (1967). "The Spread of Western Science". *Science* 156: 111-122.

Ben-David, J. (1971). *The Scientist's Role in Society: A Comparative Study*. Englewood Cliffs, N.J.: Prentice-Hall.

Bensaude-Vincent, B.; García Belmar, A., and Bertomeu Sánchez, J. R. (2003). *L'émergence d'une science des manuels: les livres de chimie en France (1789-1852)*. Paris: Éditions des archives contemporaines.

Bensaude-Vincent, B., and Rasmussen, A. (1997). *La science populaire dans la presse et l'édition: XIXe et XXe siècles*. Paris: CNRS.

Bertomeu Sánchez, J. R.; Garcia-Belmar, A., Patiniotis, M., and Lundgren, A. (2006). "Textbooks In The Scientific Periphery". *Science & Education* 15 (7-8 Special Issue): 657-880.

Bloch, M. (1954). *The Historian's Craft*. Manchester: Manchester University Press.

Chambers, D. W. (1993). "Locality and Science: Myths of Centre and Periphery". In Lafuente, A.; Elena, A., and Ortega, M. L., eds. *Mundialización de la ciencia y la cultura nacional*. Aranjuez: Doce Calles.

Chartier, R. (1984). "Culture as Appropriation: Popular Cultural Uses in Early Modern France". In Kaplan, S. L., ed. *Understanding Popular Culture*. Berlin: Mouton Publishers, pp. 229-253.

—. (1995). *Forms and Meanings: Texts, Performances, and Audiences from Codex to Computer*. Philadelphia: University of Pennsylvania Press.

Choppin, A. (1992). *Les manuels scolaires: histoire et actualité*. Paris: Hachette.

Christie, J. (1993). "Aurora, Nemesis and Clio". *British Journal for the History of Science* 26: 391-405.

Compère, M.-M. (1995). *L'histoire de l'éducation en Europe: essai comparatif sur la façon dont elle s'écrit*. Paris: INRP.

Cooter, R., and Pumphrey, S. (1994). "Separate Spheres and Public Spaces: Reflections on the History of Science Popularization and Science in Popular Culture". *History of Science* 32: 237-267.

Crawford, E. (1992). *Nationalism and Internationalism in Science, 1880-1939: Four Studies of the Nobel Population*. Cambridge: Cambridge University Press.

Cunningham, A., and Williams, P. (1993). "De-centring the 'Big Picture': The Origins of Modern Science and the Modern Origins of Science". *British Journal for the History of Science* 26: 407-432.

Curthoys, A., and Lake, M. (2005). *Connected Worlds: History in Transnational Perspective*. Canberra: Australian National University E Press.

Darnton, R. (1990). *The Kiss of Lamourette: Reflections in Cultural History*. London: Faber & Faber.

Dear, P. (2005). "What Is the History of Science the History Of? Early Modern Roots of the Ideology of Modern Science". *Isis* 96 (3): 390-406.

Detienne, M. (2002). "L'art de construire des comparables: Entre historiens et anthropologues". *Critique Internationale* 14 (1): 68-78.

Dolby, R. G. A. (1977). "The Transmission of Science". *History of Science* 15: 1-43.

Edgerton, D. (1997). "Science in the United Kingdom: A Study in the Nationalization of Science". In Krige, J., and Pestre, D., eds. *Science in the Twentieth Century*. Amsterdam: Harwood Academic, pp. 759-76.

—. (2005). "Science and the Nation: Towards New Histories of Twentieth-Century Britain". *Historical Research* 78: 96-112.

—. (2006). *The Shock of the Old: Technology and Global History science 1900*. London: Profile Books.

Elena, A.; Lafuente, A., and Ortega, M. L. (1993). *Mundialización de la ciencia y cultura nacional*. Madrid: Ediciones Doce Calles.

Espagne, M. (1999). *Les transferts culturels franco-allemands*. Paris: PUF.

Findlen, P. (2005). "The Two Cultures of Scholarship". *Isis* 96: 230-7.

Fleck, L. (1979). *Genesis and Development of a Scientific Fact*. Chicago and London: The University of Chicago Press.

Fox, R. (2003). "Centre and Periphery Revisited: The Structures of European Science". *Revue de la Maison Francaise d'Oxford* 1 (2).

—. (2006). "Fashioning the Discipline: History of Science in the European Intellectual Tradition". *Minerva* 44: 410-32.

Garber, E. (1990). "Introduction". In Garber, ed. *Beyond History of Science. Essays in Honor of Robert E. Schofield.* Bethlehem: Lehigh University Press, pp. 7-20.

Gavroglu, K.; Patiniotis, M.; Papanelopoulou, F.; Simoes, A.; Carneiro, A.; Diogo, M. P.; Bertomeu Sánchez, J. R.; García Belmar, A., and Nieto-Galan, A. (2008). "Science and Technology in the European Periphery. Some Historiographical Reflections". *History of Science* 46: 1-23.

Gellner, E. (1983). *Nations and Nationalism.* Oxford: Basil Blackwell.

Golinski, J. (1998). *Making Natural Knowledge: Constructivism and the History of Science.* Cambridge: Cambridge University Press.

Gooday, G. (2000). "Lies, Damned Lies and Declinism: Lyon Playfair, the Paris 1867 Exhibition and the Contested Rethorics of Scientific Education and Industrial Performance". In Inkster, I.; Griffin, C.; Hill, J., and Rowbotham, J., eds. *The Golden Age. Essays in British Social and Economical History, 1850-1870.* Aldershot: Ashgate, pp. 105-20.

Govoni, P. (2002). *Un pubblico per la scienza: La divulgazione scientifica nell'Italia in formazione.* Roma: Carocci editore.

Green, A. (1990). *Education and State Formation: The Rise of Education Systems in England, France and the USA.* Basingstoke and London: Macmillan.

Gregg, R. (2000). *Inside Out, Outside In: Essays in Comparative History.* Basingstoke: Macmillan.

Gross, R. A. (2000). "The Transnational Turn: Rediscovering American Studies in a Wider World". *Journal of American Studies* 34 (3): 373-93.

Hakfoort, C. (1991). "The Missing Syntheses in the Historiography of Science". *History of Science* 29: 209-216.

Halls, W. D. (1990). "Trends and Issues in Comparative Education". In Halls., ed. *Comparative Education: Contemporary Issues and Trends.* London: Jessica Kingsley Publishers-Unesco, pp. 21-65.

Harwood, J. (1993). *Styles of Scientific Thought: The German Genetics Community, 1900-1933.* Chicago: University of Chicago Press.

Harris, S. J. (1998). "Introduction: Thinking Locally, Acting Globally". *Configurations* 6 (2): 131-139.

Heilbron, J. L. (November 2002). "Science as a subject of history" [Conference at the Palau de la Generalitat]. *VII Trobada d'Història de la Ciència i de la Tècnica.* Barcelona: SCHCT.

Hess, D. J. (1995). *Science and Technology in a Multicultural World: the Cultural Politics of Facts and Artifacts*. New York: Columbia University Press.

Hiltgarner, S. (1990). "The Dominant View of Popularization: Conceptual Problems, Political Uses". *Social Studies of Science* 20: 519-39.

Hobsbawm, E., and Ranger, T., ed. (1992). *The Invention of Tradition*. Cambridge: Cambridge University Press.

Holmes, F. L. (1993). "Justus Liebig and the Construction of Organic Chemistry". In Mauskopf, S. H., ed. *Chemical Sciences in the Modern World*. Philadelphia: University of Pennsylvania Press, pp. 119-135.

Homburg, E. (2004). "Shifting Centres and Emerging Peripheries: Global Patterns in Twentieth-Century Chemistry". *Ambix* 52 (1): 3-7.

—. (2008). "Boundaries and audiences of national histories of science: insights from the history of science and technology of the Netherlands". *Nuncius* [forthcoming].

Iriye, A. (2004). "Transnational History". *Contemporary European History* 13 (2): 211-22.

Jacob, M. C. (1999). "Science Studies after Social Construction: The Turn toward the Comparative and the Global". In Bonnell, V. E., and Hunt, L., eds. *Beyond the Cultural Turn: New Directions n the Study of Society and Culture*. Berkeley: University of California Press, pp. 95-120.

Jardine, N. (1998). "The Places of Astronomy in Early-Modern Culture". *Journal for the History of Astronomy* 29 (1): 49-62.

Jeanneney, J.-N. (2007). *Google and the Myth of Universal Knowledge. A View from Europe*. Chicago: The University of Chicago Press.

Jordanova, L. (1996). "Science and National Identity". In Chartier, R., and Corsi, P., eds. *Sciences et langues en Europe*. Paris: EHESS and Centre Alexandre Koyré, pp. 221-31.

Kaiser, D. (2005). *Pedagogy and the Practice of Science. Historical and Contemporary Perspectives*. Cambridge, MA: MIT Press.

—. (2005). "Training and the Generalist's Vision in the History of Science". *Isis* 96 (2): 244-51.

Kelley, D. R., eds. (1997). *History and the Disciplines: The Reclassification of Knowledge in Early Modern Europe*. Rochester: University of Rochester Press.

Koerner, L. (1999). *Linnaeus: Nature and Nation*. Cambridge, Mass.: Harvard University Press.

Kohler, R. (2005). "A Generalist's Vision". *Isis* 96 : 224-9.

Kuhn, T. S. (1975). "Tradition mathématique et tradition expérimentale dans le développement de la physique". *Annales. Economies, sociétés, civilisations* 30 (5): 975-98.

Lindberg, D. C., and Westman, R. S. (1990). *Reappraisals of the Scientific Revolution*. Cambridge: Cambridge University Press.

Livingstone, D. N. (2003). *Putting Science in its Place: Geographies of Scientific Knowledge*. Chicago and London: The University of Chicago Press.

Lloyd, G. E. R. (1997). "The Comparative History of Pre-Modern Science: The Pitfalls and the Prizes". *Studies in History and Philosophy of Science* 28 (2): 363–68.

López Piñero, J. M. (1992). "La historia de la ciencia como disciplina". *Saber Leer* (55): 8-9.

—. (1993). "La tradición de la historiografía de la ciencia y su coyuntura actual: Los condicionantes de un congreso". In Lafuente, A.; Elena, A., and Ortega, M. L., eds. *Mundialización de la ciencia y cultura nacional*. Madrid: Ediciones Doce Calles, pp. 23-49.

Löwy, I. (2007). "The Social History of Medicine: Beyond the local". *Social History of Medicine* 20 (3): 465-81.

MacLeod, R. (1987). "On Visiting the 'Moving Metropolis': Reflections on the Architecture of Imperial Science". In Reingold, N., and Rothenberg, M., eds. *Scientific Colonialism: A Cross-Cultural Comparison*. Washington DC: Smithsonian Institution Press, pp. 217-249.

Merton, R. (1942). "Science and Technology in a Democratic Order". *Journal of Legal and Political Sociology* 1: 115-126.

—. (1973). *The Sociology of Science*. Chicago: University of Chicago Press.

Nelkin, D. (1987). *Selling Science: How the Press Covers Science and Technology*. New York: W.H. Freeman and Company.

Nieto-Galan, A. (1999). "The Images of Science in Modern Spain". In Gavroglu, K., ed. *The Sciences in the European Periphery During the Enlightenment*. Dordrecht: Kluwer Academic Publishers, pp. 73-94.

—. (2006). "The History of Science in Spain: Imperial Past, Peripheries and the Making of the Modern State". *Neusis* 15: 50-74 (in Greek).

Nowotny, H.; Scott, P., and Gibbons, M., eds. (2001). *Re-thinking Science: Knowledge and the Public in an Age of Uncertainty*. London: Polity Press.

Nye, M. J. (1986). *Science in the Provinces: Scientific Communities and Provincial Leadership in France, 1860-1930*. Berkeley: University of California Press.

—. ed. (2003). *The Cambridge History of Science. The Modern Physical and Mathematical Sciences.* Cambridge: Cambridge University Press, vol. 5.

Olby, R., Cantor, G. N., Christie, J. R. R., and Hodge, M. J. S. (1990). *Companion to the History of Modern Science.* London: Routledge.

Olesko, K. (2006). "Science Pedagogy as a Category of Historical Analysis: Past, Present, and Future". *Science and Education* 15 (7-8): 863-880.

Olesko, K. M. (1993). "Tacit Knowledge and School Formation". *Osiris* 8: 16-29.

Ophir, A., and Shapin, S. (1991). "The Place of Knowledge: A Methodological Survey". *Science in Context* 4: 3-21.

Osler, M. J. (2000). *Rethinking the Scientific Revolution.* Cambridge: Cambridge University Press.

Oudshoorn, N. and Pinch, T., eds. (2003). *How Users Matter: The Co-Construction of Users and Technologies.* Cambridge, Mass.: The MIT Press.

Palladino, P. and Worboys, M. (1993). "Science and Imperialism". *Isis* 84 (1): 91-102.

Park, K. and Daston, L., eds. (2006). *The Cambridge History of Science. Volume 3: Early Modern Science.* Cambridge: Cambridge University Press.

Pickstone, J. V. (2000). *Ways of knowing : a new history of science, technology and medicine.* Manchester: Manchester University Press.

—. (2005). "Review Article: Medical History as a Way of Life". *Social History of Medicine* 18 (2): 307-323.

—. (2007). "Working Knowledges Before and After circa 1800: Practices and Disciplines in the History of Science, Technology, and Medicine". *Isis* 98 (3): 489-516.

Polanco, X. (1990). *Naissance et développement de la science-monde: Production et reproduction des communautés scientifiques en Europe et en Amérique Latine.* Paris: Éditions La Découverte.

Pyenson, L. (2002). "Comparative History of Science". *History of Science* 40: 1-33.

—. (2002). "An End to National Science: Extension of Local Knowledge". *History of Science* 40: 251-90.

Raj, K. (2007). *Relocating Modern Science: Circulation and the Construction of Scientific Knowledge in South Asia and Europe, 1650-1900.* Basingstoke: Palgrave Macmillan.

Roberts, L. (2005). "Circulation Promises and Challenges". In *The Circulation of Knowledge and Practices: The Low Countries as an historical laboratory. Workshop.* Woudschoten, 27-28 May 2005.

Rossi, P. (1986). *I ragni e le formiche: una apologia della storia della scienza.* Bologna: Il Mulino.

Rudolph, J. L. (2008). "Historical Writing on Science Education: A View of the Landscape". *Studies in Science Education* [forthcoming].

Rose, J. (2006). "The History of Education as the History of Reading". *History of Education* 36 (4): 595-605.

Rudwick, M.; Coleman, W.; Sylla, E., and Daston, L. (1981). "Review: Critical Problems in the History of Science". *Isis* 72 (2): 267-283.

Rudwick, M. J. S. (2005). *Bursting the Limits of Time: The Reconstruction of Geohistory in the Age of Revolution.* Chicago: University of Chicago Press.

Sabra, A. I. (1987). "The Appropriation and Subsequent Naturalization of Greek Science in Medieval Islam: A Preliminary Statement". *History of Science* 25: 223-243.

Saussy, H. (2006). "Exquisite Cadavers Stitched from Fresh Nightmares: Of Memes, Hives, and Selfish Genes". In Saussy, H., ed. *Comparative Literature in an Age of Globalization.* Baltimore: Johns Hopkins University Press, pp. 3-42.

Secord, J., ed. (1993). "The Big Picture: Introduction". *British Journal for the History of Science* 26 (special issue): 387-483.

—. (2000). *Victorian Sensation: The Extraordinary Publication, Reception, and Secret Authorship of Vestiges of the Natural History of Creation.* Chicago and London: The University of Chicago Press.

—. (2004). "Knowledge in Transit". *Isis* 4 (95): 654-72.

—. (2007). "How Scientific Conversation Became Shop Talk". In Fyfe, A., and Lightman, B., eds. *Science in the Marketplace.* Chicago: Chicago University Press, pp. 23-59.

Seigel, M. (2005). "Beyond Compare: Comparative Method after the Transnational Turn". *Radical History Review* 91 (Winter): 62-90.

Sewell, W. H. (1967). "Marc Bloch and the Logic of Comparative History". *History and Theory* 6 (2): 208-218.

Shapin, S. (2005). "Hyperprofessionalism and the Crisis of Readership in the History of Science". *Isis* 96 (2): 238-243.

Shils, E. (1975). *Center and Periphery: Essays in Macrosociology.* Chicago: University of Chicago Press.

Shinn, T., and Whitley, R. (1985). *Expository Science: Forms and Functions of Popularisation.* Dordrecht: D. Reidel Publishing Company.

Simoes, A.; Carneiro, A., and Diogo, M. P., eds. (2003). *Travels of Learning: A Geography of Science in Europe*. Berlin: Springer.

Stewart, L. (1992). *The Rise of Public Science. Rhetoric, Technology and Natural Philosophy in Newtonian Britain, 1660-1750*. Cambridge: Cambridge University Press.

Stichweh, R. (1992). *Zur Entstehung des modernen Systems wissenschaftlicher Disziplinen: Physik in Deutschland*. Frankfurt: Suhrkamp.

—. (1994). "La structuration des disciplines dans les universités allemandes au XIXe siècle". *Histoire de l'éducation* 62 (mai): 55-73.

Stone, L. (1979). "The Revival of Narrative: Reflections on a New Old History". *Past and Present* 85 (November): 2-25.

Thomas, R. M. (1990). "The Nature of Comparative Education: How and Why are Education Systems Compared". In Thomas, R. M., eds. *International Comparative Education: Practices, Issues & Prospects*. Oxford: Pergamon Press, pp. 1-21.

Topham, J. R. (2000). "Scientific Publishing and the Reading of Science in Nineteenth-Century Britain: A Historiographical Survey and Guide to Sources". *Studies in History and Philosophy of Science* 4 (31): 559-612.

—. (2008). "Rethinking the History of Science Popularization/Popular Science". In Papanelopoulou, F.; Nieto-Galan, A., and Perdiguero, E., eds. *Popularizing Science and Technology in the European Periphery, 1800-2000*. Aldershot: Ashgate.

Vleuten, E. van d. and Kaijser, A. (2006). *Networking Europe: Transnational Infrastructures and the Shaping of Europe, 1850-2000*. Science History Publications: Sagamore Beach.

Waquet, F. (2003). *Parler comme un livre. L'oralité et le savoir (XVIe-XXe siècle)*. Paris: Albin Michel.

Warner, J. H. (1995). "The History of Science and the Sciences of Medicine". *Osiris* 10: 164-193.

—. (2004). "Grand Narrative and its Discontents: Medical History and the Social Tranformation of American Medicine". *Journal of Health Politics, Policy and Law* 29 (4-5): 757-780.

Werner, M. and Zimmermann, B. (2006). "Beyond Comparison: Histoire Croisée and the Challenge of Reflexivity". *History and Theory* 45 (February): 30-50.

INDEX